Advances in Porous Materials: Synthesis, Characterisations and Applications

Advances in Porous Materials: Synthesis, Characterisations and Applications

Weiqing Zhang
Yichao Wang
Zhifeng Wang

MDPI
Basel • Beijing • Wuhan • Barcelona • Belgrade • Novi Sad • Cluj • Manchester

Editors
Weiqing Zhang
Department of Research
Guangxi Medical University
Cancer Hospital
Nanning
China

Yichao Wang
School of Science
RMIT University
Melbourne
Australia

Zhifeng Wang
School of Materials Science
and Engineering
Hebei University of Technology
Tianjin
China

Editorial Office
MDPI AG
Grosspeteranlage 5
4052 Basel, Switzerland

This is a reprint of articles from the Special Issue published online in the open access journal *Materials* (ISSN 1996-1944) (available at: www.mdpi.com/journal/materials/special_issues/3NHH23093N).

For citation purposes, cite each article independently as indicated on the article page online and as indicated below:

Lastname, A.A.; Lastname, B.B. Article Title. *Journal Name* **Year**, *Volume Number*, Page Range.

ISBN 978-3-7258-2588-2 (Hbk)
ISBN 978-3-7258-2587-5 (PDF)
doi.org/10.3390/books978-3-7258-2587-5

Contents

materials

MDPI

Review

Piezoelectric Applications of Low-Dimensional Composites and Porous Materials

Xiaoqiang Luo [1,*], Qingbin Li [1,*] and Yichao Wang [2,*]

[1] College of Chemical and Environmental Engineering, Pingdingshan University, Pingdingshan 467000, China
[2] School of Engineering, Design and Built Environment, Western Sydney University, Penrith, NSW 2751, Australia
* Correspondence: lxq.5566@163.com (X.L.); lqbsbc@126.com (Q.L.); yichao.wang@westernsydney.edu.au (Y.W.)

Abstract: Low-dimensional (LD) materials, with atomically thin anisotropic structures, exhibit remarkable physical and chemical properties, prominently featuring piezoelectricity resulting from the absence of centrosymmetry. This characteristic has led to diverse applications, including sensors, actuators, and micro- and nanoelectromechanical systems. While piezoelectric effects are observed across zero-dimensional (0D), one-dimensional (1D), and two-dimensional (2D) LD materials, challenges such as effective charge separation and crystal structure imperfections limit their full potential. Addressing these issues requires innovative solutions, with the integration of LD materials with polymers, ceramics, metals, and other porous materials proving a key strategy to significantly enhance piezoelectric properties. This review comprehensively covers recent advances in synthesizing and characterizing piezoelectric composites based on LD materials and porous materials. The synergistic combination of LD materials with other substances, especially porous materials, demonstrates notable performance improvements, addressing inherent challenges. The review also explores future directions and challenges in developing these composite materials, highlighting potential applications across various technological domains.

Keywords: low-dimensional materials; piezoelectricity; composite materials; charge separation; technological applications

Citation: Luo, X.; Li, Q.; Wang, Y. Piezoelectric Applications of Low-Dimensional Composites and Porous Materials. *Materials* **2024**, *17*, 844. https://doi.org/10.3390/ma17040844

Academic Editor: Sangwook Kim

Received: 31 December 2023
Revised: 3 February 2024
Accepted: 5 February 2024
Published: 9 February 2024

1. Introduction

Piezoelectric materials have the ability to convert mechanical stimuli into electrical signals and vice versa, making them attractive candidates for use in a variety of applications, including sensors, actuators, and energy-harvesting devices [1–5]. The piezoelectric effect, first discovered in 1880, has been widely studied for its potential use in a range of technologies such as piezotronics, sensors, and energy-harvesting devices. Traditional piezoelectric materials, including ceramics and polymers, have been used for these applications due to their piezoelectric properties [6–8]. However, the limitations posed by the three-dimensional (3D) nature of these materials, particularly in micro- and nanoelectromechanical systems (MEMS/NEMS) due to size and stiffness constraints, have spurred a quest for innovative alternatives.

In recent years, there has been a growing interest in the development of low-dimensional (LD) materials, including 0D, 1D, and 2D materials, as potential candidates for piezoelectric applications due to their unique physical and chemical properties [9–11]. LD material-based piezoelectric materials have several advantages over traditional piezoelectric materials, such as piezoceramics. They are much more flexible and durable, which makes them suitable for use in a wide range of applications. They also have a higher piezoelectric coefficient, which means that they are able to generate more electrical energy from a given amount of mechanical strain. This makes them particularly suitable for use in energy-harvesting applications, where high efficiency is critical. LD materials have a high degree of flexibility, making them well suited for use in flexible devices such as soft robotics and wearable electronics. Additionally, due to their small size, LD materials can be easily integrated into miniaturized devices, making them attractive for use in micro- and nanoelectromechanical systems.

Exploring beyond traditional piezoelectric materials, the focus has now shifted to composite piezoelectric materials. These combine the strengths of traditional piezoelectric materials with unique properties from other materials, unlocking new performance levels and applications [12]. LD material-based composites can be created by combining LD materials with other materials, such as polymers, ceramics, or metals, to take advantage of the unique piezoelectric properties of both components. These composites have a range of potential applications, including sensors, actuators, and micro- and nanoelectromechanical systems. In recent years, there has been a significant amount of research focused on the development of LD material-based piezoelectric composites for use in energy harvesting and storage devices. These devices are designed to capture and store energy from ambient sources, such as vibration or temperature fluctuations, which can then be used to power small electronic devices or to charge batteries.

In addition to energy harvesting and storage, LD material-based piezoelectric composites have been explored for use in a range of other applications, including medical devices, automotive sensors, and structural health-monitoring systems. These applications often require materials that are able to withstand extreme environments, such as high temperatures, high pressures, or corrosivity. LD material-based piezoelectric composites are able to meet these requirements, making them an attractive choice for use in these applications.

Numerous review articles have extensively addressed the inherent characteristics of untouched piezoelectric materials, as well as exploring energy harvesting and flexible devices [4,13–15]. Nevertheless, there is a notable scarcity of systematic reviews on the piezoengineering of LD-based composite and porous materials. A more thorough review is essential to present the latest developments in this crucial area and underscore the prevailing trends.

This review provides a thorough overview of piezoelectric applications of low-dimensional composite and porous materials. Emphasis is placed on discussing how the structure of composite materials impacts the charge-separation, layer-dependent, and directional effects in low-dimensional materials. As a result, this review systematically summarizes the relationship between the material structure and application performance.

2. Fundamental Aspects and Evaluation Criteria of Piezoelectric Effects

2.1. Fundamentals and Methods for Characterization

When subjected to an external mechanical force or an electric field, piezoelectric materials can undergo polarization, leading to the generation of electric potential or displacement at a macroscopic level. The specified constants typically consist of two subscripts: the first one denotes the polarization direction within the piezoelectric material, while the second one indicates the direction of externally applied pressure or tension. In defining the polarization direction, subscripts 1, 2, or 3 are commonly substituted for the original X, Y, or Z directions, and the associated shear stresses are denoted by subscripts 4, 5, or 6 [3,16].

The performance indicators of piezoelectric materials are generally represented by the following parameters. Specifically, 'd' denotes the piezoelectric charge constant, measured in units of C/N or Pm/V (given that V/m = N/C). Meanwhile, 'g' signifies the piezoelectric voltage constant, with the unit Vm/N. Additionally, 'ε' represents the permittivity and 'e' corresponds to temperature-dependent factors, both measured in C/m^2. 's' means elastic compliance and the unit is m^2/N. 'K' is the electromechanical coupling factor.

2.2. Piezoelectric Material Properties: Charge and Voltage Constants, Sensitivity, and Interconnections

In the study of piezoelectric materials, the piezoelectric charge constant, represented by 'd', holds significant importance. This constant signifies the polarization generated per unit under the application of mechanical stress (T) to the piezoelectric material or, conversely, the mechanical strain experienced by the material when a unit electric field is applied. External factors such as pressure or an electric field can induce polarization in piezoelectric materials. The piezoelectric charge constant, 'd', is characterized by two subscripts, where the first subscript denotes the direction of polarization within the material and the second indicates the direction of externally induced pressure or tension. For d_{33} and d_{31}, both represent polarization in direction 3. In the case of d_{33}, the applied pressure direction aligns with direction 3, whereas, for d_{31}, the pressure is applied in direction 1 (perpendicular to the polarization direction) [17]. Another example is d_{15}, where induced polarization occurs in direction 1, and the applied force is solely a shear force. The unit of measurement for the piezoelectric charge constant is [C/N], and its value is expressed in picometers per volt ([pm/V]).

Moving forward, the piezoelectric voltage constant, denoted by 'g', takes center stage. This constant defines the electric field generated by a unit force applied to a piezoelectric material or the electric displacement caused by applying an electric field on the material's unit surface. Similar to the piezoelectric charge constant, 'g' also comprises two subscripts. The first subscript indicates the direction of the induced electric field or the generated electric displacement, while the second subscript signifies the direction in which the external force is applied.

The piezoelectric voltage constant, being a macroscopic representation of a material's characteristics, serves as a key parameter for evaluating its suitability. Commonly encountered forms of 'g' include g_{33}, g_{31}, and g_{15}. The first two, g_{33} and g_{31}, indicate that the induced electric field aligns with directions 3 and 1, respectively, while the applied external force corresponds to these directions. On the other hand, g_{15} represents a horizontal induced electric field, with the applied external force being a tangential force in direction 5. The unit of measurement for the piezoelectric voltage constant is [Vm/N].

Proceeding to the magnitude of charges (sensitivity) concerning a given strain (e_{ij}), the piezoelectric constant e_{ij} plays a crucial role. Its primary purpose is to eliminate the influence of mechanical clamping on the substrate, electrode, and surrounding unpoled film, ensuring accurate measurement using optical methods like a double-beam interferometer. The focus is often on e_{31} and e_{33}, where e_{31} reflects the piezoelectric performance of the film and e_{33} represents the piezoelectric performance of the bulk material. Bulk piezoelectric materials typically exhibit two to three times the piezoelectric constant of the film. The piezoelectric constant e_{ij} essentially reflects the material's response to changes in shape caused by external stress.

The piezoelectric constant (e_{ij} and d_{ij}) is defined as the external strain (ε_j) and stress (σ_j) coefficients of the electric polarization change (P_i, where i = x, y, z). The relationships between elastic constants, piezoelectric constants, and piezoelectric constants are articulated, emphasizing the interplay between these factors in response to external stress and electric field changes.

2.3. Methods for Measuring Piezoelectric Charge Constant

The integration of piezoelectric materials into actuators and sensors necessitates a thorough understanding of a crucial parameter—the piezoelectric charge constant. This constant, denoted as d_{ij}, represents a material's ability to generate charge under mechanical deformation and is measured in units of C/N or m/V [18,19]. Various methods are utilized to measure this coefficient, categorizing them into distinct groups. There are diverse methods for measuring piezoelectric charge constant offers and each has unique advantages and limitations, catering to specific experimental requirements and desired levels of precision in measuring the piezoelectric charge constant. Understanding these techniques is vital for accurate characterization and utilization of piezoelectric materials in diverse applications.

2.3.1. Frequency Method Measurement

This approach involves the frequency method, which is particularly useful when a complete matrix of material coefficients is required. The accuracy of the resulting value depends on precise readings of resonance frequency and other necessary values. However, this method is limited by the requirement for first-order piezoelectric materials and specific sample constructions.

2.3.2. Laser Interferometry Method (PFM)

Laser interferometry, a high-resolution technique, measures the displacement of a piezoelectric material's surface under applied voltage [4,13,20]. Although sensitive, it is costly and susceptible to external vibrations. Laser interferometry is often used to measure specific piezoelectric charge coefficients (d_{31} and d_{33}), with piezoresponse force microscopy (PFM) serving as a derivative device.

2.3.3. Quasi-Static Method

The quasi-static method provides a cost-effective alternative, comparing piezoelectric samples with a known reference sample. This method measures both d_{31} and d_{33} coefficients without requiring complex processing. The charge on the sample is measured using a charge amplifier or voltmeter with high input resistance.

2.3.4. MEMS Processing Technology (Berlincourt Method)

Microelectromechanical systems (MEMS) processing technology enables the generation of piezoelectric MEMS. The Berlincourt method utilizes a comb-drive actuator to produce dynamic forces and measures piezoelectric charges through charge-sensitive preamplifiers. This method is particularly effective for micro- and nanoscale measurements.

2.3.5. Scanning Evanescent Microwave Microscope (SEMM)

SEMM is used to measure dielectric properties related to piezoelectrics and ferroelectrics. It relies on a high-quality microwave resonator and a sharp metal tip, demonstrating high sensitivity to tip-sample separation and enabling accurate measurement of tiny piezoelectric displacements [21].

2.3.6. X-ray Diffraction Technique

X-ray diffraction is employed to observe polarization conversion, phase change, or structural deformation in piezoelectric materials. This method utilizes Bragg reflection and lattice deformation to measure piezoelectric effects. High-resolution synchrotron X-ray diffraction (HR-XRD) techniques provide exceptional resolutions [22].

2.3.7. Laser Doppler Vibrometer (LDV)

LDV exploits inverse piezoelectric characteristics by applying a voltage of a certain frequency to tungsten electrodes on a piezoelectric material film. The resulting 180-degree phase shift is measured to determine the piezoelectric response.

3. Synthesis of LD-Nanostructures and Their Impact on Piezoelectric Properties

Fabrication methods for low-dimensional piezoelectric composites differ widely, depending on the desired dimensions and specific properties. Various approaches are employed to create these specialized materials with unique piezoelectric characteristics. Techniques such as sol–gel processes, chemical vapor deposition, and template-assisted methods are commonly utilized to engineer these composites at the nanoscale [23–25]. Additionally, bottom-up assembly processes, like layer-by-layer deposition and self-assembly [26,27], contribute to the controlled formation of low-dimensional structures. These fabrication methods play a crucial role in tailoring the mechanical, electrical, and piezoelectric performance of the composites, enabling their application in diverse fields, including sensors, energy harvesting, and biomedical devices.

The intricate world of piezoelectric materials, specifically those classified as LD materials, is predominantly governed by the atomic thickness film with a non-centrosymmetric structure. This structure, as detailed in prior research, is crucial for inducing piezoelectricity. Notably, the layered material must exhibit a specific charge delocalization, as metals lack piezoelectric properties due to shielding of polarization charges by a high electron concentration.

LD materials encompass zero-dimensional, one-dimensional, and two-dimensional entities. The challenge with 0D nanoparticles lies in their size, requiring a matrix as a carrier. Moving to 1D, nanowires, nanobelts, nanotubes, and nanorods take center stage, with diameters generally not exceeding 100 nm. Widely used materials in synthesizing piezoelectric nanorods include ZnO, GaN, and AlN with wurtzite structures. The realm of 2D introduces planar structures with nanometer thicknesses, and the recent trend involves the mixed use of 1D and 2D piezoelectric nanostructures.

Upon reviewing existing research, materials can be categorized into two types: layered materials, involving interlayer coupling, and single-layer films made of bulk piezoelectric materials. Layered piezoelectric films typically rely on weak van der Waals interlayers, featuring materials such as h-BN, TMDCs (2H-MoS_2, 2H-WS_2, 2H-$MoSe_2$), and group III and IV monochalcogenides. Reduction to a single layer disrupts the centrosymmetrical structure, enhancing piezoelectric properties. Group III monochalcogenides like InSe, GaS, and GaSe display D3h point group symmetry, while Group IV monochalcogenides, including GeS, GeSe, SnS, and SnSe, exhibit a flexible folded C2V structure, demonstrating superior piezoelectric performance.

Furthermore, the directional influence on piezoelectricity in 2D materials becomes evident due to low-symmetry crystal structures, resulting in anisotropy. Taking black phosphorus (BP) as an example, its anisotropic characteristics showcase varying piezoelectric coefficients along different directions. Other materials, including rhenium disulfide (ReS_2) and rhenium diselenide ($ReSe_2$), also exhibit anisotropic piezoelectricity.

In the transition from bulk to a single layer, piezoelectric performance often improves, and materials with non-centrosymmetric 3D crystal classes present this opportunity. The lack of a center of symmetry in structures like wurtzite (ZnO, GaN, AlN, CdSe) and nanorods results in an asymmetric distribution of charges, leading to net polarization. ZnO, a representative piezoelectric material, demonstrates anisotropic properties due to its crystal structure. Additionally, the exploration of a single-phase organic–inorganic perovskite piezoelectric, trimethylchloromethyl ammonium trichloromanganese (II) ($TMCMMnCl_3$), reveals remarkable piezoelectric coefficients under bulk conditions.

Figure 1 provides a comprehensive structural model of various two-dimensional (2D) piezoelectric materials, showcasing a diversity of atomic arrangements and crystal structures. In panel a, the top view of atomically thin h-BN (hexagonal boron nitride) is depicted, illustrating hexagonal and orthogonal unit cells. The simulation using Density Functional Theory (DFT) is highlighted in blue and yellow, demonstrating the computational approach employed in understanding the material's properties. For panel b, a side view of 2H-MoS_2, a representative 2D layered piezoelectric material from the transition metal dichalcogenides family, is presented. The arrangement of atoms, including B, N, Mo, and S, is color-coded, and the arrow indicates the direction of piezoelectric polarization. Panel c introduces the side-view crystal structure of In_2Se_3, representing group III compounds, with the arrow indicating the direction of piezoelectric polarization. Panel d showcases the top and side views of a typical group IV monosulfide monolayer, offering insights into the atomic arrangement of this 2D material. Panels e and f present the top and side views of a typical two-dimensional group III–V single-layer honeycomb and an asymmetric moss monolayer structure, respectively. Finally, panel g reveals the structure of ZnO along the a-axis, highlighting the {000} and {0111} polar surfaces that are representative of the wurtzite structure. Collectively, Figure 1 serves as a visual guide to the diverse structures of 2D piezoelectric materials, providing a foundation for understanding their unique properties and potential applications in various technological domains.

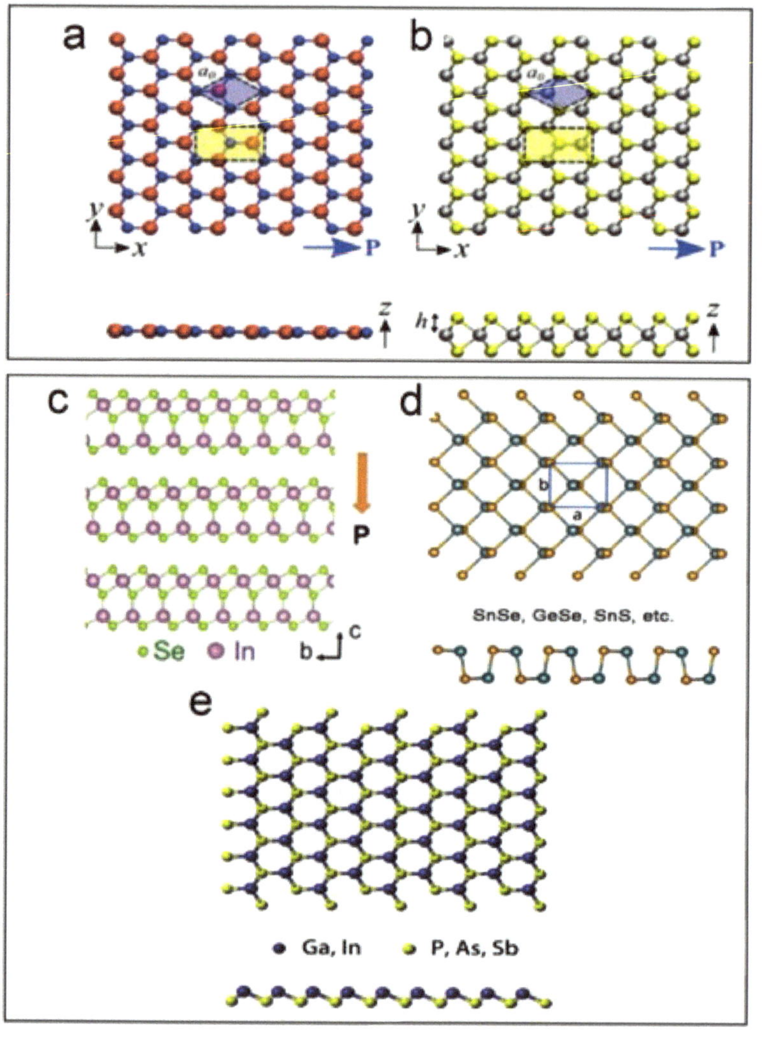

Figure 1. Structural model of two-dimensional piezoelectric materials. (**a**) Top view of atomically thin h-BN used for hexagonal and orthogonal unit cells. The DFT simulation is highlighted in blue and yellow. Reprinted with permission from ref. [28]. Copyright (2012) American Chemical Society, (**b**) Side view 2H-MoS_2, which is representative of 2D layered piezoelectric materials of typical transition metal dichalcogenides. B, N, Mo, and S atoms are red, blue, silver, and yellow, respectively. The arrow indicates the direction of piezoelectric polarization. (**c**) Side-view crystal structure of In_2Se_3 as representative of group III compounds. The direction of piezoelectric polarization is indicated by the arrow (adapted from ref [29], copyright American Chemical Society, 2017) (**d**) Top view and side view of a typical group IV monosulfide monolayer. (**e**) The top and side views of the typical two-dimensional group III–V single-layer honeycomb [30].

4. Composite-Based Piezoelectric Materials

In the realm of enhancing piezoelectric properties, the technique of compositing low-dimensional materials with other compounds emerges as a powerful and efficient approach. This method capitalizes on the synergy of different components within the heterostructure, leading to elevated polarization levels and improved piezoelectric characteristics. The resultant heterostructures often exhibit superior energy-harvesting performance compared to individual components, owing to electrostatic fields and complex interfaces, resulting in nonzero charges and heightened moduli for enhanced energy-harvesting devices.

A significant aspect of this approach involves the conversion of nonpiezoelectric materials into piezoelectric ones through the formation of LD material-based composites. The combination of two entities in a composite induces piezoelectricity by disrupting the original centrosymmetric crystal structures. A notable example involves the enhancement of piezoelectric energy-harvesting properties in polarized poly (vinylidene fluoride–trifluoroethylene) [PVDF–TrFE] and graphene oxide (GO) bilayer films. The bilayer film demonstrates superior energy-harvesting performance compared to unipolar PVDF–TrFE films, showcasing the potential for diverse applications. Researchers such as Bhavanasi et al. and Rodrigues et al. have explored methods to induce piezoelectricity through compositing. Bhavanasi's work involves enhancing the energy-harvesting properties of PVDF–TrFE and GO bilayer films, demonstrating improved performance with higher voltage and power output. Rodrigues, on the other hand, deposited a single-layer graphene onto SiO_2 grating substrates, resulting in observable piezoelectric activities. These examples underscore the versatility of compositing LD materials for tailored piezoelectric applications. Mahmud et al. took a distinctive approach by integrating 1D and 2D ZnO nanostructures on a single substrate, deviating from traditional piezoelectric nanogenerators [31]. This innovative combination aimed to enhance the performance of piezoelectric nanogenerators by merging the advantages of 1D nanowires and 2D nanorods. In another study, Jae Cheol Shin reported that a 10% loading of nano-brick enhanced the remanent polarization from 0.04 to 6.10 $\mu C\ cm^{-2}$ for PVDF, and the dielectric constant at 1 kHz increased by more than 5 and 6 times at 10% ($\varepsilon_r = 13$) and 20% ($\varepsilon_r = 15.4$) loading, respectively, compared to the pure PVDF matrix with an original dielectric constant of 2.5 [32].

The method of enhancing piezoelectric properties through compositing goes beyond engineering transformations. Instead, it focuses on combining different types of piezoelectric materials to amplify their collective properties. Heterostructures, particularly those involving PVDF, serve diverse applications, including actuators, energy storage devices, optoelectronics, and biomedical applications. The exploration of asymmetric layered structures within heterostructures facilitates the generation of electrostatic fields during deformation, contributing to the broadening scope of application scenarios. Therefore, LD materials and their composites present a promising avenue for advancing piezoelectric properties and diversifying applications. This approach holds substantial potential for shaping the future development of piezoelectric materials and their multifaceted applications.

5. Applications

Nanogenerators made of piezoelectric materials can effectively convert tiny kinetic energy into electrical energy, so they are widely used in medical monitoring sensors, environmental monitoring, wearable devices, energy generators, etc. Since, in 1880, the brothers Pierre Curie and Jacques Curie discovered the piezoelectric effect, many applications based on the piezoelectric effect have been created, especially the development of military technology experienced during the First World War and the Second World War. There have been many typical cases such as ultrasonic submarine detectors, echolocation devices, piezoelectric igniters, etc. The development of devices is also progressing with the development of new materials, as piezoelectric materials change from the original lithium niobate ($LiNbO_3$), lead titanate ($PbTiO_3$), quartz, and other materials to ceramics, lead-free piezoceramics, III–V and II–VI semiconductors, and polymers. The morphology of piezoelectric materials has gradually shifted from bulk form to 2D atomic-thickness film form. These innovations have promoted the tremendous development of devices. In recent decades, the main applications of piezoelectric materials are high voltage and power sources, actuators, frequency standards, piezoelectric motors, photovoltaics, and even the surgery field. In recent years, the research on piezoelectric materials has mainly focused on 2D. The synthesis of piezoelectric material films and the direction of ultra-thin materials have been the main directions of research on piezoelectric materials. And they also show huge advantages. The first is that the piezoelectric coefficient is higher than the bulk form. The second is that 2D material film has more electrical characteristics, and the third is that a 2D material provides the basis for the fabrication of flexible devices. The subsequent investigation summarizes the latest applications of 2D piezoelectric materials in recent years. The application of piezoelectric materials is mainly divided into two aspects: one is the mutual conversion of mechanical energy and electric energy, and the other is the production of photoelectric devices related to composites.

5.1. Mechanical or Strain Sensor

A mechanical or strain sensor piezoelectric device converts the surrounding mechanical stress or mechanical vibration due to mechanical stress into a piezoelectric response. On the contrary, it is possible to use an AC power source to make piezoelectric devices produce ideal vibrations. Examples of the mutual transformation of mechanical vibration or strain and piezoelectric response are mature, but, in recent years, as the thickness of piezoelectric materials tends to be at the atomic level, this has made flexible piezoelectric sensor fabrication possible. At present, the most novel sensor is a wireless wearable self-powered flexible electronic device. These wearable devices provide a wide range of applications for the development of intelligence and medical care, such as human health monitoring. Additionally, there are other application fields, for example, nanogenerators, resonators, sonar, texture energy harvesters, micromachined ultrasonic transducers, abd strain energy harvesters. Wearable devices generally require a flexible substrate. PVDF is used as a piezoelectric sub-material, and it is also a flexible polymer, which is an ideal substrate material. Mahmud et al. reported 1D/2D ZnO hybrid nanostructures grown on the same substrate (shown in Figure 2a). The output open-circuit voltage and short-circuit current had almost the same mechanical force in each cycle under the same measurement conditions, at a force of 5 N and a frequency of 5 Hz. The frequency average peak output open-circuit voltage and short-circuit current reach 10.18 V and 15.9 μA [31]. Mina et al. combined the piezoelectric properties of polyvinylidene fluoride (PVDF) nanofibers and the planner nanofillers graphene oxide and graphene; compared with PVDF powder, the electroactive phase (β-phase) of PVDF nanofiber felt increased by about 49% [33]. Mohit et al. dipped the 2D-$SnSe_2$ nanosheet on Whatman filter paper; after frying and annealing, they produced packed $SnSe_2$-decorated paper. They stacked this paper using Ag paste, and, finally, the sensor could monitor a human breath at rate of 3.2–3.5 s/breath (Figure 2b) [34]. Dai et al. reported real-time health monitoring by using seven-layer α-In_2Se_3; these flexible/wearable devices' outputs could reach 0.363 V, with a current responsivity of 598.1 pA for 1% strain (Figure 3a) [35]. Zhang et al. demonstrated one ammonia (NH_3) sensor driven by a novel flexible piezoelectric nanogenerator; the generator was fabricated using semiconductor MoS_2 flakes [36].

There are many novel nanogenerators fabricated. As the earliest nanogenerator, Wang et al. used the two-dimensional odd number of layers of MoS_2 to fabricate nanowire- and nanofilm-based nanogenerators [40]. Yiin et al., chose the piezoelectric polymer PVDF to fabricate piezoelectric fibers (Figure 2c) onto the surface of the monolayer and bilayer CVD-grown graphene. The graphene piezoelectric fiber generator (GPFG) showed highly transparent properties and achievable output voltage/current values of 2 V/200 nA [37]. Mengjun et al., explored a high-output flexible lead-free piezoelectric nanogenerator (PENG); they dropped inorganic piezoelectric $0.91K_{0.48}Na_{0.52}NbO_3$–$0.04Bi_{0.5}Na_{0.5}ZrO_3$–$0.05AgSbO_3$–0.2%$Fe_2O_3$ (KNN–BNZ–AS–Fe) particles on the PDMS matrix; this piezoelectric nanogenerator achieved an ultrahigh piezoelectric coefficient (d_{33}) of 500 pC N-1 [41]. Gyoung-Ja et al. fabricated a piezoelectric energy harvester, using two-dimensional (2D) piezoelectric hexagonal boron nitride nanoflakes (h-BN NFs) deposited onto a flexible plastic substrate (polyimide, 125 μm). This harvester converted a piezoelectric voltage of ~9 V, a current of ~200 nA, and an effective output power of ~0.3 μW [42]. In Figure 3b, Kar et al. reported one self-cleaning piezoelectric energy harvester; this harvester combined inorganic–organic 2D SnO_2 nanosheet-/PVDF-based piezoelectric nanogenerators (PSNG). The output density could reach up to 4900 $W{\cdot}m^{-3}$ with an efficiency ca. 16.3% due to pressure gently imparted by a human finger [43]. Song et al., reported a piezoelectric nanogenerator/strain sensor; hexagonal PbI_2 nanosheets were separated on the surface of polyethylene terephthalate substrate. The peak value of piezoelectric device open-circuit voltage, short-circuit current, and loading power were 29.4 mV, 20 pA, and 0.12 pW [44].

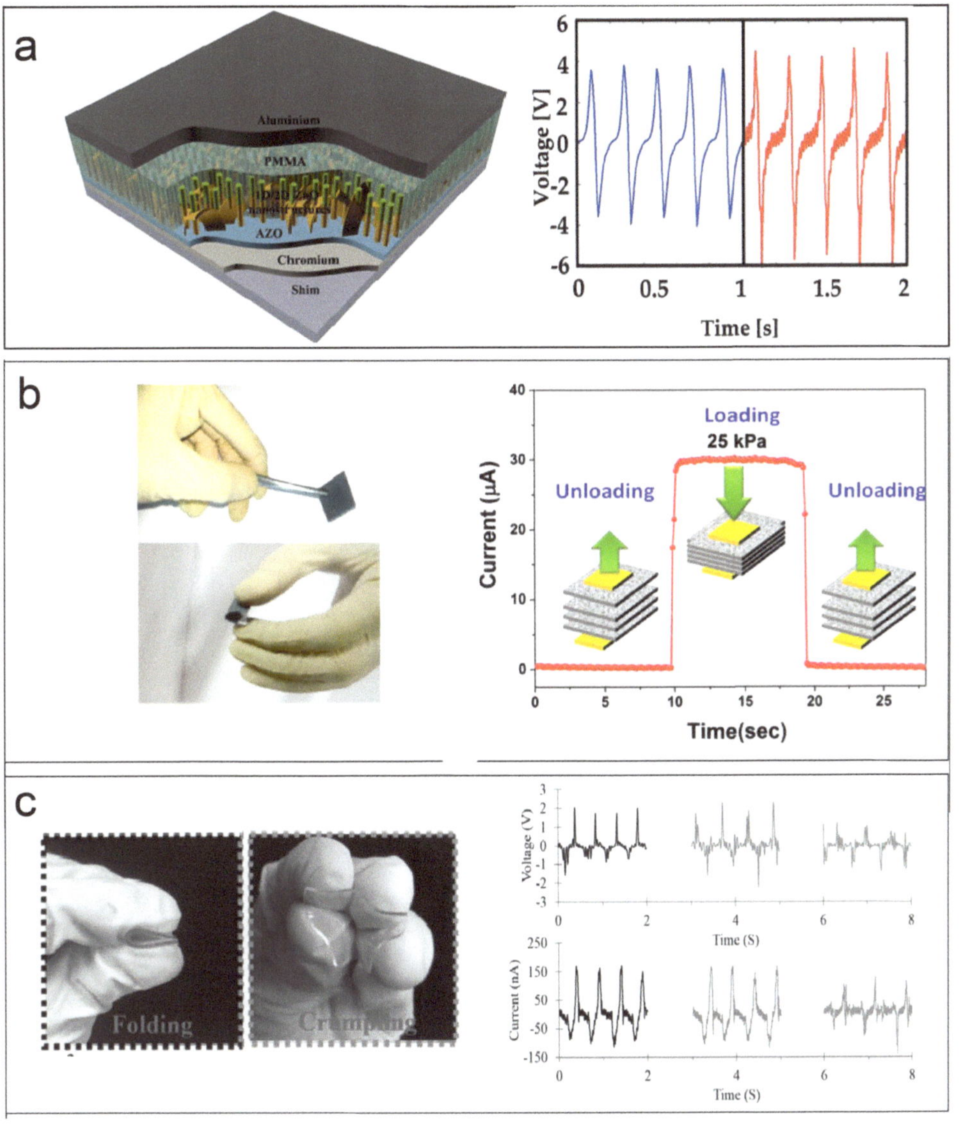

Figure 2. (**a**) 3D schematic of hybrid zinc oxide nanostructures and the output performance [31]. (**b**) The paper-based sensor fabricated by few-layer $SnSe_2$ nanosheets and a single pulse of the time-resolved response with loading and unloading mechanism [34]. (**c**) Photographs of a graphene-piezoelectric fiber generator under different deformation types, and the corresponding output voltages and currents obtained [37].

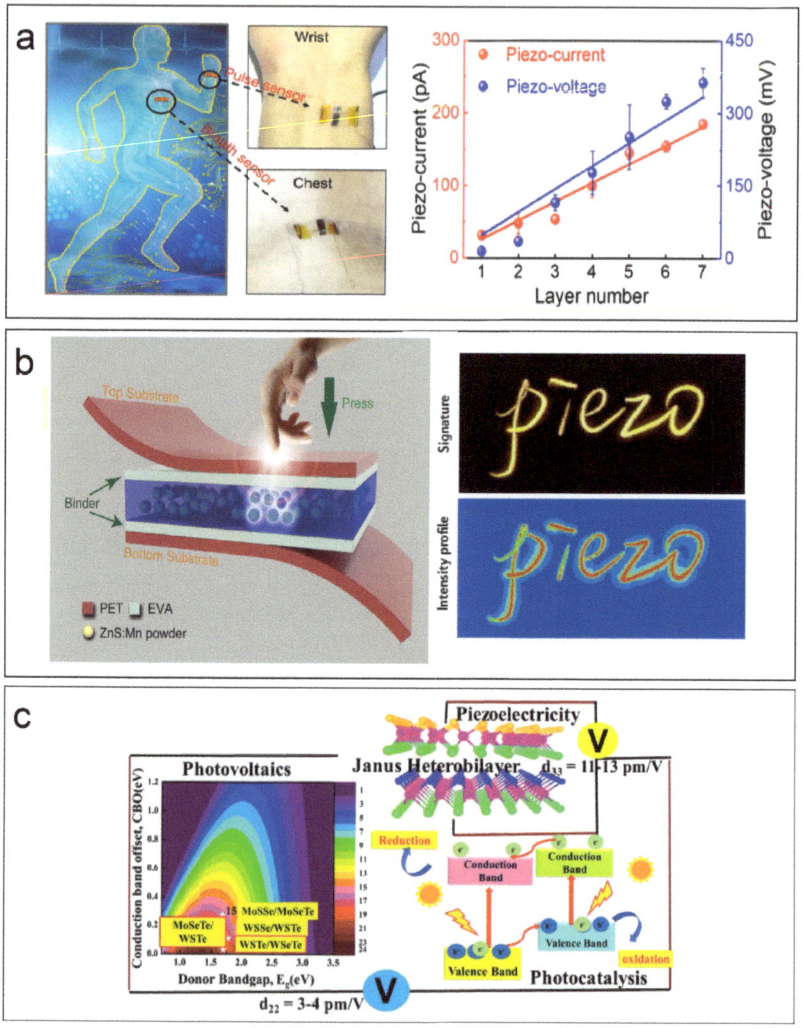

Figure 3. (**a**) Self-powered piezoelectric sensors can monitor physiological signals, pulse and respiration, when attached to the skin of the wrist (top right) and chest (bottom right), along with the piezoelectric current and voltage outputs [35]. (**b**) Pressure sensor matrix (PSM) device schematic configuration. The structure of the device with the characterization of ZnS:Mn particles. And visualization of the dynamic pressure distribution generated by writing "Piezo" by hand [38]. (**c**) The piezoelectricity, photocatalytic performance, and carrier mobility of Janus hybrid bilayers; the power conversion efficiency of 2D ultra-thin exciton solar cells composed of some heterobilayers is between 15 and 20%. The piezoelectric coefficient of the bilayer (d_{33} = 13.91 pm/V) is close to the piezoelectric coefficient of multilayer/bulk [39].

Nanogenerators are the main application method for converting mechanical energy into electrical energy. Furthermore, the transducers use the reverse mechanism; they are applied in many fields such as ultrasonic medical imaging, ultrasonic communication, ultrasonic range-finding, and handwriting input systems [45]. Park et al., reported honeycomb-shaped 1–3 connectivity piezoelectric micropillar arrays, with a crystal type having a size of 42 μm and an aspect ratio of 5 [46].

5.2. Photoelectric Sensor (No-Single-Mechanical or Strain Sensor)

In addition, there are non-single-mechanical piezoelectric sensors. This kind of sensor is generally heterogeneously combined with another characteristic material semiconductor material, so that the characteristics of the piezoelectric material, such as the high carrier concentration, can be used. Wi et al. developed a vertically stacked heterostructure photovoltaic device composed of indium tin oxide/n-MoS_2/plasma-doped MoS_2/Au, with a high short-ring photocurrent density of 20.9 mA/cm^2 [47]. In Figure 3b, Wang et al. introduced one personalized handwriting application, wherein a flexible sensor records the habits associated with handwriting graphics and signatures. The realization of these features was based on the mechanical luminescence (ML) (ZMPs) of ZnS:Mn particles. Through the conversion process between mechanical stress and visible light emission, the single-point pressure and the two-dimensional planar distribution in the range of 0.6–50 MPa were instantaneously performed [38]. Rawat et al. formed a combined heterobilayer composed of two-sided monolayers, which was helpful to study the electronic, optical, piezoelectric, transmission, photocatalytic, and photovoltaic properties of "two-sided monolayers" relative to two-sided monolayers. Upgrading these heterobilayers can cause a large visible-light coefficient ($\alpha \approx$ 105 cm^{-1} (Figure 3c) [39]. Danget et al. designed one graphene field-effect transistor; the transistor mechanism was that the piezoelectric effect of ZnO NRs under static or dynamic pressure modulated the channel conductivity (σ) and caused a positive displacement of 0.25%/kPa at the Dirac point [48].

5.3. Thermal Applications

Thermoelectric materials use waste heat to generate electricity and will play an important role in global sustainable energy solutions. LD materials provide a new way for high-performance thermoelectric performance due to their unique electron-and-hole state density. As a typical representative of piezoelectric materials, MoS_2 has very high Seebeck coefficients, on the order of ~10 mV/K, meaning that it is a representative material with the potential to make thermoelectric devices. Huang et al., used 1T phase MoS_2 to show superior thermoelectric properties: the room temperature power factor could reach 73.1 μW m^{-1} K^{-2}, which was much higher performance than the original graphene or single-walled carbon nanotubes [49]. Wang et al. reported a hybrid nanogenerator that included an electromagnetic generator (EMG), triboelectric nanogenerator (TENG), and thermoelectric generator, which collected both mechanical energy and thermal energy in one process [50]; thus, it was a hybridized nanogenerator for simultaneously scavenging mechanical and thermal energies by electromagnetic–triboelectric–thermoelectric effects [50]. In addition, Aly et al. tried to use a one-dimensional phononic crystal (PnC) that contained a piezoelectric material as a defect layer. Using the control pass band generated in the middle of the band gap, the position of the pass band within the band gap was adjusted by temperature changes. In this way, the application of an acoustic switch and a temperature sensor could be carried out [51] to determine thermal properties of a one-dimensional piezoelectric phononic crystal. Piezoelectric thermoelectric materials, as a new application of piezoelectric materials in recent years, have great potential as a new application field. There is a certain correlation between mechanical energy and thermal energy because thermoelectricity will also be accompanied by machinery in the future. This transformation will become more developed.

5.4. Biomedical Applications

The development of piezoelectric materials in the medical field mainly depends on the great progress in microelectronics, such as the shrinking of electronic devices and the diversification of functions. The development of implantable medical products is mainly to improve the quality of human life, but also to help overcome human defects. Implantable medical electronics (IMEs) are implanted into the human body as diagnostic and therapeutic tools (in applications such as cardioverter defibrillators, sensors, cochlear implants, pacemakers, artificial retinas and stimulators (for nerves, brain, and bones), deep brain stimulation, piezoelectric acoustic sensors for biomimetic artificial hair cells, and tissue engineering). These implanted electronic devices provide diagnostic tools (such as monitoring of temperature, heart rate, and blood pressure) for several diseases in the body while supporting their treatment. At the same time, this can also help us to better understand the complex operating mechanisms in the human body, as seen in the developments in many biomedical fields in recent years [52]. Lee et al., studied the development of piezoelectric viruses, peptides, and other polymeric biomaterials. Their self-assembled thin film of phage exhibited piezoelectric strengths of up to 7.8 pm/V. Higher energy output could be obtained by arranging ordered 23 phage membranes in series or parallel [53]. Kapusetti et al., observed the piezoelectric scaffold placed on a predefined damage site, wherein the scaffold bore the functional load of the subject. The stent converted functional stress into electrical signals through piezoelectric phenomena. The generated synchronous electrical stimulation could regulate the Ca^{2+} channel, thereby enhancing the synthesis of various molecules and promoting the rapid regeneration of damaged bone tissue [54]. Electrical stimulation of cells and tissues is an important way to combine devices with living substances. Traditionally, standard stimulation methods are usually accompanied by destructiveness. Therefore, piezoelectric materials achieve indirect electrical stimulation. This possibility is a completely new solution for all biomedical research. As shown in Figure 4a, Marino et al., used piezoelectric nanomaterials to wirelessly deliver indirect electrical stimulation to neurons. By remotely activating the nanosensor, it was possible to induce Ca^{2+} and Na^{+} transients in neurons by activating the voltage membrane channel [55]. Tang et al., introduced one graphene/BT($BaTiO_3$)/PMMA biopiezoelectric composite. They incorporated $BaTiO_3$ (BT) particles into biological materials, such as polymethyl methacrylate (PMMA) bone cement. The G/BT/PMMA biopiezoelectric composite material is not cytotoxic, and graphene can also promote cell adhesion and proliferation on the surface of the composite material (shown in Figure 4b). Polarized biopiezoelectric composite materials can improve cell morphology and promote cell proliferation [56]. Cheng et al., reported a blood pressure (BP) self-powered monitor with a piezoelectric thin film; the monitor had good linearity (R2$\frac{1}{4}$ 0.971), the sensitivity was up to 14.32 mV/mmHg, and the maximum instantaneous power output in the body was 40 nW. This medical sensor integrated the coupling functions of energy harvesting and biomedical sensing and could effectively detect changes in cardiac blood flow (Figure 4c) [57]. The development of piezoelectric materials in the medical field is mainly based on the use of miniaturized generators as energy sources to perform various functional medical behaviors and on the stimulation of tissues by weak electricity caused by the interactions during tissue movement. As a brand-new solution to solve medical problems, piezoelectricity has huge development potential in the medical field.

Overall, the applications of piezoelectric materials can be divided into three categories. One, with the most numerous applications, is applications that convert the mechanical energy from vibration into electrical energy, such as many nano-generators, resonators, sonar, texture energy-harvesting wireless power supplies for wearable devices, etc. The second is the inverse application of the piezoelectric effect; the application is mainly electrical energy that is converted to mechanical energy, including resonators, ink injectors, piezoelectric motors, piezo-surgery, etc. The third is the heterogeneous combination of piezoelectric materials and other semiconductor properties. These possess the combination of piezoelectric properties and other electronic properties, and the two combined properties are used to make sensor devices, such as some transistors. However, in general, the future of piezoelectric devices will develop in the direction of flexible wearable devices. These are built based on flexible substrates, so PVDF is a key material. The development of this trend is also in line with the intelligent development of the Internet of Things.

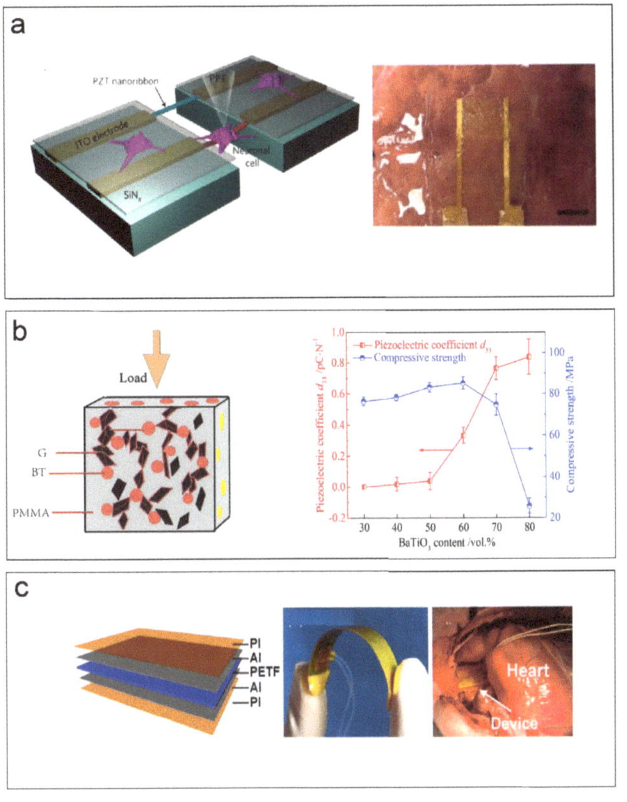

Figure 4. (**a**) Schematic diagram of the sensor device for suspending piezoelectric nanobelts and culturing PC12 cells. The mechanical deflection of the battery is caused by a glass pipette (PPT). PZT nanobelts can detect the deflection of cells and convert them into electrical signals. Then, the electrical signal is collected by indium tin oxide (ITO) electrodes, which are electrically isolated due to the SiN_x coating [55]. Scale bar: 1 cm. (**b**) The mechanism of adding graphene to increase the piezoelectric coefficient of the G/BT/PMMA biopiezoelectric composite material, and the compressive strength of the BT/PMMA biopiezoelectric composite material's piezoelectric coefficient d_{33} [56]. (**c**) The schematic diagram showing a device with a multilayer film structure. In turn, the photos of the device to the right show its flexibility (die > 1) and the device wrapped in a pig's heart [57].

6. Conclusions and Outlook

Piezoelectricity, a valuable phenomenon utilized in diverse applications such as sensing, thermal, biomedical, and energy-harvesting applications, is the focus of this review, which assesses various piezoelectric materials based on their properties, strengths/weaknesses, applications, and ongoing state-of-the-art research. The discussion delves into the materials science perspective of piezoelectricity mechanisms, categorizing materials into different composites, with an emphasis on the linear response in multiple directions. While ceramics traditionally exhibit superior piezoelectric coefficients and polymers offer flexibility and lightweight properties that are suitable for applications like wearable electronics, composite piezoelectric materials emerge as advantageous due to a well-balanced set of properties. However, challenges persist, notably in the efficiency of energy conversion and the need to match resonant frequencies with ambient vibrations for effective energy harvesting. Despite advancements, there is room for improvement in material properties, particularly in lowering resonant frequencies and addressing performance degradation at elevated temperatures. The review suggests avenues for enhancement, such as developing databases for existing composites to facilitate the creation of novel materials and the exploration of alterations in material architecture. Improved piezoelectric properties could lead to innovative applications, including self-powered ocean observation equipment and biocompatible nanocomposites powering cardiac pacemakers, ultimately positioning piezoelectric materials as crucial components in sensing and energy-harvesting technologies through composite optimization.

Funding: This research received no external funding.

Conflicts of Interest: The authors declare no conflicts of interest.

References

1. Sherrell, P.C.; Fronzi, M.; Shepelin, N.A.; Corletto, A.; Winkler, D.A.; Ford, M.; Shapter, J.G.; Ellis, A.V. A bright future for engineering piezoelectric 2D crystals. *Chem. Soc. Rev.* **2022**, *51*, 650–671. [CrossRef]
2. Wu, X.; Li, Z.; Hu, J.; Wang, S.; Wang, Y.; Lin, P.; Zhou, H.; Zhao, W.W. Metallointercalated-DNA Nanotubes as Functional Light Antenna for Organic Photoelectrochemical Transistor Biosensor with Minimum Background. *Anal. Chem.* **2023**, *95*, 11800–11806. [CrossRef]
3. Wang, Y.; Guo, X.; Li, L.H.; Zhang, J.; Li, G.K.; Zavabeti, A.; Li, Y. Enhanced Piezoelectric Properties Enabled by Engineered Low-Dimensional Nanomaterials. *ACS Appl. Nano Mater.* **2022**, *5*, 12126–12142. [CrossRef]
4. Guo, X.; Nguyen, C.K.; Syed, N.; Ravindran, A.; Islam, M.A.; Filleter, T.; Cao, K.; Wang, Y.; Mazumder, A.; Xu, C.; et al. Multi-Functional Atomically Thin Oxides from Bismuth Liquid Metal. *Adv. Funct. Mater.* **2023**, 2307348. [CrossRef]
5. Wang, Y.; Vu, L.M.; Lu, T.; Xu, C.; Liu, Y.; Ou, J.Z.; Li, Y. Piezoelectric Responses of Mechanically Exfoliated Two-Dimensional SnS_2 Nanosheets. *ACS Appl. Mater. Interfaces* **2020**, *12*, 51662–51668. [CrossRef]
6. Oueslati, A.; Aydi, A.J. Impedance and modulus studies of $Na_{0.9}Ba_{0.1}Nb_{0.9}(Sn_{0.5}Ti_{0.5})_{0.1}O_3$ ceramic. *Adv. Dielectr.* **2021**, *11*, 2150013. [CrossRef]
7. Sharma, S.; Kumar, R.; Vaish, R. Universal converse flexoelectricity in dielectric materials via varying electric field direction. *Int. J. Smart Nano Mater.* **2021**, *12*, 107–128. [CrossRef]
8. Zhao, Z.; Zhu, J.; Chen, W. Size-dependent vibrations and waves in piezoelectric nanostructures: A literature review. *Int. J. Smart Nano Mater.* **2022**, *13*, 391–431. [CrossRef]
9. Cui, C.; Xue, F.; Hu, W.J.; Li, L.J. Two-dimensional materials with piezoelectric and ferroelectric functionalities. *NPJ 2D Mater. Appl.* **2018**, *2*, 18. [CrossRef]
10. Li, R.; Cheng, Y.; Huang, W. Recent Progress of Janus 2D Transition Metal Chalcogenides: From Theory to Experiments. *Small* **2018**, *14*, 1802091. [CrossRef] [PubMed]
11. Chen, Q.; Chatzigeorgiou, G.; Meraghni, F.; Javili, A. Homogenization of size-dependent multiphysics behavior of nanostructured piezoelectric composites with energetic surfaces. *Eur. J. Mech. A Solids* **2022**, *96*, 104731. [CrossRef]
12. Chen, Q.; Chatzigeorgiou, G.; Meraghni, F. Extended mean-field homogenization of unidirectional piezoelectric nanocomposites with generalized Gurtin-Murdoch interfaces. *Compos. Struct.* **2023**, *307*, 116639. [CrossRef]
13. Guo, X.; Nguyen, C.K.; Mazumder, A.; Wang, Y.; Syed, N.; Gaspera, E.D.; Daeneke, T.; Walia, S.; Ippolito, S.J.; Sabri, Y.; et al. Gas sensors based on the oxide skin of liquid indium. *Nanoscale* **2023**, *15*, 4972–4981. [CrossRef] [PubMed]
14. Wang, Y.; Zavabeti, A.; Yao, Q.; Tran, T.L.C.; Yang, W.; Kong, L.; Cahill, D. Nanobionics-Driven Synthesis of Molybdenum Oxide Nanosheets with Tunable Plasmonic Resonances in Visible Light Regions. *ACS Appl. Mater. Interfaces* **2022**, *14*, 55285–55294. [CrossRef]
15. Wang, Y.; Ren, B.; Ou, J.Z.; Xu, K.; Yang, C.; Li, Y.; Zhang, H. Nanobionics-Driven Synthesis of Molybdenum Oxide Nanosheets with Tunable Plasmonic Resonances in Visible Light Regions. Engineering two-dimensional metal oxides and chalcogenides for enhanced electro- and photocatalysis. *Sci. Bull.* **2021**, *66*, 1228–1252. [CrossRef]
16. Akbari, M.K.; Verpoort, F.; Hu, J.; Zhuiykov, S. Acoustic-Activated Se Crystalline Nanodomains at Atomically-Thin Liquid-Metal Piezoelectric Heterointerfaces for Synergistic CO_2 Conversion. *ACS Appl. Mater. Interfaces* **2023**, *15*, 39716–39731. [CrossRef] [PubMed]
17. Wu, T.; Zhang, H. Piezoelectricity in Two-Dimensional Materials. *Angew. Chem. Int. Ed.* **2015**, *54*, 4432–4434. [CrossRef]
18. Apte, A.; Mozaffari, K.; Samghabadi, F.S.; Hachtel, J.A.; Chang, L.; Susarla, S.; Idrobo, J.C.; Moore, D.C.; Glavin, N.R.; Litvinov, D.; et al. 2D Electrets of Ultrathin MoO_2 with Apparent Piezoelectricity. *Adv. Mater.* **2020**, *32*, 2000006. [CrossRef]
19. Zhang, J.; Meguid, S.A. Piezoelectricity of 2D nanomaterials: Characterization, properties, and applications. *Semicond. Sci. Technol.* **2017**, *32*, 043006. [CrossRef]
20. Guo, X.; Wang, Y.; Elbourne, A.; Mazumder, A.; Nguyen, C.K.; Krishnamurthi, V.; Yu, J.; Sherrell, P.C.; Daeneke, T.; Walia, S.; et al. Doped 2D SnS materials derived from liquid metal-solution for tunable optoelectronic devices. *Nanoscale* **2022**, *14*, 6802–6810. [CrossRef]
21. Gao, C.; Hu, B.; Li, X.; Liu, C.; Murakami, M.; Chang, K.S.; Long, C.J.; Wuttig, M.; Takeuchi, I. Measurement of the magnetoelectric coefficient using a scanning evanescent microwave microscope. *Appl. Phys. Lett.* **2005**, *87*, 153505. [CrossRef]

22. Thery, V.; Bayart, A.; Blach, J.-F.; Roussel, P.; Saitzek, S. Effective piezoelectric coefficient measurement of $BaTiO_3$ thin films using the X-ray diffraction technique under electric field available in a standard laboratory. *Appl. Surf. Sci.* **2015**, *351*, 480–486. [CrossRef]
23. Li, Y.; Goei, R.; Ong, A.J.; Zou, Y.; Dayan, A.S.; Rahmany, S.; Etgar, L.; Tok, A.I.Y. Atomic layer deposition of piezoelectric materials: A timely review. *Mater. Today Energy* **2024**, *39*, 101457. [CrossRef]
24. Kholkin, A.L.; Yarmarkin, V.K.; Wu, A.; Avdeev, M.; Vilarinho, P.M.; Baptista, J.L. PZT-based piezoelectric composites via a modified sol–gel route. *J. Eur. Ceram. Soc.* **2001**, *21*, 1535–1538. [CrossRef]
25. Ou, C.; Sanchez-Jimenez, P.E.; Datta, A.; Boughey, F.L.; Whiter, R.A.; Sahonta, S.-L.; Kar-Narayan, S. Template-Assisted Hydrothermal Growth of Aligned Zinc Oxide Nanowires for Piezoelectric Energy Harvesting Applications. *ACS Appl. Mater. Interfaces* **2016**, *8*, 13678–13683. [CrossRef] [PubMed]
26. Zhang, Z.; Li, X.; Peng, Z.; Yan, X.; Liu, S.; Hong, Y.; Shan, Y.; Xu, X.; Jin, L.; Liu, B.; et al. Active self-assembly of piezoelectric biomolecular films via synergistic nanoconfinement and in-situ poling. *Nat. Commun.* **2023**, *14*, 4094. [CrossRef]
27. Lee, S.; Yeom, B.; Kim, Y.; Cho, J. Layer-by-layer assembly for ultrathin energy-harvesting films: Piezoelectric and triboelectric nanocomposite films. *Nano Energy* **2019**, *56*, 1–15. [CrossRef]
28. Duerloo, K.A.N.; Ong, M.T.; Reed, E.J. Intrinsic Piezoelectricity in Two-Dimensional Materials. *J. Phys. Chem. Lett.* **2012**, *3*, 2871–2876. [CrossRef]
29. Zhou, Y.; Wu, D.; Zhu, Y.; Cho, Y.; He, Q.; Yang, X.; Herrera, K.; Chu, Z.; Han, Y.; Downer, M.C.; et al. Out-of-Plane Piezoelectricity and Ferroelectricity in Layered α-In_2Se_3 Nanoflakes. *Nano Lett.* **2017**, *17*, 5508–5513. [CrossRef]
30. Lin, P.; Pan, C.; Wang, Z.L. Two-dimensional nanomaterials for novel piezotronics and piezophototronics. *Mater. Today Nano* **2018**, *4*, 17–31. [CrossRef]
31. Mahmud, A.; Khan, A.A.; Voss, P.; Das, T.; Abdel-Rahman, E.; Ban, D. A High Performance and Consolidated Piezoelectric Energy Harvester Based on 1D/2D Hybrid Zinc Oxide Nanostructures. *Adv. Mater. Interfaces* **2018**, *5*, 1801167. [CrossRef]
32. Mane, S.; Teli, A.; Deonikar, V.; Patil, D.; Shin, J.C. Hydrothermally prepared nano bricks of potassium sodium niobate for enhancing thermal and electrical properties of Poly(vinylidene fluoride). *Mater. Lett.* **2023**, *338*, 134040. [CrossRef]
33. Abbasipour, M.; Khajavi, R.; Yousefi, A.A.; Yazdanshenas, M.E.; Razaghian, F.J. The piezoelectric response of electrospun PVDF nanofibers with graphene oxide, graphene, and halloysite nanofillers: A comparative study. *Mater. Sci. Mater. Electron.* **2017**, *28*, 15942–15952. [CrossRef]
34. Tannarana, M.; Solanki, G.K.; Bhakhar, S.A.; Patel, K.D.; Pathak, V.M.; Pataniya, P.M. 2D-$SnSe_2$ Nanosheet Functionalized Piezo-resistive Flexible Sensor for Pressure and Human Breath Monitoring. *ACS Sustain. Chem. Eng.* **2020**, *8*, 7741–7749. [CrossRef]
35. Dai, M.; Wang, Z.; Wang, F.; Qiu, Y.; Zhang, J.; Xu, C.-Y.; Zhai, T.; Cao, W.; Fu, Y.; Jia, D.; et al. Two-Dimensional van der Waals Materials with Aligned In-Plane Polarization and Large Piezoelectric Effect for Self-Powered Piezoelectric Sensors. *Nano Lett.* **2019**, *19*, 5410–5416. [CrossRef] [PubMed]
36. Zhang, D.; Yang, Z.; Li, P.; Pang, M.; Xue, Q. Flexible self-powered high-performance ammonia sensor based on Au-decorated MoSe2 nanoflowers driven by single layer MoS_2-flake piezoelectric nanogenerator. *Nano Energy* **2019**, *65*, 103974. [CrossRef]
37. Fuh, Y.K.; Kuo, C.C.; Huang, Z.M.; Li, S.C.; Liu, E.R. A Transparent and Flexible Graphene-Piezoelectric Fiber Generator. *Small* **2016**, *12*, 1875–1881. [CrossRef] [PubMed]
38. Wang, X.; Zhang, H.; Yu, R.; Dong, L.; Peng, D.; Zhang, A.; Zhang, Y.; Liu, H.; Pan, C.; Wang, Z.L. Dynamic Pressure Mapping of Personalized Handwriting by a Flexible Sensor Matrix Based on the Mechanoluminescence Process. *Adv. Mater.* **2015**, *27*, 2324–2331. [CrossRef] [PubMed]
39. Rawat, A.; Mohanta, M.K.; Jena, N.; Dimple; Ahammed, R.; De Sarkar, A.J. Nanoscale Interfaces of Janus Monolayers of Transition Metal Dichalcogenides for 2D Photovoltaic and Piezoelectric Applications. *J. Phys. Chem. C* **2020**, *124*, 10385–10397. [CrossRef]
40. Zhou, Y.; Liu, W.; Huang, X.; Zhang, A.; Zhang, Y.; Wang, Z.L. Theoretical study on two-dimensional MoS2 piezoelectric nanogenerators. *Nano Res.* **2016**, *9*, 800–807. [CrossRef]
41. Wu, M.; Zheng, T.; Zheng, H.; Li, J.; Wang, W.; Zhu, M.; Li, F.; Yue, G.; Gu, Y.; Wu, J.J. High-performance piezoelectric-energy-harvester and self-powered mechanosensing using lead-free potassium–sodium niobate flexible piezoelectric composites. *Mater. Chem. A* **2018**, *6*, 16439–16449. [CrossRef]
42. Lee, G.-J.; Lee, M.-K.; Park, J.-J.; Hyeon, D.Y.; Jeong, C.K.; Park, K.-I. Piezoelectric Energy Harvesting from Two-Dimensional Boron Nitride Nanoflakes. *ACS Appl. Mater. Interfaces* **2019**, *11*, 37920–37926. [CrossRef]
43. Kar, E.; Bose, N.; Dutta, B.; Banerjee, S.; Mukherjee, N.; Mukherjee, S. 2D SnO_2 nanosheet/PVDF composite based flexible, self-cleaning piezoelectric energy harvester. *Energy Convers. Manag.* **2019**, *184*, 600–608. [CrossRef]
44. Song, H.; Karakurt, I.; Wei, M.; Liu, N.; Chu, Y.; Zhong, J.; Lin, L. Lead iodide nanosheets for piezoelectric energy conversion and strain sensing. *Nano Energy* **2018**, *49*, 7–13. [CrossRef]
45. Jung, J.; Kim, S.; Lee, W.; Choi, H.J. Fabrication of a two-dimensional piezoelectric micromachined ultrasonic transducer array using a top-crossover-to-bottom structure and metal bridge connections. *Micromech. Microeng.* **2013**, *23*, 125037. [CrossRef]
46. Park, J.M.; Shin, D.S.; Han, J.S.; Oh, J.W.; Park, S.C.; Kim, Y.D.; Jang, J.M.; Lee, W.S.; Park, S.J. Design, fabrication of honeycomb-shaped 1–3 connectivity piezoelectric micropillar arrays for 2D ultrasound transducer application. *Ceram. Int.* **2020**, *46*, 12023–12030. [CrossRef]

47. Wi, S.; Kim, H.; Chen, M.; Nam, H.; Guo, L.J.; Meyhofer, E.; Liang, X. Enhancement of Photovoltaic Response in Multilayer MoS_2 Induced by Plasma Doping. *ACS Nano* **2014**, *8*, 5270–5281. [CrossRef]
48. Dang, V.Q.; Kim, D.-I.; Duy, L.T.; Kim, B.-Y.; Hwang, B.-U.; Jang, M.; Shin, K.-S.; Kim, S.-W.; Lee, N.-E. Piezoelectric coupling in a field-effect transistor with a nanohybrid channel of ZnO nanorods grown vertically on graphene. *Nanoscale* **2014**, *6*, 15144–15150. [CrossRef]
49. Huang, H.; Cui, Y.; Li, Q.; Dun, C.; Zhou, W.; Huang, W.; Chen, L.; Hewitt, C.A.; Carroll, D.L. Metallic 1T phase MoS_2 nanosheets for high-performance thermoelectric energy harvesting. *Nano Energy* **2016**, *26*, 172–179. [CrossRef]
50. Wang, X.; Wang, Z.L.; Yang, Y. Hybridized nanogenerator for simultaneously scavenging mechanical and thermal energies by electromagnetic-triboelectric-thermoelectric effects. *Nano Energy* **2016**, *26*, 164–171. [CrossRef]
51. Aly, A.H.; Nagaty, A.; Mehaney, A. Thermal properties of one-dimensional piezoelectric phononic crystal. *Eur. Phys. J. B* **2018**, *91*, 251. [CrossRef]
52. Ali, F.; Raza, W.; Li, X.; Gul, H.; Kim, K.-H. Piezoelectric energy harvesters for biomedical applications. *Nano Energy* **2019**, *57*, 879–902. [CrossRef]
53. Lee, B.Y.; Zhang, J.; Zueger, C.; Chung, W.-J.; Yoo, S.Y.; Wang, E.; Meyer, J.; Ramesh, R.; Lee, S.-W. Virus-based piezoelectric energy generation. *Nat. Nanotechnol.* **2012**, *7*, 351–356. [CrossRef] [PubMed]
54. More, N.; Kapusetti, G. Piezoelectric material—A promising approach for bone and cartilage regeneration. *Med. Hypotheses* **2017**, *108*, 10–16. [CrossRef]
55. Marino, A.; Genchi, G.G.; Sinibaldi, E.; Ciofani, G. Piezoelectric Effects of Materials on Bio-Interfaces. *ACS Appl. Mater. Interfaces* **2017**, *9*, 17663–17680. [CrossRef]
56. Tang, Y.; Chen, L.; Duan, Z.; Zhao, K.; Wu, Z. Graphene/barium titanate/polymethyl methacrylate bio-piezoelectric composites for biomedical application. *Ceram. Int.* **2020**, *46*, 6567–6574. [CrossRef]
57. Cheng, X.; Xue, X.; Ma, Y.; Han, M.; Zhang, W.; Xu, Z.; Zhang, H.; Zhang, H. Implantable and self-powered blood pressure monitoring based on a piezoelectric thinfilm: Simulated, in vitro and in vivo studies. *Nano Energy* **2016**, *22*, 453–460. [CrossRef]

materials

MDPI

Review

Recent Advances in Electrospun Membranes for Radiative Cooling

Dongxue Zhang [1,†], Haiyan Zhang [1,†], Zhiguang Xu [2,*] and Yan Zhao [1,*]

1 College of Textile and Clothing Engineering, Soochow University, Suzhou 215123, China; 20204215025@stu.suda.edu.cn (D.Z.); 20204015004@stu.suda.edu.cn (H.Z.)

2 China-Australia Institute for Advanced Materials and Manufacturing, Jiaxing University, Jiaxing 314001, China

* Correspondence: zhiguang.xu@zjxu.edu.cn (Z.X.); yanzhao@suda.edu.cn (Y.Z.)

† These authors contributed equally to this work.

Abstract: Radiative cooling is an approach that maximizes the thermal emission through the atmospheric window in order to dissipate heat, while minimizing the absorption of incoming atmospheric radiation, to realize a net cooling effect without consuming energy. Electrospun membranes are made of ultra-thin fibers with high porosity and surface area, which makes them suitable for radiative cooling applications. Many studies have investigated the use of electrospun membranes for radiative cooling, but a comprehensive review that summarizes the research progress in this area is still lacking. In this review, we first summarize the basic principles of radiative cooling and its significance in achieving sustainable cooling. We then introduce the concept of radiative cooling of electrospun membranes and discuss the selection criteria for materials. Furthermore, we examine recent advancements in the structural design of electrospun membranes for improved cooling performance, including optimization of geometric parameters, incorporation of highly reflective nanoparticles, and designing multilayer structure. Additionally, we discuss dual-mode temperature regulation, which aims to adapt to a wider range of temperature conditions. Finally, we provide perspectives for the development of electrospun membranes for efficient radiative cooling. This review will provide a valuable resource for researchers working in the field of radiative cooling, as well as for engineers and designers interested in commercializing and developing new applications for these materials.

Keywords: radiative cooling; electrospun membrane; electrospinning; thermal management

Citation: Zhang, D.; Zhang, H.; Xu, Z.; Zhao, Y. Recent Advances in Electrospun Membranes for Radiative Cooling. *Materials* **2023**, *16*, 3677. https://doi.org/10.3390/ma16103677

Academic Editors: Zhifeng Wang, Weiqing Zhang and Yichao Wang

Received: 19 April 2023
Revised: 8 May 2023
Accepted: 9 May 2023
Published: 11 May 2023

1. Introduction

Maintaining a comfortable living environment is a crucial aspect of modern life, but it comes at a significant cost to our planet. The energy used for effective cooling and heating accounts for about 32–33% of the global energy use [1–3]. With the rise of global temperatures due to climate change, the need for cooling is becoming even more critical. At present, space cooling is widely used in summer, mainly relying on traditional vapor compression-based technologies, which are energy-intensive and take up 17% of the world's total electricity. Existing cooling methods have been leading to environmental issues including ozone depletion, the greenhouse effect, and air pollution. Research conducted by the Intergovernmental Panel on Climate change (IPCC) indicates that the global surface temperature has increased by 0.74 ± 0.18 °C over the past century [4,5]. Given these reasons, it is imperative to develop alternative cooling strategies, methods, and technologies.

Radiative cooling technology reflects solar irradiance and radiates heat through the "atmospheric window" to achieve cooling [6–8]. Unlike most cooling methods that need energy input to remove heat, radiative cooling is a way to achieve passive cooling without consuming energy [9–13]. The concept was first proposed by Arago in 1828 [14,15], and has since attracted attention from scientists worldwide. Radiative cooling can help reduce energy and environmental problems by saving energy and electricity [5,16–19]. For instance, for personal thermal management, wearing radiative cooling clothing can help individuals maintain their personal thermal comfort at higher ambient temperatures, reducing the need for air conditioning and saving energy in the process. By allowing for cooling setpoints of air conditioners to be higher while still maintaining the same level of personal thermal comfort, radiative cooling clothing can lead to significant energy savings. In fact, studies have shown that with an increase of 1–4 °C in the setpoint temperature, an energy saving of 7–45% can be achieved, depending on the specific conditions [20].

So far, radiative cooling materials have proven to be highly versatile and have been developed for a variety of applications including energy-efficient buildings [21–26], personal thermal management [20,27–30], cooling solar cells [31–34], enhancing dew yield in water harvesting [35,36], and generating temperature differential for thermoelectric generators [37–40]. In general, to achieve a cooling effect that is lower than the ambient temperature, even for the case under direct sunlight, radiative cooling materials need to possess two key characteristics: strong emission through the atmospheric window and the ability to reflect most of the sunlight. There are various preparation methods available for radiative cooling materials, including vacuum evaporation [17,41], micro–nano processing technology [42,43], direct coating [44], electrospinning [28], and extrusion molding [27,45]. Among these methods, electrospinning is the only technology that can produce continuous ultra-fine fibers with diameters ranging from several micrometers to several hundreds of nanometers, by applying an appropriate electric field on the viscose polymeric fluid [46]. This technique is applicable to almost all soluble or fusible polymers. The fiber diameter, morphology, alignment, porosity, and surface area of electrospun fibers can be easily adjusted by optimizing a series of parameters including equipment configuration, spinning conditions, material ratio, and solvent type [47,48]. The specific wavelength scattering properties of fiber-shaped materials make them an ideal candidate for radiative cooling applications. As a result, electrospun membranes have been extensively researched in recent years due to their excellent radiative cooling capabilities, for various applications including personal thermal management [49–56], passive cooling of buildings [57–59], temperature control of electronic devices [60,61], cooling down food and agricultural products [62], and automobile sun protection clothing [63–65].

To date, radiative cooling has been widely studied and a number of literature reviews have been published on this topic, which covers a wide range of aspects related to the design, fabrication, and application of radiative cooling materials [3,15,66–83]. However, despite the extensive coverage of radiative cooling in the literature and a few review papers that cover fiber-based materials obtained by different spinning techniques for radiative cooling applications [84], there is still a lack of comprehensive reviews that specifically address the research progress of radiative cooling electrospun membranes. In this review, we first briefly summarize the fundamental principles of radiative cooling and its significance in achieving sustainable cooling. We then introduce the concept of radiative cooling of electrospun membranes and discuss the selection criteria for materials used in this process. Furthermore, we examine recent advancements in the structural design of electrospun membranes for improved cooling performance, including optimization of geometric parameters, incorporation of highly reflective nanoparticles, and designing multilayer structure. Additionally, we discuss dual-mode temperature regulation, which aims to adapt to a wider range of temperature conditions. Finally, we provide perspectives for the development of electrospun membranes for efficient radiative cooling.

2. Basic Principles of Radiative Cooling

Heat is a form of energy that always transfers spontaneously from a region of high temperature to another region of lower temperature. In fact, the outer space outside the Earth's atmosphere can be considered as a huge natural cold storage, with its temperature close to absolute zero. From the point of view of thermal radiation, the outer space can be regarded as a blackbody with zero absolute temperature, and all objects with temperatures higher than absolute zero will emit heat to the outer space via infrared radiation [69], through which heat is released from objects on the Earth into the cold universe via thermal radiation [85].

Although the radiative exothermic and endothermic processes of any object occur at the same time, the object can still cool down as long as the heat radiation is more than the heat absorbed from the atmosphere. This is because after receiving solar radiation on the Earth's surface, the energy is mainly concentrated in the wavelength range of 2.5–50 μm to emit thermal radiation. Most of the energy is absorbed by the atmosphere, which is composed of ozone, water vapor, carbon dioxide, and suspended particles that absorb, scatter, and emit electromagnetic waves, where water vapor and carbon dioxide make the most contributions in the thermal radiation of the atmosphere [85]. On one hand, the atmosphere has a highly transparent window ranging from 8 μm to 13 μm (Figure 1a), where the atmosphere's radiative emission is very weak [66]. Outside this window, the atmosphere is highly emissive. On the other hand, the peak thermal radiation of a black body defined by Planck's law at around 300 K coincidentally falls within this atmospheric window. This feature allows for a terrestrial body to cool down by removing heat in the form of thermal emission through the atmospheric window, which is the passive radiative cooling mechanism.

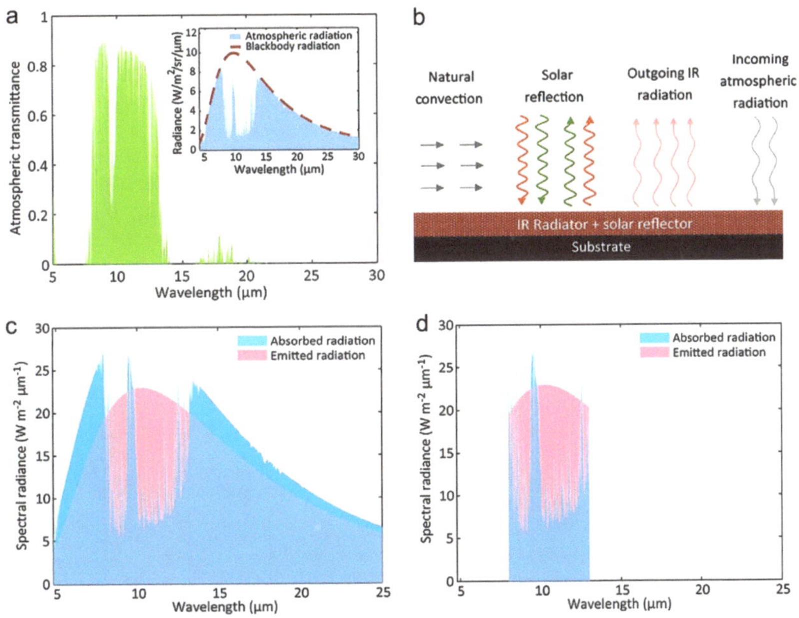

Figure 1. (**a**) A modeled atmospheric transmittance (Inset is the downward atmospheric radiation) [66]. (**b**) Schematics of the thermal exchange of an IR radiator with the environment through radiative and nonradiative processes [66]. Emitted radiation and absorbed radiation at 16 °C for (**c**) broadband and (**d**) selective radiators, respectively [66].

For practical use, the cooling property of a radiator is highly dependent on factors including the incoming atmospheric radiation, the solar radiation during daytime, and the convective or conductive heat gain (Figure 1b). To achieve high cooling efficiency, the thermal emission through the atmospheric window must be maximized, while the incoming atmospheric radiation, the nonradiative heat gain from the surrounding and absorption of solar light must be minimized [66]. Moreover, the spectral emission profiles of the radiative materials significantly affect the performance of radiative cooling. There are two types of radiative materials. The first type is the broadband radiative materials, which are similar to blackbody and have emissivity within the entire emission wavelength range of the atmosphere (Figure 1c). The second type is the selective radiative materials that have high emissivity in 8–13 μm wavelength range, but very low emissivity outside the 8–13 μm window (Figure 1d). For broadband radiative materials, it is noted that they may receive more radiative power than they emit at the ambient temperature, which may prevent them from cooling well below the ambient temperature. In comparison, the selective radiative materials do not absorb radiative power outside the atmospheric window (while in the 8–13 μm window, the atmosphere is highly emissive), but strongly emit within the 8–13 μm wavelength range, which ensures that the nonradiative or solar heat gain can be overcome and thus it is possible to realize effective cooling effects below the ambient temperature. However, for applications where the aim is to decrease the temperature of a device producing a lot of heat, rather than to cool the device below the ambient temperature, the broadband materials can be more useful than the selective radiative materials.

3. Radiative Cooling of Electrospun Membranes

Electrospinning is a well-established technique that enables the production of submicro- and nanofibers with great precision and control. In the electrospinning process (Figure 2a) [86], as the electric field is applied, the droplet at the tip of the needle elongates into a cone shape known as the Taylor cone. When the electric field exceeds the surface tension of the solution, a jet is formed, and the solvent evaporates as the jet travels from the needle tip to the collector, leading to the formation of an ultrafine fiber. The structure and morphology of the fibers can be adjusted by controlling the parameters of the electrospinning process, including the electric field strength, polymer concentration, solution viscosity, flow rate of the polymer solution, and the distance between the spinneret and collector. For instance, by increasing the polymer concentration, the diameter of the resulting fibers can be increased, while decreasing the concentration can produce smaller diameter fibers. By increasing the electric field strength, the fibers can be stretched, resulting in thinner fibers. Similarly, by changing the distance between the spinneret and collector, the fibers can be oriented in different directions, which can affect their mechanical and optical properties.

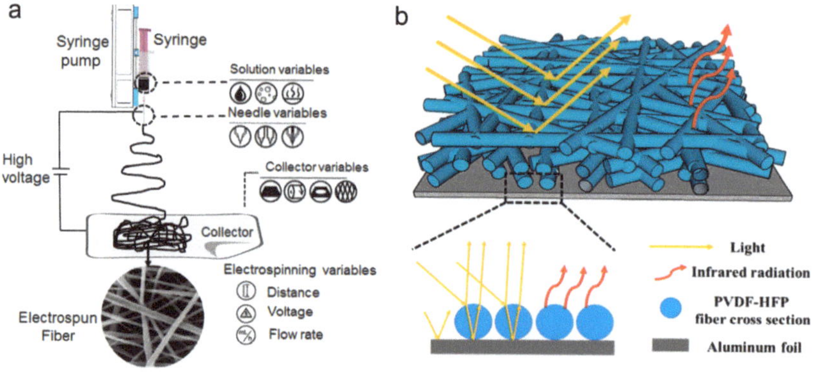

Figure 2. Schematic illustration of (**a**) the electrospinning process [86] and (**b**) the radiative cooling principle of electrospun membranes. Reprinted with permission from [60]. Copyright 2021 Elsevier.

Electrospun membranes are a promising type of material for radiative cooling applications due to their high porosity, large surface area, and tunable optical properties. The large surface area allows for a greater amount of heat transfer through radiation, as more surface area is available for the material to emit radiation. Electrospun membranes can also be designed to have specific optical properties, such as high reflectivity to the solar light by adjusting fiber diameter and pore size, and high emissivity in the 8–13 μm window by selecting high emissivity materials. This allows them to reflect sunlight and emit thermal radiation efficiently (Figure 2b) [60], which further enhances their cooling performance. More importantly, the electrospun membranes are characterized by an in-plane anisotropic structure with fibers arranged mostly in a planar fashion. It has been confirmed that such an anisotropic structure provides higher diffuse reflectance when illuminated from the perpendicular direction than isotropic alignment [87].

In terms of other properties, electrospun membranes for radiative cooling need to be strong enough to maintain their shape and structure over time, without becoming brittle or losing their elasticity. In addition, thermal stability is also important for electrospun membranes to maintain their structural integrity and cooling performance even under high temperatures. This is crucial for applications in hot and sunny climates where radiative cooling can be most effective. Electrospun membranes also have the potential for reusability, as they can be easily maintained and reused in radiative cooling systems. However, the actual reusability of these membranes depends on various factors such as the type of polymer used and the specific conditions of their use.

For electrospun membranes used for radiative cooling, besides the general characterizations such as their mechanical, thermal, and wetting properties, spectroscopic characterization of their emissivity, reflectivity, transmissivity, and absorptivity at specific wavelength ranges is usually conducted. Some tests that are often specifically tailored for electrospun membranes, such as measuring the fiber diameter, the porosity, and the orientation degree of fibers, are also performed. Moreover, the radiative cooling performance can be directly evaluated by the thermal measurement of a radiative cooling surface. For the thermal measurement, the membrane is enclosed by insulation foam, and a convection shield is used to prevent non-radiative heat transfer (conduction and convection heat loss) between the ambient environment and the radiative cooling surface (at sub ambient temperature). Typically, materials with high transmittance within the atmospheric window, such as low-/high-density polyethylene, are used as the convection shield.

In addition, it is noted that the thermal properties of electrospun membranes play an important role in passive radiative cooling applications. Due to their porous structure, electrospun membranes usually exhibit quite low thermal conductivity [88,89], which is one of the main limitations of them when they are used to decrease the temperature of hot substrates, because they are not able to efficiently extract and dissipate excess heat from the substrates. However, for the case of sub ambient cooling, environmental heat gain significantly affects the cooling performance at sub ambient temperatures, and thus it is important to isolate the heat gain from the environment. Therefore, good thermal insulation against ambient air is advantageous for achieving sub ambient cooling. For instance, Zhong et al. [58] prepared a thermal insulating membrane consisting of coaxially electrospun hierarchically hollow microfibers. The membrane shows a high solar reflectivity of 94% and also a high infrared thermal emissivity of 0.94, while it has a lower thermal conductivity than air, which helps to effectively shield the ambient thermal gain. It is worth noting that the influence of the incorporation of thermal insulating function is negligible to the radiative cooling performance. Under direct sunlight radiation with an intensity of 900 W/m^2, the membrane shows a temperature drop of about 9 °C.

In the following sections, the associated factors affecting the cooling performance of electrospun membranes, including the selection of materials and structural design, as well as the dual-mode temperature regulation, are discussed.

4. Selection of Materials

Based on the above discussion, it is obvious that the maximization of radiative emission through the 8–13 μm window and the minimization of heat gain (especially the solar radiation during the daytime) are two necessary approaches to achieving high cooling efficiency [66]. Therefore, to enhance radiative cooling, it would be useful to choose materials with high emissivity within the atmospheric window. For the case of cooling during the daytime, on the other hand, it is highly important to improve the reflectivity to solar light and make it cooperate with the high emissivity to achieve high cooling performance.

4.1. High Emissivity Materials

As discussed above, the radiative cooling performance is highly dependent on the emission profile of the two types of radiative cooling materials. Both types of radiative materials should have low absorptivity to the solar light. Regarding the cooling performance, the use of broadband materials is to maximize the net cooling power, while the use of selective ones is to maximize the temperature difference between the radiator and the environment [15]. Many materials inherently emit highly across the entire IR wavelength range and the emissivity is almost uniform, and thus most radiative materials belong to the type of broadband materials [15].

Polymers are commonly used as base materials for radiative cooling due to their process simplicity, low cost, and wide applicability. The IR emission of polymers depends on the vibrations of their functional groups [90]. As indicated in Figure 3, the fingerprint region (6.7–16.7 μm) corresponds to the atmospheric window and can be used as a main selection metric to identify highly efficient radiative cooling polymers [15]. In the fingerprint region, groups such as C-O, C-F, C-N, and C-Cl have relatively strong absorption due to the bending vibration, and thus polymers having these groups can be selected for high-efficient radiative cooling. For the preparation of radiative cooling electrospun membranes, polymers such as poly(vinylidene fluoride-co-hexafluoropropylene) (PVDF-HFP) [59,61,91], polyethylene oxide (PEO) [92], cellulose acetate (CA) [93,94], and polyvinylidene fluoride (PVDF) [54,64,65,95] have been widely used.

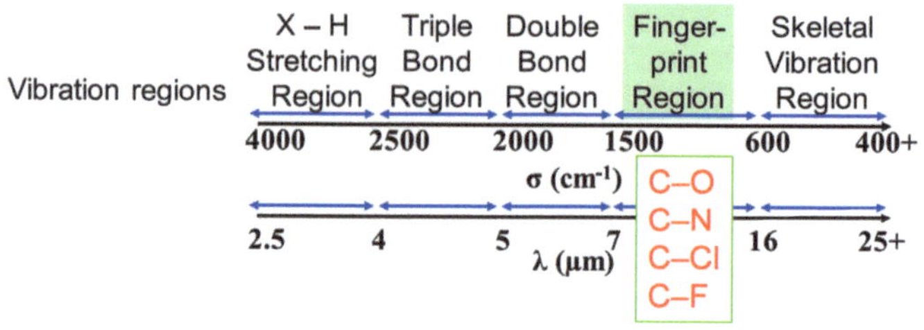

Figure 3. Vibration regions of functional groups in polymers. Reprinted with permission from [15]. Copyright 2021 Elsevier.

4.2. High Reflectivity Materials

It is known that most polymer films (with high IR emissivity) are transparent, and they do not reflect sunlight. To solve this problem, metal substrates or inorganic nanoparticles (Al_2O_3, SiO_2, ZnO, etc.) are often added for reflection. Some of these materials can not only strongly reflect visible light but also have high IR emission. For instance, Al_2O_3 has almost no absorption in the visible-near-infrared wavelength range (0.4–2 μm), and so it is difficult to produce heat under direct solar radiation, but it has strong phonon polarization in the mid-infrared regime (8–20 μm) and thus good emissivity [96,97]. SiO_2 is also transparent to visible light, while it is capable of effectively emitting infrared light [11,60]. This occurs because SiO_2 has a high refractive index for infrared light, which causes the light to be trapped and absorbed within the material. ZnO also possesses a high refractive index of approximately $n \approx 2$, and also exhibits virtually no absorption in the visible to mid-infrared wavelength range of 0.4–16 μm [98,99]. TiO_2 exhibits hyperspectral reflectivity in the wavelength range of visible and near-infrared [100], but it has strong UV absorption and has been rarely used for radiative cooling applications [57].

5. Structural Design

According to the scattering theory established by Rayleigh [101] and Mie [102], materials can strongly scatter electromagnetic waves with wavelengths close to their size. The apertures in the micron to nanometer scale can reflect the energy of sunlight in the visible band, preventing an increase in object temperature. Additionally, these apertures can enhance the emissivity of polymers in the infrared window band, leading to improved radiative cooling performance. In this section, we summarize the approaches employed to enhance the radiative cooling performance through structural design on fibers themselves and the entire electrospun membranes, including the optimization of geometric parameters, the incorporation of highly reflective nanoparticles, and the design of multilayer structures.

5.1. Optimization of Geometric Parameters

By changing the process parameters, the diameter of the electrospun fibers can be facilely adjusted from a few nanometers to a few microns. The influence of nanofiber diameter on cooling efficiency has been studied by using both scattering theory and experiments. The results indicate that the diameter of the fibers plays a critical role in the effectiveness of radiative cooling. For instance, Li et al. [92] prepared a PEO electrospun membrane and used Mie theory to calculate the change in the scattering efficiency of PEO nanofibers with the fiber diameter across the solar spectrum. The results showed that nanofibers with diameter in the range of 500–1200 nm can strongly scatter sunlight, especially in the 0.3–1.2 μm wavelength range. The PEO electrospun membrane exhibits a high reflectivity of >96.3% to the sunlight (0.3–2.5 μm) and a relatively high emissivity of 78% in the 8–13 μm wavelength range.

Besides the high solar light reflectivity, high IR transmittance is important for the radiative heat dissipation of the human body. Kim et al. [95] optimized the fiber diameter for achieving high scattering at a shorter wavelength (~2500 nm), and also a relatively high transmittance at a longer wavelength (~20 μm), so as to simultaneously realize a high reflectivity to solar light (wavelength range 250–2500 nm) and a high transmittance to human body radiation (wavelength range 3–20 μm) (Figure 4a). They demonstrated that PVDF electrospun membrane with an average fiber diameter of ~600 nm has a remarkable radiative cooling ability with solar and NIR reflectance higher than 90%, to significantly block the sunlight energy influx, and an IR transmittance of ~50% for effective radiative heat dissipation of human body. On a simulated skin, the membrane shows a cooling effect of ~12 °C compared to normal textile materials.

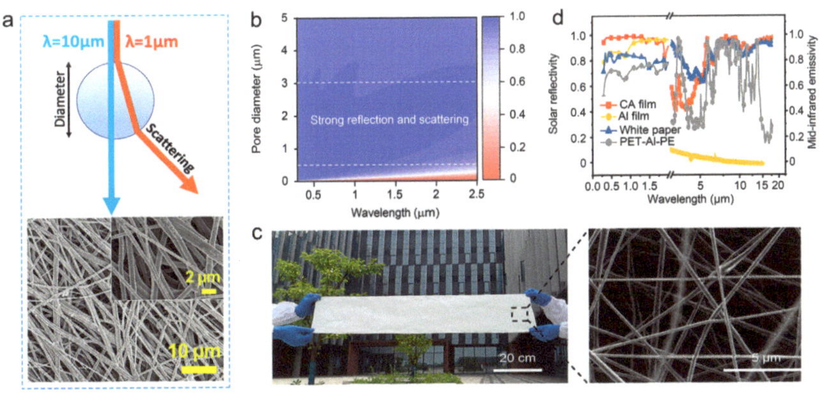

Figure 4. (**a**) Cross-sectional illustration of the optical scattering of different wavelengths across a fiber, and SEM image of PVDF fibers with an average diameter of 600 nm. Reprinted with permission from [95]. Copyright 2021 American Chemical Society. (**b**) Theoretical analysis of the effect of pore size on the scattering of the incident sunlight. On the right side of the graph, the color bar represents reflectivity [62]. (**c**) Digital photo and SEM image of the CA membrane [62]. (**d**) Mid-infrared emissivity and solar reflectivity spectra of the CA membrane [62].

In addition to the fiber diameter, the size of the pores in the electrospun membrane also affects the reflection of sunlight. Li et al. [62] theoretically analyzed the effect of the pore size on the scattering of the incident sunlight, revealing that a porous structure with multiscale pores having size in the range from 500 nm to 3 μm can strongly scatter and reflect the incident sunlight (300 nm to 2.5 μm wavelength range) (Figure 4b). According to this theoretical analysis, they prepared cellulose acetate (CA) electrospun membrane that has multiscale pores having sizes in the same range as that in the theoretical analysis (Figure 4c). It was found that the porous structure and the inherent molecular vibration of CA endow the membrane with a reflectivity as high as 0.974 for the solar spectrum (AM 1.5 G) and a broadband mid-infrared emissivity of 0.92 with an emissive peak within the atmospheric transparent window (Figure 4d). Under direct sunlight irradiation, the CA membrane achieved a cooling temperature of ~12 °C and a cooling power of ~110 W/m^2. Cheng et al. [103] prepared hierarchically porous PVDF-HFP fibers with interconnected nanospheres structure, that possess large roughness and high specific surface area, via vapor-induced phase separation in the process of electrospinning. The PVDF-HFP membrane shows an average solar reflectance of ~93.7% and an IR emittance of ~91.9%, yielding a temperature drop of ~19.8 °C compared to the bare skin simulator under solar intensity of ~950 W/m^2.

5.2. Incorporation of Highly Reflective Nanoparticles

According to Mie theory [102], dielectric and semiconductor nanoparticles with high-refractive index, in a specific size range, may exhibit strong resonant light scattering in the visible spectral range. Recent research has focused on utilizing this phenomenon to improve the radiative cooling performance of electrospun membranes by introducing such micro/nano particles with a size comparable to the solar wavelength. The particles are usually incorporated into the electrospun fibers by blending them into the spinning solutions. For instance, Jayathilaka et al. [57] studied the radiative cooling performance of electrospun fibers doped with Al_2O_3, SiO_2, and TiO_2 particles. It was found that, when the fibers doped with Al_2O_3 and SiO_2 particles were stacked together and the final membrane thickness was controlled at 1.1 mm, the membrane could cool as high as ~22.3 °C below the ambient temperature. Jing et al. [63] made an electrospun membrane consisting of polyvinylidene fluoride/alumina (PVDF/Al_2O_3) ultrafine fibers having a diameter ranging from 0.5 μm to 2.5 μm for daytime radiative cooling below the ambient temperature, based on the high emissivity of PVDF in the transparent atmospheric window, strong sunlight reflectivity of Al_2O_3 nanoparticles, as well as the scattering of solar light by fibers that have diameter equivalent to the wavelength of the solar light. The membrane shows an atmospheric window emissivity of 0.95, a solar reflectivity of 0.97, a net radiative cooling power of 82.7 W/m^2, and a temperature drop of about 4.0 °C under direct sunlight. Xiao et al. [56] fabricated an infrared-radiation-enhanced nanofiber membrane (NFM) composed of polyamide 6 (PA6) nanofibers and randomly distributed SiO_2 submicron spheres. Under a clear sky, the NFM membrane results in temperatures that are about 0.4–1.7 °C lower than those of commercial textiles when they are covered by dry or wet hands, and temperatures that are about 1.0–2.5 °C lower than the ambient temperature when tested in a closed device with thermal conduction and convection being isolated. For an electrospun membrane composed of PVDF electrospun fibers and randomly distributed SiO_2 microspheres [65], an average solar reflectivity of 92.3% and an average transmittance of 0.86 in the atmospheric window were obtained. When the membrane was irradiated under sunlight (700.8 W/m^2), its temperature was 5.8 °C lower than the ambient temperature. Another electrospun membrane consisting of PVDF-HFP fibers incorporated with SiO_2 nanoparticles shows a reflectivity exceeding 0.97 in the solar band and emissivity over 0.94 in the atmospheric window, leading to a cooling effect of 15.9 °C under direct sunlight [104].

In addition to the solution blending method, particles can be combined into the electrospun membranes by post-treatment. For instance, Wang et al. [64] prepared PVDF/tetraethyl orthosilicate (TEOS) electrospun membrane with numerous nanopores in the fibers resulted from the solvent-evaporation-induced phase separation, and then SiO_2 microspheres (diameter 6–14 μm) were introduced into the membrane by simple emulsion deposition. The composite membrane shows an average IR emissivity of higher than 0.96, a solar light reflectivity of ~97%, an average cooling power of 61 W/m^2, and a temperature drop of 6 °C under direct sunlight radiation (with a peak solar intensity of 1000 W/m^2). Hu et al. [59] prepared polyvinylidene fluoride–hexafluoropropylene (PVDF-HFP) electrospun membrane, followed by the electrospraying of SiO_2 nanoparticles, to realize a high solar reflectivity of 98.5% and an average mid-infrared emissivity of higher than 95%. Under sunlight, the average temperature decrease of the membrane was 11.6 °C compared with the surrounding air. Similarly, Meng et al. [105] prepared bead (SiO_2 nanoparticles)-on-string (PVDF-HFP fibers) structured membrane by electrospinning and electrostatic spraying. The membrane achieves average solar reflectance of 97.8% and average atmospheric window emittance of 96.6%, leading to sub-ambient temperature drops of 11.5 and 4.1 °C in daytime and nighttime outdoor conditions, respectively.

5.3. Designing Multilayer Structures

By incorporating different materials with complementary properties, the resulting radiative cooling material can not only improve its cooling performance, but also compensate for any shortcomings associated with a single material. The multilayer structure design strategy has led to the improvement of radiative cooling and also improved the compatibility of radiative cooling function with practical performance requirements such as moisture permeation and waterproofness. For instance, Zhang et al. [53] prepared an electrospun membrane composed of a bottom CA/Al_2O_3 layer (near the skin) and a top PA6/SiO_2 layer nanofiber membrane (Figure 5a). For the top layer, the pore size (300–500 nm) and fiber diameter (100 nm) make it reflect in the ultraviolet-visible region, and the addition of SiO_2 increases its reflectivity in the near-infrared region. In the bottom layer, the pore size of 900–1500 nm and the addition of Al_2O_3 further improve the reflectivity in the near-infrared region. Regarding the IR radiative cooling, the CA in the bottom layer shows a desirable selective emissivity within the atmospheric window, while in the top layer, the PA6 is an IR-transparent material, and it allows thermal radiation emitted by the bottom CA/Al_2O_3 layer to transmit effectively. Overall, the whole membrane achieves a high reflectivity of 99.16% in the 0.3–0.76 μm wavelength range and 88.60% in the 0.76–2.5 μm wavelength range, as well as a selective emissivity of 78.13% in the 8–13 μm window.

In addition to the combination of two or more layers of electrospun membranes, an additional layer that is not prepared by electrospinning can also be used to enhance the cooling performance. For instance, Song et al. [51] used nanoporous PE membrane (PENM, pore size 100–1000 nm) to form a bilayer membrane together with nylon 6 electrospun membrane (PANF) (Figure 5b). The nanoporous PE membrane was used as the outer layer, as it has good IR transmittance and its abundant nanopores results in a strong reflection to visible light (380–760 nm) (Figure 5c). In addition, they also proposed a three-layer film composed of a middle layer of Si_3N_4@PVDF electrospun fibers, an outer layer of nanoporous PE membrane, and an inner layer of hydrophilic-modified PE membrane (mPE) [52]. The middle Si_3N_4@PVDF layer serves to emit IR, because PVDF has strong absorptions at 8.1, 8.5, 9.3, 11.4, and 11.9 μm, and the major absorption of Si_3N_4 is at 10.0 μm. The middle layer also functions to reflect visible light (PVDF fiber diameter 200–600 nm, Si_3N_4 particles 200–1000 nm). The outer and inner PE layers further enhance the reflectance to visible light. The purpose of hydrophilic modification of inner PE layer is to facilitate the moisture permeation. This design results in an IR emittance of 87.31% and a sunlight reflectance of 93.28%. Under direct sunlight radiation, the three-layer film shows a high cooling performance, in which the temperature of the skin is 7.7–10.8 °C lower compared to common textiles.

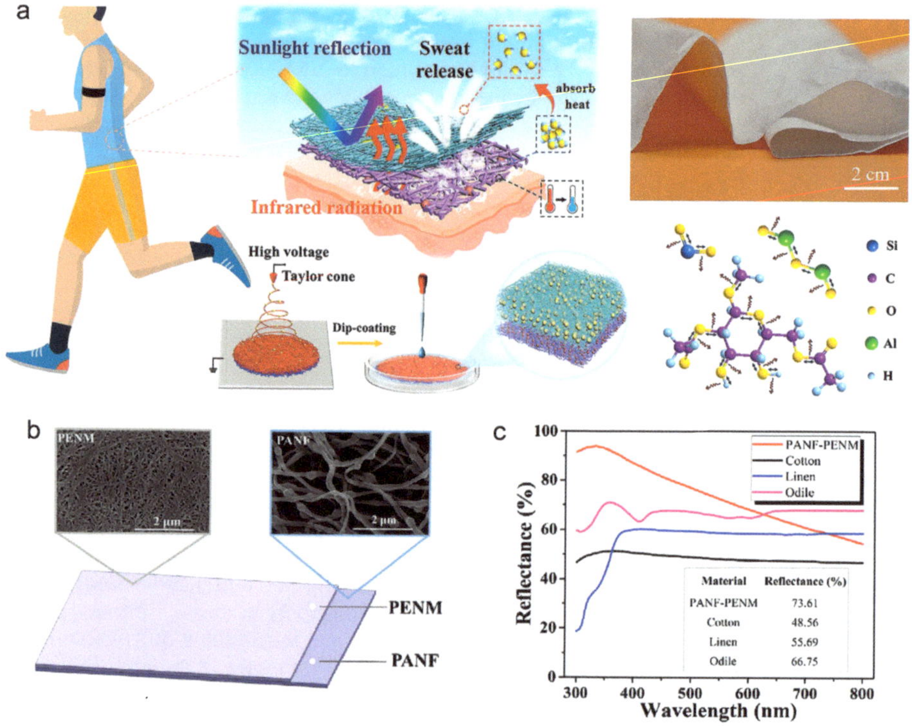

Figure 5. (**a**) Schematic illustration of the preparation process of the bilayer membrane, its photograph, and the schematic illustration of the IR emission by bond vibration of CA, Al_2O_3, and SiO_2. Reprinted with permission from [53]. Copyright 2022 American Chemical Society. (**b**) Structural design and SEM images of the PANF-PENM membrane. Reprinted with permission from [51]. Copyright 2018 American Chemical Society. (**c**) Reflectance curves of PANF-PENM membrane and other commercial textiles. Reprinted with permission from [51]. Copyright 2018 American Chemical Society.

Moreover, Song et al. [28] further combined nanoporous PE membrane with electrospun nylon (PA) and PVDF membranes to form a three-layer film (PA/PVDF/PE). The composite film can induce a temperature drop of 4.5–6.5 °C for the human body under direct sunlight radiation. Besides the nanoporous PE membrane, porous PTFE was also used to compensate for PVDF/SiO_2 electrospun membrane which has a relatively low reflectance to UV light [54]. Under a clear sky, the PVDF/SiO_2-PTFE bilayer film achieves radiative cooling performance of ~9.2 °C temperature drop compared to the bare simulated human skin.

Compared to the approach of optimizing the geometric parameters, the methods of incorporating highly reflective nanoparticles and designing multi-layer structures require more intricate fabrication processes, which inevitably increase the cost and complexity of producing the membrane. Despite the added expense and complexity, research results in the literature consistently demonstrate that these methods can provide higher cooling efficiency than single-material or single-layer approaches. Therefore, for applications that require high-performance cooling capabilities, the benefits of these advanced techniques may justify the additional cost and complexity involved in their implementation.

6. Dual-Mode Temperature Regulation

Besides the high reflectivity that is often utilized to enhance the radiative cooling performance, additional heating functions can be added without affecting the radiative performance so as to allow radiative cooling membranes to adapt to a wider range of temperature conditions, enhancing their versatility and usefulness in various applications. For instance, Xiang et al. [61] reported an electrospun membrane with dual-functional radiative cooling and solar heating abilities for daytime thermal management. PVDF-HFP electrospun fibers were employed for radiative cooling purpose, and polypyrrole (PPy) was spray-coated onto one side of the PVDF-HFP fibrous membrane for heating purposes. During hot sunny days, the side without PPy (white) can face towards the sun to realize a sub-ambient cooling effect with temperature drop of ~4.5 °C. On the contrary, during cold sunny days, the side with PPy (black) can be turned towards the sun to absorb the sunlight and thus to get a heating temperature of ~35.8 °C (with solar intensity of ~850 W/m^2).

In another example, a Janus membrane with dual functions of cooling and heating was prepared by consequently spray-coating ZnO nanosheet, carbon nanotube (CNT), silver nanowire (AgNW), and PDMS membrane on the top of the PVDF electrospun membrane [106]. For the cooling mode, the dual-functional membrane exhibited a radiative emissivity of 89.2% and a very low sunlight absorptivity of only 9.4%, and as a result, temperature drops of 8.2–12.6, 9.0–14.0, and 10.9 °C were realized for a substrate, a closed space, and a semi closed space, respectively. However, for the heating mode, a low thermal radiation emissivity of 10.5% and a high sunlight absorptivity of 74.1% were achieved. For a substrate, a closed space, and a semi closed space, the temperature rise was 3.8–4.6, 4.0–4.8, and 12.5 °C, respectively.

To achieve the solar heating function, the above two examples coated additional sunlight-absorbing materials on one side of the electrospun membranes, and it is required to manually flip the membranes for obtaining the cooling-heating switch. Very recently, Li et al. [107] utilized the changes in the optical reflectivity induced by stretching to realize the reversible switching between radiative cooling and solar heating for electrospun polyurethane membrane, where the inherent nature of electrospun membranes plays a critical role. In the radiative cooling mode (0% strain), the membrane shows a high and angular-independent reflectance of 95.6% in the 0.25–2.5 μm wavelength range, an IR emissivity of 93.3% in the 8–13 μm atmospheric window, a temperature drop of 10 °C at midday, and a corresponding cooling power of 118.25 W/m^2. Under an elongation strain of 80%, the membrane is in the solar heating mode, and it shows a reflectivity of 61.1%, resulting in a net temperature increase of 9.5 °C above ambient of an absorbing substrate and an equivalent power of 220.34 W/m^2.

Compared with other membranes with single temperature regulation mode and limited temperature regulation range, dual-mode membranes prepared by electrospinning have great potential in practical applications. The dual-mode temperature regulation characteristic can prolong the buffer time for objects in the face of huge and abrupt weather changes, so as to make them more suitable for diverse and complex environments.

7. Conclusions and Perspectives

In conclusion, we introduced the basic principles of radiative cooling, the radiative cooling concept of electrospun membranes, and the selection criteria for materials, and summarized the recent advancements in the structural design of electrospun membranes for improved cooling performance, as well as the dual-mode temperature regulation that aims to adapt to a wider range of temperature conditions. Based on the studies reviewed, it can be concluded that the morphology and composition of electrospun membranes can be tailored to optimize their radiative cooling performance, making them a promising candidate for sustainable and energy-efficient radiative cooling technologies. The optimization of geometric parameters, incorporation of highly reflective nanoparticles, and designing multilayer structures have all been effective strategies for enhancing their cooling performance. Moreover, the development of dual-mode temperature regulation has expanded the applicability of electrospun membranes for cooling in a wider range of temperature conditions. Within the scope of this review, the following perspectives were attained:

(1) Based on the diversity of material selection and structural design, it is important to carefully analyze the intended use of the radiative cooling electrospun membranes in order to optimize the material and structure for maximum effectiveness. This will allow us to create radiative cooling electrospun membranes that are compatible with a wide range of applications and offer excellent cooling performance, helping to meet the growing demand for sustainable and energy-efficient solutions.

(2) For accurately comparing the cooling performance of different electrospun membranes, it is very important to standardize the self-built test platform used in the testing process. This involves using the same equipment and providing clear information on the thickness of the membranes being tested, under identical environmental conditions whenever possible. In addition to these standardization measures, it is also important to consider various factors when explaining the cooling performance of a membrane. These factors include the cooling power, temperature difference, sunlight reflectivity, and infrared emissivity. By combining and analyzing these metrics, a more comprehensive and accurate understanding of a membrane's cooling performance can be obtained.

(3) The cooling effect of electrospun membrane is an important factor to consider, but it should not be the sole focus at the expense of other practical properties. Breathability, wear-resistance, skin-friendliness, mechanical strength, and durability are also crucial characteristics that cannot be overlooked. Only when radiative cooling membranes are developed to be compatible with a variety of practical properties can they truly have value for practical application and successful commercialization in the market. Therefore, it is important to strike a balance between cooling and other practical properties in order to produce high-quality and functional electrospun membranes.

Overall, electrospun membranes are advantageous in their versatility in terms of their composition and structure. They show promise as a material for radiative cooling applications. However, there are also challenges associated with the use of electrospun membranes for radiative cooling. One of the main challenges is their mechanical durability. The membranes can be susceptible to abrasion, stretching, and other types of damage. Additionally, the fabrication process for electrospun membranes can be complex and expensive, which can limit their scalability for large-scale applications. Further research is needed to address these challenges, particularly regarding the improvement of their mechanical durability and scalability. In addition, enhancing their radiative cooling performance is also necessary to fully harness their potential for this application.

Author Contributions: Conceptualization, Y.Z. and Z.X.; investigation, D.Z. and H.Z.; writing—original draft preparation, D.Z. and H.Z.; writing—review and editing, Y.Z. and Z.X.; supervision, Y.Z. All authors have read and agreed to the published version of the manuscript.

Funding: This research received no external funding.

Institutional Review Board Statement: Not applicable.

Informed Consent Statement: Not applicable.

Data Availability Statement: Not applicable.

Conflicts of Interest: The authors declare no conflict of interest.

References

1. Zhao, H.; Sun, Q.; Zhou, J.; Deng, X.; Cui, J. Switchable cavitation in silicone coatings for energy-saving cooling and heating. *Adv. Mater.* **2020**, *32*, e2000870. [CrossRef] [PubMed]
2. Zeyghami, M.; Goswami, D.Y.; Stefanakos, E. A review of clear sky radiative cooling developments and applications in renewable power systems and passive building cooling. *Sol. Energy Mater. Sol. Cells* **2018**, *178*, 115–128. [CrossRef]
3. Zhao, B.; Hu, M.; Ao, X.; Chen, N.; Pei, G. Radiative cooling: A review of fundamentals, materials, applications, and prospects. *Appl. Energy* **2019**, *236*, 489–513. [CrossRef]
4. Hansen, J.E.; Ruedy, R.; Sato, M.; Lo, K. Global surface temperature change. *Rev. Geophys.* **2010**, *48*, RG4004. [CrossRef]

5. Hsu, P.-C.; Liu, X.; Liu, C.; Xie, X.; Lee, H.R.; Welch, A.J.; Zhao, T.; Cui, Y. Personal thermal management by metallic nanowire-coated textile. *Nano Lett.* **2015**, *15*, 365–371. [CrossRef]
6. Chen, Z.; Zhu, L.; Raman, A.; Fan, S. Radiative cooling to deep sub-freezing temperatures through a 24-h day–night cycle. *Nat. Commun.* **2016**, *7*, 13729. [CrossRef]
7. Cheng, Z.; Wang, F.; Wang, H.; Liang, H.; Ma, L. Effect of embedded polydisperse glass microspheres on radiative cooling of a coating. *Int. J. Therm. Sci.* **2019**, *140*, 358–367. [CrossRef]
8. Cheng, Z.; Han, H.; Wang, F.; Yan, Y.; Shi, X.; Liang, H.; Zhang, X.; Shuai, Y. Efficient radiative cooling coating with biomimetic human skin wrinkle structure. *Nano Energy* **2021**, *89*, 106377. [CrossRef]
9. You, P.; Li, X.; Huang, Y.; Ma, X.; Pu, M.; Guo, Y.; Luo, X. High-performance multilayer radiative cooling films designed with flexible hybrid optimization strategy. *Materials* **2020**, *13*, 2885. [CrossRef]
10. Son, S.; Jeon, S.; Chae, D.; Lee, S.Y.; Liu, Y.; Lim, H.; Oh, S.J.; Lee, H. Colored emitters with silica-embedded perovskite nanocrystals for efficient daytime radiative cooling. *Nano Energy* **2021**, *79*, 105461. [CrossRef]
11. Zhai, Y.; Ma, Y.; David, S.N.; Zhao, D.; Lou, R.; Tan, G.; Yang, R.; Yin, X. Scalable-manufactured randomized glass-polymer hybrid metamaterial for daytime radiative cooling. *Science* **2017**, *355*, 1062–1066. [CrossRef] [PubMed]
12. Leroy, A.; Bhatia, B.; Kelsall, C.C.; Castillejo-Cuberos, A.; Di Capua, H.M.; Zhao, L.; Zhang, L.; Guzman, A.M.; Wang, E.N. High-performance subambient radiative cooling enabled by optically selective and thermally insulating polyethylene aerogel. *Sci. Adv.* **2019**, *5*, eaat9480. [CrossRef] [PubMed]
13. Mandal, J.; Fu, Y.; Overvig, A.C.; Jia, M.; Sun, K.; Shi, N.N.; Zhou, H.; Xiao, X.; Yu, N.; Yang, Y. Hierarchically porous polymer coatings for highly efficient passive daytime radiative cooling. *Science* **2018**, *362*, 315–319. [CrossRef] [PubMed]
14. Orel, B.; Gunde, M.K.; Krainer, A. Radiative cooling efficiency of white pigmented paints. *Sol. Energy* **1993**, *50*, 477–482. [CrossRef]
15. Yu, X.; Chan, J.; Chen, C. Review of radiative cooling materials: Performance evaluation and design approaches. *Nano Energy* **2021**, *88*, 106259. [CrossRef]
16. Sadineni, S.B.; Madala, S.; Boehm, R.F. Passive building energy savings: A review of building envelope components. *Renew. Sustain. Energy Rev.* **2011**, *15*, 3617–3631. [CrossRef]
17. Raman, A.P.; Anoma, M.A.; Zhu, L.; Rephaeli, E.; Fan, S. Passive radiative cooling below ambient air temperature under direct sunlight. *Nature* **2014**, *515*, 540–544. [CrossRef] [PubMed]
18. Cai, L.; Song, A.Y.; Wu, P.; Hsu, P.-C.; Peng, Y.; Chen, J.; Liu, C.; Catrysse, P.B.; Liu, Y.; Yang, A.; et al. Warming up human body by nanoporous metallized polyethylene textile. *Nat. Commun.* **2017**, *8*, 496. [CrossRef]
19. Yu, X.; Li, Y.; Wang, X.; Si, Y.; Yu, J.; Ding, B. Thermoconductive, moisture-permeable, and superhydrophobic nanofibrous membranes with interpenetrated boron nitride network for personal cooling fabrics. *ACS Appl. Mater. Interfaces* **2020**, *12*, 32078–32089. [CrossRef]
20. Hsu, P.-C.; Song, A.Y.; Catrysse, P.B.; Liu, C.; Peng, Y.; Xie, J.; Fan, S.; Cui, Y. Radiative human body cooling by nanoporous polyethylene textile. *Science* **2016**, *353*, 1019–1023. [CrossRef]
21. Yu, X.; Chen, C. Coupling spectral-dependent radiative cooling with building energy simulation. *Build. Environ.* **2021**, *197*, 107841. [CrossRef]
22. Lu, X.; Xu, P.; Wang, H.; Yang, T.; Hou, J. Cooling potential and applications prospects of passive radiative cooling in buildings: The current state-of-the-art. *Renew. Sustain. Energy Rev.* **2016**, *65*, 1079–1097. [CrossRef]
23. Zhao, D.; Yin, X.; Xu, J.; Tan, G.; Yang, R. Radiative sky cooling-assisted thermoelectric cooling system for building applications. *Energy* **2019**, *190*, 116322. [CrossRef]
24. Ma, M.; Zhang, K.; Chen, L.; Tang, S. Analysis of the impact of a novel cool roof on cooling performance for a low-rise prefabricated building in China. *Build. Serv. Eng. Res. Technol.* **2020**, *42*, 26–44. [CrossRef]
25. Li, X.; Xie, W.; Sui, C.; Hsu, P.-C. Multispectral thermal management designs for net-zero energy buildings. *ACS Mater. Lett.* **2020**, *2*, 1624–1643. [CrossRef]
26. Zhao, D.; Aili, A.; Yin, X.; Tan, G.; Yang, R. Roof-integrated radiative air-cooling system to achieve cooler attic for building energy saving. *Energy Build.* **2019**, *203*, 109453. [CrossRef]
27. Peng, Y.; Chen, J.; Song, A.Y.; Catrysse, P.B.; Hsu, P.-C.; Cai, L.; Liu, B.; Zhu, Y.; Zhou, G.; Wu, D.S.; et al. Nanoporous polyethylene microfibres for large-scale radiative cooling fabric. *Nat. Sustain.* **2018**, *1*, 105–112. [CrossRef]
28. Song, Y.-N.; Li, Y.; Yan, D.-X.; Lei, J.; Li, Z.-M. Novel passive cooling composite textile for both outdoor and indoor personal thermal management. *Compos. Part A Appl. Sci. Manuf.* **2020**, *130*, 105738. [CrossRef]
29. Zeng, S.; Pian, S.; Su, M.; Wang, Z.; Wu, M.; Liu, X.; Chen, M.; Xiang, Y.; Wu, J.; Zhang, M.; et al. Hierarchical-morphology metafabric for scalable passive daytime radiative cooling. *Science* **2021**, *373*, 692–696. [CrossRef]
30. Kim, G.; Park, K.; Hwang, K.; Choi, C.; Zheng, Z.; Seydel, R.; Coza, A.; Jin, S. Black textile with bottom metallized surface having enhanced radiative cooling under solar irradiation. *Nano Energy* **2021**, *82*, 105715. [CrossRef]
31. Zhao, B.; Hu, M.; Ao, X.; Xuan, Q.; Pei, G. Spectrally selective approaches for passive cooling of solar cells: A review. *Appl. Energy* **2020**, *262*, 114548. [CrossRef]
32. Wang, Z.; Kortge, D.; Zhu, J.; Zhou, Z.; Torsina, H.; Lee, C.; Bermel, P. Lightweight, passive radiative cooling to enhance concentrating photovoltaics. *Joule* **2020**, *4*, 2702–2717. [CrossRef]

33. Slauch, I.M.; Deceglie, M.G.; Silverman, T.J.; Ferry, V.E. Model for characterization and optimization of spectrally selective structures to reduce the operating temperature and improve the energy yield of photovoltaic modules. *ACS Appl. Energy Mater.* **2019**, *2*, 3614–3623. [CrossRef]
34. Li, W.; Shi, Y.; Chen, K.; Zhu, L.; Fan, S. A comprehensive photonic approach for solar cell cooling. *ACS Photonics* **2017**, *4*, 774–782. [CrossRef]
35. Liu, C.; Fan, J.; Bao, H. Hydrophilic radiative cooler for direct water condensation in humid weather. *Sol. Energy Mater. Sol. Cells* **2020**, *216*, 110700. [CrossRef]
36. Xu, J.; Zhang, J.; Fu, B.; Song, C.; Shang, W.; Tao, P.; Deng, T. All-day freshwater harvesting through combined solar-driven interfacial desalination and passive radiative cooling. *ACS Appl. Mater. Interfaces* **2020**, *12*, 47612–47622. [CrossRef]
37. Mu, E.; Wu, Z.; Wu, Z.; Chen, X.; Liu, Y.; Fu, X.; Hu, Z. A novel self-powering ultrathin teg device based on micro/nano emitter for radiative cooling. *Nano Energy* **2019**, *55*, 494–500. [CrossRef]
38. Ishii, S.; Dao, T.D.; Nagao, T. Radiative cooling for continuous thermoelectric power generation in day and night. *Appl. Phys. Lett.* **2020**, *117*, 013901. [CrossRef]
39. Xia, Z.; Zhang, Z.; Meng, Z.; Ding, L.; Yu, Z. Thermoelectric generator using space cold source. *ACS Appl. Mater. Interfaces* **2019**, *11*, 33941–33945. [CrossRef]
40. Deppe, T.; Munday, J.N. Nighttime photovoltaic cells: Electrical power generation by optically coupling with deep space. *ACS Photonics* **2019**, *7*, 33941–33945. [CrossRef]
41. Hu, M.; Pei, G.; Wang, Q.; Li, J.; Wang, Y.; Ji, J. Field test and preliminary analysis of a combined diurnal solar heating and nocturnal radiative cooling system. *Appl. Energy* **2016**, *179*, 899–908. [CrossRef]
42. Zhu, L.; Raman, A.P.; Fan, S. Radiative cooling of solar absorbers using a visibly transparent photonic crystal thermal blackbody. *Proc. Natl. Acad. Sci. USA* **2015**, *112*, 12282–12287. [CrossRef] [PubMed]
43. Zou, C.; Ren, G.; Hossain, M.; Nirantar, S.; Withayachumnankul, W.; Ahmed, T.; Bhaskaran, M.; Sriram, S.; Gu, M.; Fumeaux, C. Metal-loaded dielectric resonator metasurfaces for radiative cooling. *Adv. Opt. Mater.* **2017**, *5*, 1700460. [CrossRef]
44. Huang, Z.; Ruan, X. Nanoparticle embedded double-layer coating for daytime radiative cooling. *Int. J. Heat Mass Transf.* **2017**, *104*, 890–896. [CrossRef]
45. Cai, L.; Peng, Y.; Xu, J.; Zhou, C.; Zhou, C.; Wu, P.; Lin, D.; Fan, S.; Cui, Y. Temperature regulation in colored infrared-transparent polyethylene textiles. *Joule* **2019**, *3*, 1478–1486. [CrossRef]
46. Wu, Y.; Chen, C.; Jia, Y.; Wu, J.; Huang, Y.; Wang, L. Review on electrospun ultrafine phase change fibers (pcfs) for thermal energy storage. *Appl. Energy* **2018**, *210*, 167–181. [CrossRef]
47. Shi, S.; Si, Y.; Han, Y.; Wu, T.; Iqbal, M.I.; Fei, B.; Li, R.K.Y.; Hu, J.; Qu, J. Recent progress in protective membranes fabricated via electrospinning: Advanced materials, biomimetic structures, and functional applications. *Adv. Mater.* **2022**, *34*, 2107938. [CrossRef]
48. Bognitzki, M.; Frese, T.; Steinhart, M.; Greiner, A.; Wendorff, J.H.; Schaper, A.; Hellwig, M. Preparation of fibers with nanoscaled morphologies: Electrospinning of polymer blends. *Polym. Eng. Sci.* **2001**, *41*, 982–989. [CrossRef]
49. Tong, J.K.; Huang, X.; Boriskina, S.V.; Loomis, J.; Xu, Y.; Chen, G. Infrared-transparent visible-opaque fabrics for wearable personal thermal management. *ACS Photonics* **2015**, *2*, 769–778. [CrossRef]
50. Miao, D.; Cheng, N.; Wang, X.; Yu, J.; Ding, B. Integration of janus wettability and heat conduction in hierarchically designed textiles for all-day personal radiative cooling. *Nano Lett.* **2022**, *22*, 680–687. [CrossRef]
51. Song, Y.-N.; Ma, R.-J.; Xu, L.; Huang, H.-D.; Yan, D.-X.; Xu, J.-Z.; Zhong, G.-J.; Lei, J.; Li, Z.-M. Wearable polyethylene/polyamide composite fabric for passive human body cooling. *ACS Appl. Mater. Interfaces* **2018**, *10*, 41637–41644. [CrossRef] [PubMed]
52. Song, Y.-N.; Lei, M.-Q.; Deng, L.-F.; Lei, J.; Li, Z.-M. Hybrid metamaterial textiles for passive personal cooling indoors and outdoors. *ACS Appl. Polym. Mater.* **2020**, *2*, 4379–4386. [CrossRef]
53. Zhang, X.; Yang, W.; Shao, Z.; Li, Y.; Su, Y.; Zhang, Q.; Hou, C.; Wang, H. A moisture-wicking passive radiative cooling hierarchical metafabric. *ACS Nano* **2022**, *16*, 2188–2197. [CrossRef]
54. Yang, C.; Sun, X.; Hu, H.; Zhang, K.; Ni, Y.; Shang, S.; Liu, Y. Scalable fabrication of pvdf/SiO_2-ptfe fiber membrane for effective daytime radiative cooling. *Mater. Lett.* **2022**, *320*, 132372. [CrossRef]
55. Hu, R.; Wang, N.; Hou, L.; Liu, J.; Cui, Z.; Zhang, C.; Zhao, Y. Bilayer nanoporous polyethylene membrane with anisotropic wettability for rapid water transportation/evaporation and radiative cooling. *ACS Appl. Mater. Interfaces* **2022**, *14*, 9833–9843. [CrossRef]
56. Xiao, R.; Hou, C.; Yang, W.; Su, Y.; Li, Y.; Zhang, Q.; Gao, P.; Wang, H. Infrared-radiation-enhanced nanofiber membrane for sky radiative cooling of the human body. *ACS Appl. Mater. Interfaces* **2019**, *11*, 44673–44681. [CrossRef]
57. Jayathilaka, W.A.D.M.; Chinnappan, A.; Ghosh, R.; Baskar, C.; Ramakrishna, S. Highly efficient polystyrene/metal oxide fiber composites for passive radiative cooling. *Adv. Eng. Mater.* **2021**, *24*, 2100694. [CrossRef]
58. Zhong, H.; Li, Y.; Zhang, P.; Gao, S.; Liu, B.; Wang, Y.; Meng, T.; Zhou, Y.; Hou, H.; Xue, C.; et al. Hierarchically hollow microfibers as a scalable and effective thermal insulating cooler for buildings. *ACS Nano* **2021**, *15*, 10076–10083. [CrossRef]
59. Hu, L.-C.; Xue, C.-H.; Liu, B.-Y.; Guo, X.-J.; Wang, J.-H.; Deng, F.-Q. Scalable superhydrophobic flexible nanofiber film for passive daytime radiative cooling. *ACS Appl. Polym. Mater.* **2022**, *4*, 3343–3351. [CrossRef]
60. Song, W.-Z.; Wang, X.-X.; Qiu, H.-J.; Wang, N.; Yu, M.; Fan, Z.; Ramakrishna, S.; Hu, H.; Long, Y.-Z. Single electrode piezoelectric nanogenerator for intelligent passive daytime radiative cooling. *Nano Energy* **2021**, *82*, 105695. [CrossRef]

61. Xiang, B.; Zhang, R.; Zeng, X.; Luo, Y.; Luo, Z. An easy-to-prepare flexible dual-mode fiber membrane for daytime outdoor thermal management. *Adv. Fiber Mater.* **2022**, *4*, 1058–1068. [CrossRef]
62. Li, J.; Liang, Y.; Li, W.; Xu, N.; Zhu, B.; Wu, Z.; Wang, X.; Fan, S.; Wang, M.; Zhu, J. Protecting ice from melting under sunlight via radiative cooling. *Sci. Adv.* **2022**, *8*, eabj9756. [CrossRef] [PubMed]
63. Jing, W.; Zhang, S.; Zhang, W.; Chen, Z.; Zhang, C.; Wu, D.; Gao, Y.; Zhu, H. Scalable and flexible electrospun film for daytime subambient radiative cooling. *ACS Appl. Mater. Interfaces* **2021**, *13*, 29558–29566. [CrossRef] [PubMed]
64. Wang, X.; Liu, X.; Li, Z.; Zhang, H.; Yang, Z.; Zhou, H.; Fan, T. Scalable flexible hybrid membranes with photonic structures for daytime radiative cooling. *Adv. Funct. Mater.* **2019**, *30*, 1907562. [CrossRef]
65. Li, M.; Zhang, M.; Mahar, F.K.; Wei, L.; Wang, Z.; Wang, X.; Wei, K. Fabrication of fibrous nanofiber membranes for passive radiation cooling. *J. Mater. Sci.* **2022**, *57*, 16080–16090. [CrossRef]
66. Hossain, M.M.; Gu, M. Radiative cooling: Principles, progress, and potentials. *Adv. Sci.* **2016**, *3*, 1500360. [CrossRef]
67. Li, W.; Fan, S. Radiative Cooling: Harvesting the Coldness of the Universe. *Opt. Photonics News* **2019**, *30*, 32–39. [CrossRef]
68. Zhao, D.; Aili, A.; Zhai, Y.; Xu, S.; Tan, G.; Yin, X.; Yang, R. Radiative sky cooling: Fundamental principles, materials, and applications. *Appl. Phys. Rev.* **2019**, *6*, 021306. [CrossRef]
69. Yin, X.; Yang, R.; Tan, G.; Fan, S. Terrestrial radiative cooling: Using the cold universe as a renewable and sustainable energy source. *Science* **2020**, *370*, 786–791. [CrossRef]
70. Liu, J.; Zhang, Y.; Li, S.; Valenzuela, C.; Shi, S.; Jiang, C.; Wu, S.; Ye, L.; Wang, L.; Zhou, Z. Emerging materials and engineering strategies for performance advance of radiative sky cooling technology. *Chem. Eng. J.* **2023**, *453*, 139739. [CrossRef]
71. Farooq, A.S.; Zhang, P.; Gao, Y.; Gulfam, R. Emerging radiative materials and prospective applications of radiative sky cooling—A review. *Renew. Sustain. Energy Rev.* **2021**, *144*, 110910. [CrossRef]
72. Liu, J.; Zhang, J.; Tang, H.; Zhou, Z.; Zhang, D.; Ye, L.; Zhao, D. Recent advances in the development of radiative sky cooling inspired from solar thermal harvesting. *Nano Energy* **2021**, *81*, 105611. [CrossRef]
73. Ulpiani, G.; Ranzi, G.; Feng, J.; Santamouris, M. Expanding the applicability of daytime radiative cooling: Technological developments and limitations. *Energy Build.* **2021**, *243*, 110990. [CrossRef]
74. Zhang, J.; Yuan, J.; Liu, J.; Zhou, Z.; Sui, J.; Xing, J.; Zuo, J. Cover shields for sub-ambient radiative cooling: A literature review. *Renew. Sustain. Energy Rev.* **2021**, *143*, 110959. [CrossRef]
75. Ulpiani, G.; Ranzi, G.; Shah, K.W.; Feng, J.; Santamouris, M. On the energy modulation of daytime radiative coolers: A review on infrared emissivity dynamic switch against overcooling. *Sol. Energy* **2020**, *209*, 278–301. [CrossRef]
76. Liu, J.; Zhou, Z.; Zhang, J.; Feng, W.; Zuo, J. Advances and challenges in commercializing radiative cooling. *Mater. Today Phys.* **2019**, *11*, 100161. [CrossRef]
77. Lin, K.-T.; Han, J.; Li, K.; Guo, C.; Lin, H.; Jia, B. Radiative cooling: Fundamental physics, atmospheric influences, materials and structural engineering, applications and beyond. *Nano Energy* **2021**, *80*, 105517. [CrossRef]
78. Chen, J.; Lu, L. Development of radiative cooling and its integration with buildings: A comprehensive review. *Sol. Energy* **2020**, *212*, 125–151. [CrossRef]
79. Li, W.; Li, Y.; Shah, K.W. A materials perspective on radiative cooling structures for buildings. *Sol. Energy* **2020**, *207*, 247–269. [CrossRef]
80. Pirvaram, A.; Talebzadeh, N.; Leung, S.N.; O'Brien, P.G. Radiative cooling for buildings: A review of techno-enviro-economics and life-cycle assessment methods. *Renew. Sustain. Energy Rev.* **2022**, *162*, 112415. [CrossRef]
81. Zhang, Q.; Wang, S.; Wang, X.; Jiang, Y.; Li, J.; Xu, W.; Zhu, B.; Zhu, J. Recent progress in daytime radiative cooling: Advanced material designs and applications. *Small Methods* **2022**, *6*, 2101379. [CrossRef] [PubMed]
82. Ding, Z.; Li, X.; Fan, X.; Xu, M.; Zhao, J.; Li, Y.; Xu, H. A review of the development of colored radiative cooling surfaces. *Carbon Capture Sci. Technol.* **2022**, *4*, 100066. [CrossRef]
83. Qi, M.; Wu, T.; Wang, Z.; Wang, Z.; He, B.; Zhang, H.; Liu, Y.; Xin, J.; Zhou, T.; Zhou, X.; et al. Progress in metafibers for sustainable radiative cooling and prospects of achieving thermally drawn metafibers. *Adv. Energy Sustain. Res.* **2021**, *3*, 2100168. [CrossRef]
84. Li, X.; Ding, Z.; Kong, L.; Fan, X.; Li, Y.; Zhao, J.; Pan, L.; Wiersma, D.S.; Pattelli, L.; Xu, H. Recent progress in organic-based radiative cooling materials: Fabrication methods and thermal management properties. *Mater. Adv.* **2023**, *4*, 804–822. [CrossRef]
85. Lim, X. The super-cool materials that send heat to space. *Nature* **2019**, *577*, 18–20. [CrossRef]
86. Fadil, F.; Affandi, N.D.N.; Misnon, M.I.; Bonnia, N.N.; Harun, A.M.; Alam, M.K. Review on electrospun nanofiber-applied products. *Polymers* **2021**, *13*, 2087. [CrossRef]
87. Utel, F.; Cortese, L.; Wiersma, D.S.; Pattelli, L. Optimized White Reflectance in Photonic-Network Structures. *Adv. Opt. Mater.* **2019**, *7*, 1900043. [CrossRef]
88. Tao, D.; Li, X.; Dong, Y.; Zhu, Y.; Yuan, Y.; Ni, Q.; Fu, Y.; Fu, S. Super-low thermal conductivity fibrous nanocomposite membrane of hollow silica/polyacrylonitrile. *Compos. Sci. Technol.* **2020**, *188*, 107992. [CrossRef]
89. Gbewonyo, S.; Carpenter, A.W.; Gause, C.B.; Mucha, N.R.; Zhang, L. Low thermal conductivity carbon fibrous composite nanomaterial enabled by multi-scale porous structure. *Mater. Des.* **2017**, *134*, 218–225. [CrossRef]
90. Aili, A.; Wei, Z.; Chen, Y.; Zhao, D.; Yang, R.; Yin, X. Selection of polymers with functional groups for daytime radiative cooling. *Mater. Today Phys.* **2019**, *10*, 100127. [CrossRef]
91. Sun, Y.; Ji, Y.; Javed, M.; Li, X.; Fan, Z.; Wang, Y.; Cai, Z.; Xu, B. Preparation of passive daytime cooling fabric with the synergistic effect of radiative cooling and evaporative cooling. *Adv. Mater. Technol.* **2021**, *7*, 2100803. [CrossRef]

92. Li, D.; Liu, X.; Li, W.; Lin, Z.; Zhu, B.; Li, Z.; Li, J.; Li, B.; Fan, S.; Xie, J.; et al. Scalable and hierarchically designed polymer film as a selective thermal emitter for high-performance all-day radiative cooling. *Nat. Nanotechnol.* **2020**, *16*, 153–158. [CrossRef] [PubMed]
93. Wei, W.; Zhu, Y.; Li, Q.; Cheng, Z.; Yao, Y.; Zhao, Q.; Zhang, P.; Liu, X.; Chen, Z.; Xu, F.; et al. An Al_2O_3-cellulose acetate-coated textile for human body cooling. *Sol. Energy Mater. Sol. Cells* **2020**, *211*, 110525. [CrossRef]
94. Wang, X.; Zhang, Q.; Wang, S.; Jin, C.; Zhu, B.; Su, Y.; Dong, X.; Liang, J.; Lu, Z.; Zhou, L.; et al. Sub-ambient full-color passive radiative cooling under sunlight based on efficient quantum-dot photoluminescence. *Sci. Bull.* **2022**, *67*, 1874–1881. [CrossRef] [PubMed]
95. Kim, G.; Park, K.; Hwang, K.-J.; Jin, S. Highly sunlight reflective and infrared semi-transparent nanomesh textiles. *ACS Nano* **2021**, *15*, 15962–15971. [CrossRef] [PubMed]
96. Fu, Y.; Yang, J.; Su, Y.; Du, W.; Ma, Y. Daytime passive radiative cooler using porous alumina. *Sol. Energy Mater. Sol. Cells* **2019**, *191*, 50–54. [CrossRef]
97. Dong, S.; Wu, Q.; Zhang, W.; Xia, G.; Yang, L.; Cui, J. Slippery passive radiative cooling supramolecular siloxane coatings. *ACS Appl. Mater. Interfaces* **2022**, *14*, 4571–4578. [CrossRef]
98. Srikant, V.; Clarke, D.R. On the optical band gap of zinc oxide. *J. Appl. Phys.* **1998**, *83*, 5447–5451. [CrossRef]
99. Cai, L.; Song, A.Y.; Li, W.; Hsu, P.; Lin, D.; Catrysse, P.B.; Liu, Y.; Peng, Y.; Chen, J.; Wang, H.; et al. Spectrally selective nanocomposite textile for outdoor personal cooling. *Adv. Mater.* **2018**, *30*, e1802152. [CrossRef]
100. Xue, X.; Qiu, M.; Li, Y.; Zhang, Q.M.; Li, S.; Yang, Z.; Feng, C.; Zhang, W.; Dai, J.; Lei, D.; et al. Creating an eco-friendly building coating with smart subambient radiative cooling. *Adv. Mater.* **2020**, *32*, e1906751. [CrossRef]
101. RAYLEIGH. Scattering of light by resonating molecules. *Nature* **1919**, *104*, 276. [CrossRef]
102. Mie, G. Eiträge zur optik trüber medien, speziell kolloidaler metallösungen. *Ann. Der Phys.* **1908**, *330*, 377–445. [CrossRef]
103. Cheng, N.; Miao, D.; Wang, C.; Lin, Y.; Babar, A.A.; Wang, X.; Wang, Z.; Yu, J.; Ding, B. Nanosphere-structured hierarchically porous pvdf-hfp fabric for passive daytime radiative cooling via one-step water vapor-induced phase separation. *Chem. Eng. J.* **2023**, *460*, 141581. [CrossRef]
104. Zhang, Y.; Yu, J. Scalable and high-performance radiative cooling fabrics through an electrospinning method. *ACS Appl. Mater. Interfaces* **2022**, *14*, 45707–45715. [CrossRef] [PubMed]
105. Meng, X.; Chen, Z.; Qian, C.; Song, Z.; Wang, L.; Li, Q.; Chen, X. Hierarchical superhydrophobic poly(vinylidene fluoride-*co*-hexafluoropropylene) membrane with a bead (SiO_2 nanoparticles)-on-string (nanofibers) structure for all-day passive radiative cooling. *ACS Appl. Mater. Interfaces* **2023**, *15*, 2256–2266. [CrossRef]
106. Song, Y.-N.; Lei, M.-Q.; Han, D.-L.; Huang, Y.-C.; Wang, S.-P.; Shi, J.-Y.; Li, Y.; Xu, L.; Lei, J.; Li, Z.-M. Multifunctional membrane for thermal management applications. *ACS Appl. Mater. Interfaces* **2021**, *13*, 19301–19311. [CrossRef]
107. Li, X.; Ding, Z.; Lio, G.E.; Zhao, J.; Xu, H.; Pattelli, L.; Pan, L.; Li, Y. Strain-adjustable reflectivity of polyurethane nanofiber membrane for thermal management applications. *Chem. Eng. J.* **2023**, *461*, 142095. [CrossRef]

materials

MDPI

Article

Cu and Ni Co-Doped Porous Si Nanowire Networks as High-Performance Anode Materials for Lithium-Ion Batteries

Can Mi [1,2,3,†], Chang Luo [1,†], Zigang Wang [1], Yongguang Zhang [1,*], Shenbo Yang [4] and Zhifeng Wang [1,2,3,*]

1 School of Materials Science and Engineering, Hebei University of Technology, Tianjin 300401, China
2 Key Laboratory for New Type of Functional Materials in Hebei Province, Hebei University of Technology, Tianjin 300401, China
3 Collaborative Innovation Center for Vehicle Lightweighting, Hebei University of Technology, Tianjin 300401, China
4 Hongzhiwei Technology (Shanghai) Co., Ltd., Shanghai 201206, China
* Correspondence: yongguangzhang@hebut.edu.cn (Y.Z.); wangzf@hebut.edu.cn (Z.W.); Tel.: +86-22-60202006 (Z.W.)
† These authors contributed equally to this work.

Abstract: Due to its extremely high theoretical mass specific capacity, silicon is considered to be the most promising anode material for lithium-ion batteries (LIBs). However, serious volume expansion and poor conductivity limit its commercial application. Herein, dealloying treatments of spray dryed Al-Si-Cu-Ni particles are performed to obtain a Cu/Ni co-doped Si-based anode material with a porous nanowire network structure. The porous structure enables the material to adapt to the volume changes in the cycle process. Moreover, the density functional theory (DFT) calculations show that the co-doping of Cu and Ni can improve the capture ability towards Li, which can accelerate the electron migration rate of the material. Based on the above advantages, the as-prepared material presents excellent electrochemical performance, delivering a reversible capacity of 1092.4 mAh g^{-1} after 100 cycles at 100 mA g^{-1}. Even after 500 cycles, it still retains 818.7 mAh g^{-1} at 500 mA g^{-1}. This study is expected to provide ideas for the preparation and optimization of Si-based anodes with good electrochemical performance.

Keywords: dealloying; doping; porous; lithium-ion battery; anode

Citation: Mi, C.; Luo, C.; Wang, Z.; Zhang, Y.; Yang, S.; Wang, Z. Cu and Ni Co-Doped Porous Si Nanowire Networks as High-Performance Anode Materials for Lithium-Ion Batteries. *Materials* **2023**, *16*, 6980. https://doi.org/10.3390/ma16216980

Academic Editor: Christian M. Julien

Received: 5 September 2023
Revised: 29 October 2023
Accepted: 30 October 2023
Published: 31 October 2023

1. Introduction

With the rapid development of the economy, portable electronic products and electric vehicles reveal increasingly high requirements for energy storage devices [1–3]. Lithium-ion batteries (LIBs) have received widespread attention as potential energy storage devices. However, due to the poor theoretical capacity of graphite electrodes (372 mA h g^{-1}), the energy density of lithium-ion batteries is low, which cannot meet the market demand [4,5]. Therefore, it is an urgent task to develop high-capacity anode materials. Silicon (Si) has extremely high theoretical mass specific capacity (~4200 mA h g^{-1}), low potential (<0.4 V vs. Li^+/Li) and is abundant on Earth (the second most abundant element in the Earth's crust) [6], making it the most promising anode material for the next generation of LIBs [7,8]. However, the two main drawbacks during lithiation/delithiation processes are significant volume changes (up to 300–400%) [9–12] and low intrinsic electrical conductivity (10^{-5} to 10^{-3} S cm^{-1}) [13–15], which lead to dramatic structural damage and capacity loss that hinder the practical application of Si-based anodes.

To improve the structural stability and conductivity of Si, researchers from various countries have adopted various strategies, such as reducing the particle size of Si from the micrometer level to the nanometer level. In this situation, various forms of Si nanomaterials, including Si nanowires [16,17], Si nanotubes [18,19], and hollow Si nanospheres [20,21] with reasonable pore space, were synthesized to suppress cracking and fragmentation.

Wang et al. prepared a coral-like network of Si branches with a one-dimensional nanowire structure that can accommodate significant volume changes during fast charging [22]. Due to the need for precision equipment and high synthesis costs in the synthesis process of existing Si nanomaterials, their production scale and practical application are limited [23]. Another way to improve the structural stability and conductivity of the Si anode is through carbon coating [24–26] or graphite mixing [27,28]. However, the addition of a large amount of carbon reduces the theoretical capacity and initial capacity of the Si-based anode [29]. In addition, another effective method to improve the conductivity and ion mobility of Si materials is to dope different elements [30], including boron (B) [31,32] and phosphorus (P) [33] into Si materials. B was doped by Chen et al. into a Si anode to obtain good initial Coulombic efficiency and long cycle stability [31]. In recent years, researchers have utilized different transition metals for doping/alloying/composite with Si anodes, which improve the structural stability and conductivity of materials and significantly improve their overall electrochemical performance [34].

Inspired by the above progress, Cu and Ni co-doped porous Si nanowire networks were obtained by dealloying spray dried Al-Si-Cu-Ni microspheres in this work. Dealloying refers to an etching process in which fewer valuable components are etched away from the precursor, and the remaining parts are self-assembled into various nano-/micro-porous structures. Some structural features of the precursor can be inherited and retained by the dealloying inheritance effect during the etching process [35,36]. At the same time, Al-Si-based powder synthesized by the spray drying method is a common spraying material used to repair aluminum, magnesium and their alloy parts. Both the spray drying method and the dealloying method involved in this paper are suitable for large-scale production [37,38]. Herein, it was found that the addition of Cu and Ni can improve the conductivity of Si-based anodes, enhance the electron transport rate in electrochemical reactions and the cycle stability of the battery. At 100 mAh g^{-1}, a reversible capacity of 1092.4 mA g^{-1} can be maintained for 100 cycles. In addition, according to the density functional theory (DFT) calculations, it was revealed that the addition of copper and nickel promoted the adsorption of lithium atoms, and the adsorption energy was the highest when copper and nickel were co-doped at the same time, thus improved the electrochemical performance. This work has a positive impact on the development of a Si-based anode with low cost and high-performance.

2. Materials and Methods

2.1. Material Synthesis

The sample preparation process is shown in Figure 1. The initial precursor is Al-Si-Cu-Ni microsphere particles synthesized by spray drying method [39–41], which is obtained from China New Metal Materials Technology Co., Ltd., Beijing, China. The fabrication process of the final products mainly includes two chemical dealloying steps with/without ultrasonic oscillation. Firstly, the precursor particles were divided into five parts, and 1 g of Al-Si-Cu-Ni particles were dealloyed in 500 mL HF solution (0.02 M, 0.05 M, 0.1 M) for 4.5 h or in 300 mL HCl liquid (2.5% and 5%) for 5 days using five process parameters, respectively, to obtain the first step of dealloying products. The detailed conditions of the first step of dealloying are listed in Table 1. The related dealloying reaction formula for this step is shown as follows, revealing the etching of Al and the production of H_2.

$$2Al + 6HCl = 2AlCl_3 + 3H_2 \uparrow \quad (1)$$

$$2Al + 6HF = 2AlF_3 + 3H_2 \uparrow \quad (2)$$

The second step of dealloying treatment was performed on 100 mg of the first step dealloying product in 50 mL of 5% HCl, 0.1 M HF, or 0.05 M HF, respectively, for 20 min, accompanied by ultrasonic oscillation. The detailed conditions of the second dealloying step are listed in Table 2.

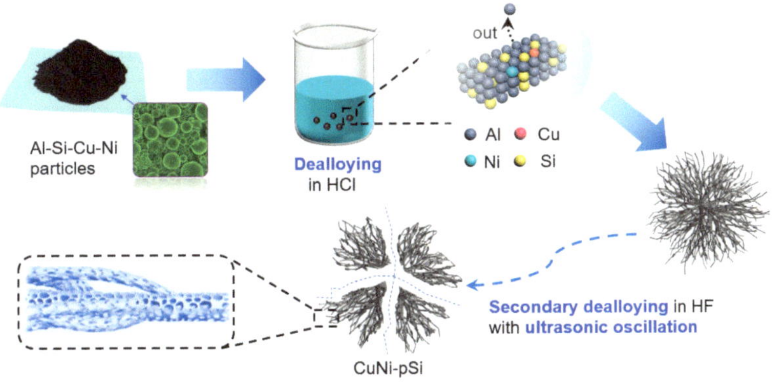

Figure 1. Schematic revealing the preparation process of CuNi-pSi.

Table 1. Treatment conditions of the first step of dealloying process.

Precursor	Corrosive Solutions	Corrosion Time	Sample Name
Al-Si-Cu-Ni particles 1.0 g	0.02 M HF, 500 mL	4.5 h	Sample-1
	0.05 M HF, 500 mL		Sample-2
	0.1 M HF, 500 mL		Sample-3
	2.5% HCl, 300 mL	5 days	Sample-4
	5% HCl, 300 mL		Sample-5

Table 2. Treatment conditions of the second step of dealloying process.

Precursor	Corrosive Solutions	Corrosion Time	Sample Name
Sample-5 100 mg	5% HCl, 50 mL	20 min	Sample-6
	0.1 M HF, 50 mL		Sample-7
	0.05 M HF, 50 mL		CuNi-pSi

All dealloyed samples (CuNi-pSi) were washed three times with anhydrous ethanol and then dried in a vacuum oven at 60 °C overnight for further characterization. The contrast material of porous Si (pSi) was also prepared by the same process as CuNi-pSi through dealloying of $Al_{90}Si_{10}$ microsphere particles.

2.2. Material Characterization

The Cu Kα X-ray diffractometer (XRD, D8 Discover, Bruker, Karlsruhe, Germany) was used to study the phase structure of the samples at 2θ, in the range of 10–90°. The morphology of the samples was observed by scanning electron microscopy (SEM, JSM-6700F, JEOL, Tokyo, Japan) and transmission electron microscopy (TEM, JEM 2100F, JEOL, Tokyo, Japan). The Brunauer Emmett Teller (BET) method and the Barrett Joyner hallenda (BJH) method were used to analyze the specific surface area and pore size distribution of the materials, respectively. To further detect the elemental composition and valence states of the sample, X-ray photoelectron spectroscopy (XPS, V-Sorb 2800P, Beijing, China) was used. Raman spectra were measured in the wavenumber range of 100~2000 cm^{-1}.

2.3. Electrochemical Measurements

The active materials (CuNi-pSi), conductive agent and binder were mixed at a weight ratio of 7:2:1, and then deionized water was added for grinding to make a uniform slurry and coated on the copper foil. After drying at 60 °C for 12 h, the original tablets were cut into

10 mm, and the active substance loaded on each electrode was about 1.1 mg. The electrolyte is 1 M $LiPF_6$ in the mixture of ethylene carbonate (EC) and dimethyl carbonate (DMC) (volume ratio 1:1), and lithium foil is used as the counter electrode. The charge/discharge measurements of the assembled battery were carried out by the NEWARE CT-4000 battery testing system at room temperature. The cyclic voltammetry (CV) curves and EIS spectra were measured on the electrochemical workstation (Princeton: Versa STAT 4).

2.4. DFT Calculations

Based on DFT, all our calculations are performed using DS-PAW software (HZWTECH, version 2022A, Shanghai, China) integrated into the Device Studio program [42,43]. The pure Si surface was constructed by the 2×2 supercell of Si (111) of three layers of atoms. The surface of CuNi-pSi is obtained by replacing Si atoms with Cu and Ni atoms on the surface of Si. In order to avoid the interaction between adjacent configurations, it is necessary to apply a vacuum region of 15 Å. In addition, the energy cutoff was chosen to be 500 eV, and the Monkhorst-Pack of k-points with a $3 \times 3 \times 1$ grid was used to sample the Brillouin zone. The convergence criteria for energy and force calculations are set at 10^{-6} EV and 0.05 EV Å.

3. Results and Discussion

Figure S1a shows the morphology of the Al-Si-Cu-Ni particles with a symmetrical and smooth spherical structure (average diameter of about 10 μm). Figure S1b–f reveals a backscatter electron image (BSEM) and corresponding EDS mapping of the precursor particles, indicating a uniform distribution of Al, Si, Cu, and Ni elements. The atomic ratio of Al:Si:Cu:Ni is about 89.53:10.26:0.08:0.13 based on EDS results (Figure S1g). The SEM images of the samples after the first dealloying step are shown in Figure S2. It can be seen from Figure S2a–d that the corrosion toward sample-1 to sample-4 is insufficient, and rare pores can be formed in these processes. In stark contrast to these samples, Figures S2e and 2a present an obviously porous morphology, implying that the current etching conditions (sample-5) are effective for the formation of porous networks. Because the precursor belongs to a typical hypoeutectic Al–Si based alloy, it mainly contains network eutectic and a large amount of Al. When it is etched under suitable conditions, the aluminum phase in the precursor will be removed, leaving only the eutectic network skeleton. In order to obtain a higher specific surface area, the second step of the dealloying treatment was performed (Figure 2b–e), accompanied by ultrasonic treatment. In this way, most of the Al elements in eutectic nanowires are further etched to create porous nanowires (Figure 2f). At the same time, through ultrasonic vibration, the microspheres collapse into porous network fragments, thus eliminating the internal uncorroded zone. It can be found that among the three different second-step etching processes (Figures S3 and 2e), the experimental parameter of 0.05 M HF and the resulting porous morphology is the optimum. In this situation, narrow ligaments and uniform network structures can be formed. The large number of pores during the cycling process can effectively adapt to the volume changes in the electrode [44,45], which is conducive to electrolyte permeation and rapid ion transport, while the nanowire network is conducive to rapid electron transfer [46]. The SEM images of pSi are shown in Figure S4, showing a similar morphological characteristic as CuNi-pSi.

Figure 2g uncovers the HR–TEM image of the sample after two dealloying processes, showing the crystal plane spacing of 0.31 nm, which is consistent with the Si (1–11) plane [47]. The selected regional electron diffraction (SAED) pattern (Figure 2h) [48] further reveals the Si [−112] zone axis [31]. The EDS mapping shown in Figure 2i–m presents the uniform distribution of Si, Cu, Ni, and Al elements. The atomic ratio of Al:Si:Cu:Ni of the local nanowire based on TEM mapping is about 3.38:95.69:0.46:0.47 (Figure S5) and the residual Al content is similar to that of pSi (Figure S6). Trace doping of Cu and Ni elements may improve electronic conductivity and reaction kinetics during the charge–discharge cycle [49,50], while a small amount of Al can increase conductivity, thereby increasing the lithiation kinetics [51].

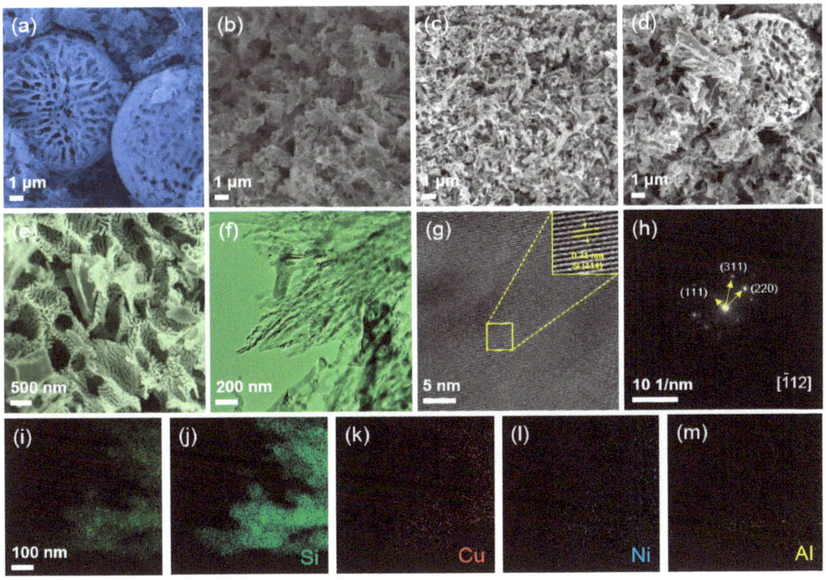

Figure 2. SEM images of the Al-Si-Cu-Ni precursor after the first step of dealloying (Sample-5) (**a**) and the second step of dealloying: (**b**) Sample-6, (**c**) Sample-7, and (**d**,**e**) CuNi-pSi; (**f**) TEM image, (**g**) HRTEM image, and (**h**) SAED patterns of the CuNi-pSi material; (**i**–**m**) elemental mapping of the CuNi-pSi product: (**i**) Mix, (**j**) Si, (**k**) Cu, (**l**) Ni, and (**m**) Al.

The X-ray diffraction (XRD) patterns of the dealloyed CuNi-pSi are displayed in Figure 3a. All the peaks can be indexed to cubic Si (JPCDS 27-1402) [33]. Compared with peak intensities of pSi, that of CuNi-pSi is higher, indicating that the crystallinity of CuNi-pSi is higher. The porous structure of the CuNi-pSi sample was further disclosed by the N_2 adsorption–desorption isotherm curves (Figure 3b). The type III curve with an H3-type hysteresis loop is revealed, indicating the presence of a certain content of mesopores in the product [52]. The specific surface area of the material is about 25.08 $m^2\ g^{-1}$, and the pore size is mainly between 1.7 and 10 nm (Figure 3c). These pores are able to facilitate the permeation of electrolytes and allow for more complete lithiation reactions. Raman spectra of the CuNi-pSi sample are shown in Figure 3d. The dominating peaks at 495 cm^{-1} and the two weak peaks at 280 cm^{-1} and 925 cm^{-1} are typical signals of Si. Compared to the standard silicon peak of 510 cm^{-1}, the peak of CuNi-pSi shifts to 495 cm^{-1}. This shift is due to the stress generation surrounding the Si network after the co-doping of Cu and Ni, resulting in a structural disorder in Si [31]. The elemental valence state of the CuNi-pSi is further investigated by XPS. A typical XPS survey spectrum confirms the presence of Si, Cu, Ni and Al (Figure 3e). Si 2p XPS spectra are displayed in Figure 3f. The peak at 98.74 eV belongs to the Si−Si bond, while weak peaks with the binding energy of 102.84 eV can be ascribed to the Si−O bond [53], indicating the presence of trace Si oxides on the sample surface. However, there is no obvious XRD peak of silicon oxide in Figure 3a, meaning that the content of silicon oxide is very small [54]. The XPS spectra of Ni 2p, Cu 2p, O 1s and Al 2p are shown in Figure S7a–d, respectively. The minimum element proportion threshold at which the XPS device can detect an element signal is about 5%. Because the ratio of Cu and Ni elements in CuNi-pSi is less than 5%, there are weak characteristic peaks of Cu 2p and Ni 2p in the XPS fine spectrum. Moreover, the residual Al exists in the form of Al^{3+} due to unavoidable surface oxidation, while O presents an obvious M–O bond (M refers to Si, Al, Ni and Cu in this work).

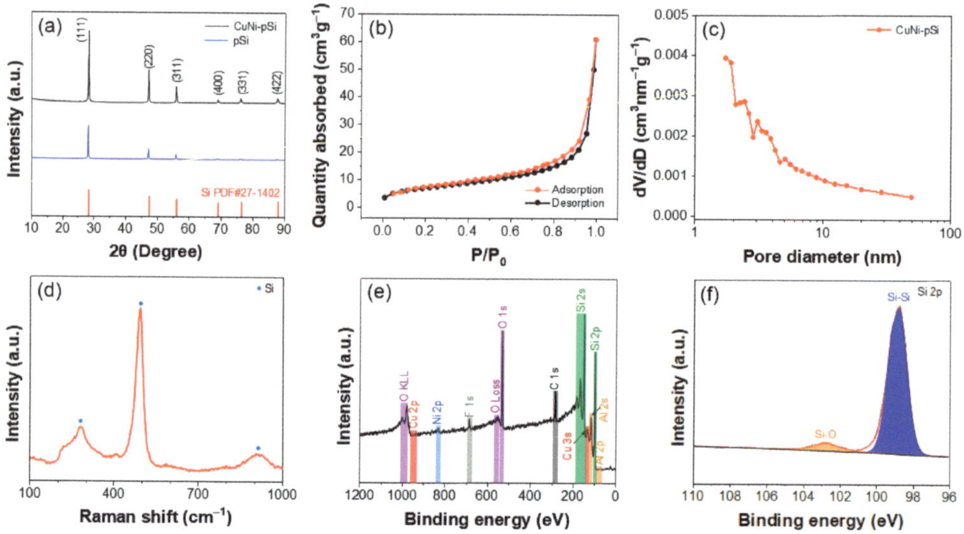

Figure 3. (**a**) XRD patterns of CuNi-pSi and pSi. N_2 adsorption-desorption isotherms of CuNi-pSi (**b**) and corresponding pore-size distribution (**c**). (**d**) Raman spectra of CuNi-pSi, red line: the measured curve. XPS spectra of CuNi-pSi: (**e**) XPS survey spectra, black line: the measured curve, color areas: the peaks of different elements; (**f**) high-resolution core-level spectra of Si 2p, black line: base line, red line: the fitting curve, color areas: different peaks of silicon.

To better elucidate the effect of trace doping of Cu and Ni on electrochemical performance, pSi (Figure S4a,b) was also obtained by the same preparation process and tested as a control sample of CuNi-pSi. Thus, a series of electrochemical tests were conducted. Figure 4a shows the CV curves of the CuNi-pSi electrode in the first five cycles. In the initial discharge process, two broad cathode peaks appeared at 1.55 V and 0.68 V due to the formation of the solid electrolyte interface (SEI). In the charging process, the anode peaks located at around 0.43 V and 0.53 V correspond to the Li separation process from lithium silicide. In the subsequent cycle, the two broad cathode peaks disappeared, indicating that the SEI layer generated on the electrode surface quickly reached stability [55,56]. The cathode peak at around 0.18 V can be attributed to the lithiation of Si. The reversible discharging–charging process of Si can be expressed as the following reaction [2]:

$$Li_xSi \leftrightarrows Si + xLi^+ + xe^- \quad (3)$$

Meanwhile, due to the activation process of Si, an increase in CV peak intensity was observed in the initial few cycles, which can also be observed in previously reported Si-based anode materials [29]. In addition, the curve shows that when the scanning voltage (vs. Li^+/Li) exceeds 2 V, the capacity provided by the battery accounts for a very small proportion of the total capacity. Comparing the CV curves of CuNi-pSi and pSi (Figure S8), the peak positions are basically consistent, and the CuNi-pSi displays a larger integrated area than the pSi, indicating that CuNi-pSi has better electrochemical activity and excellent reversible lithium storage capacity.

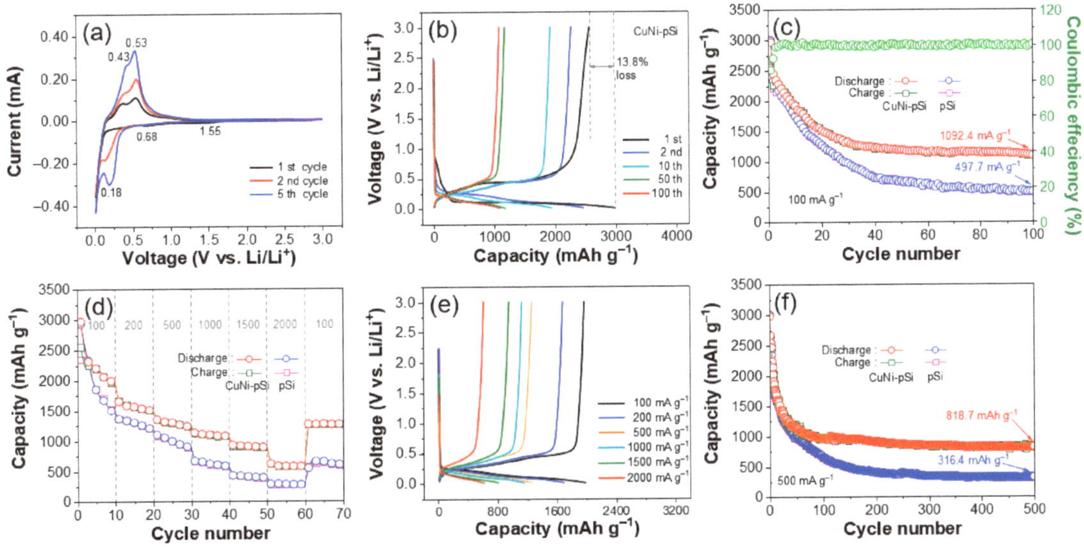

Figure 4. (**a**) CV curves of the CuNi-pSi at 0.1 mV s^{-1}. (**b**) Charge–discharge curves of the CuNi-pSi anode at the current density of 100 mA g^{-1}. (**c**) Cycling performance of the two anodes at the current density of 100 mA g^{-1}. (**d**) Rate performance of the two anodes at various current densities. (**e**) Charge–discharge profiles of the CuNi-pSi anode at various current densities. (**f**) Cycling performance of the two anodes at the current density of 500 mA g^{-1} (the batteries are firstly activated at 100 mA g^{-1} for three cycles).

Figure 4b shows the constant current charge–discharge curves of CuNi-pSi at 100 mA g^{-1}. All voltage platforms in the discharge and charging curves correspond well to the cathode and anode peaks in the CV curve. The discharge plateau in the first cycle is relatively low, which is caused by the first lithiation of Si, representing the stable formation of the SEI layer. The initial discharge capacity of the CuNi-pSi electrode is 2972 mAh g^{-1}, while the charging capacity is 2560 mAh g^{-1}, which is equivalent to a capacity loss of 13.8%. The main reason for the capacity loss in the first cycle is the formation of the SEI layer. As the number of cycles increases, the curve gradually moves to the left, indicating a gradual decrease in capacity. The discharge–charge platform is well maintained after 100 cycles, indicating that the electrochemical process of CuNi-pSi is relatively stable.

The cyclic performance of sample-1 to sample-7 at a current density of 100 mA g^{-1} is shown in Figure S9. It can be clearly seen that after the first dealloying step, sample-5 exhibits the best electrochemical properties among the first five samples, which is related to its good porous structure. In view of the advantages of the three-dimensional structure and electrochemical properties, it is reasonable to use sample-5 as a precursor in the second step of corrosion. The cyclic performance of CuNi-pSi and pSi at 100 mA g^{-1} is shown in Figure 4c. It can be seen that the capacity of CuNi-pSi rapidly decays during the first 30 cycles and gradually stabilizes thereafter. During the initial cycles, the formation of SEI membranes leads to the consumption and loss of some active lithium, resulting in a decrease in capacity. The capacity fade may be recovered through a newly reported reserve lithium-ion battery strategy [57]. Moreover, the porous nanowire network structure provides room to adapt to the volume expansion of Si and eventually reach a stable state. At 100 mA g^{-1}, CuNi-pSi provides a reversible capacity of 1092.4 mA g^{-1} after 100 cycles, while pSi rapidly decays to below 1000 mA g^{-1} after only 30 cycles. It is evident that the CuNi-pSi porous nanowire networks doped with Cu and Ni exhibit more stable cycling performance compared with sample-1 to sample-7 and pSi. One possible reason is that the slow electrochemical kinetics of Si leads to a large amount of lithium detained in the Si anode [58], while the addition of Cu and Ni accelerates the electrochemical kinetics of the Si-based anode.

Figure 4d shows the rate performance of the two anodes at current densities from 100 to 2000 mA g^{-1}. The reversible capacities of the CuNi-pSi anode at 100, 200, 500, 1000, 1500 and 2000 mA g^{-1} are 2203.54, 1569.75, 1301.36, 1098.33, 906.19 and 575.05 mAh g^{-1}, respectively. As the current density restores to 100 mA g^{-1}, the discharge capacity returns to 1264 mAh g^{-1}, indicating that CuNi-pSi has good reversibility. Accordingly, the pSi anode maintains a capacity of 274 mAh g^{-1} at 2000 mA g^{-1}. When the current density is restored to 100 mA g^{-1}, only 602.15 mAh g^{-1} reversible capacity can be maintained. Moreover, CuNi-pSi has a higher capacity compared to pSi at different current densities, uncovering that CuNi-pSi has a better rate performance and reversibility. The comparison of the rate capability of the CuNi-pSi and pSi anode can be more clearly observed in the charge–discharge curve at different current densities shown in Figures 4e and S8b. It is obvious that as the current density gradually increases from 100 mA g^{-1} to 2000 mA g^{-1}, the curves gradually shift to the left, and the reversible capacity decreases with the increase in current density. The distance between the charge and discharge curves of CuNi-pSi at different current densities is smaller than that of pSi, indicating that CuNi-pSi has better rate performance.

Figure 4f shows the cycling performance of CuNi-pSi and pSi anodes. The batteries are firstly activated at 100 mA g^{-1} for three cycles and then cycled at 500 mA g^{-1} for four hundred and ninety-seven cycles. CuNi-pSi maintains a reversible capacity of 818 mAh g^{-1} after 500 cycles, while pSi experiences severe capacity decay, maintaining only 316.4 mAh g^{-1} after 500 cycles. The excellent cycling performance of CuNi-pSi can be attributed to the better conductivity of the Si-based anode due to the addition of Cu and Ni. A small amount of Cu and Ni atoms are embedded in Si, suppressing the volume and morphology changes in the silicon anode, reducing the drastic volume changes in silicon, preventing the pulverization of silicon anodes, and maintaining the nanowire network structure [59].

To further understand the electrochemical behavior of the two anodes, the electrode morphology after cycling for 100 cycles at 100 mA g^{-1} was characterized. Cross-sectional and typical top-view SEM images of the CuNi-pSi electrode after cycling are shown in Figure 5a,c. During cycling, due to changes in electrode volume, some porous Si nanowires gradually merged together, but the overall structural integrity of the electrode remained good without fracture. The thickness of the active materials is about 12.3 nm. Figure 5b,d presents the cross-sectional and top-view SEM images of the pSi electrode after cycling. Obvious cracks can be observed, and the thickness of the active materials is about 29.7 μm, which is more than twice as thick as that of CuNi-pSi. This proves that the CuNi-pSi electrode can adapt to the volume change in the process of cycling and can keep a relatively low expansion rate. Figure 5e shows the EIS diagram and equivalent circuit (inset) of two fresh samples. The EIS curve shows a semicircle in the high-frequency region and a slope line in the low-frequency region, corresponding to the charge transfer resistance and ion diffusion resistance, respectively [60]. The resistance of CuNi-pSi is smaller than that of pSi, which means that the addition of copper and nickel increases the conductivity of the anode. The EIS diagram and equivalent circuit of the sample after cycling for 100 cycles are shown in Figure 5f. Compared with the pSi anode, the resistance of the CuNi-pSi anode increases less after cycling. This is because the co-doping of Cu and Ni can reduce the huge stress caused by the volume change during the cycling process to ensure the integrity of the electrode so that the anode presents better cycle stability. Furthermore, during the cycling process, the expansion rate of Al after lithium absorption (94%) is lower than that of Si (280%), which can effectively prevent the lattice expansion of Si and improve its performance as an anode material for lithium-ion batteries [51]. As shown in Figure 5g, in order to investigate the usability of the Si anode, the battery was connected to a light bulb, which emitted bright light and did not significantly darken after 30 min, representing the enormous application prospects of the CuNi-pSi anode.

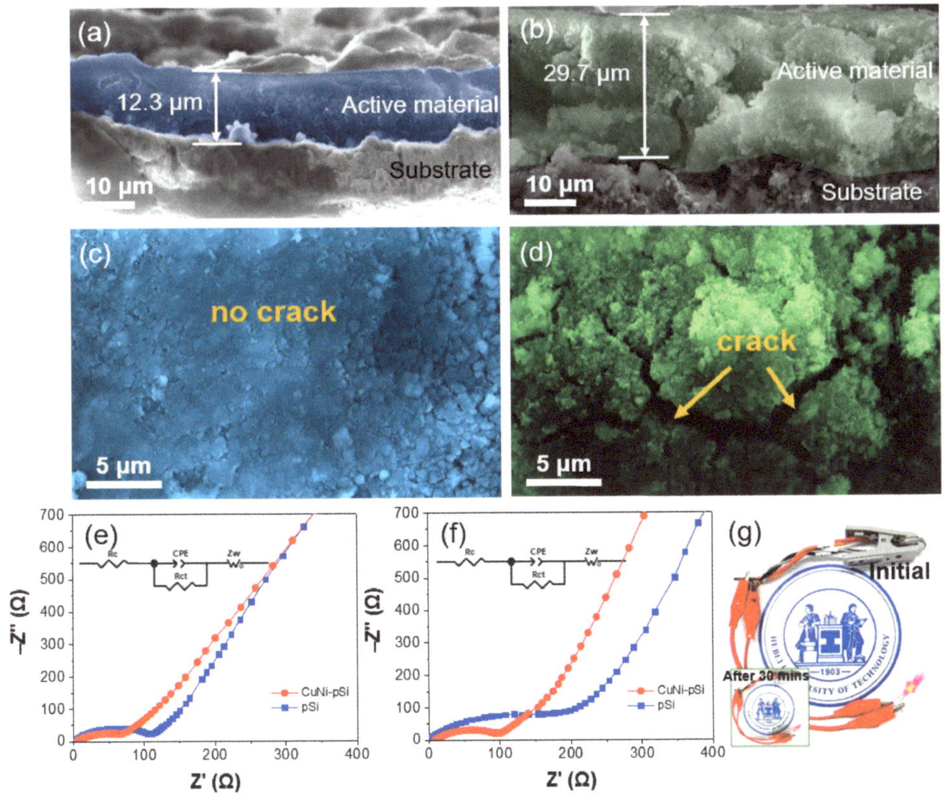

Figure 5. Cross-sectional and typical top-view SEM images of (**a**,**c**) CuNi-pSi and (**b**,**d**) pSi after cycling at the current density of 100 mA g^{-1} for 100 cycles. EIS spectra equivalent circuit (inset) of the CuNi-pSi and pSi anodes: (**e**) Fresh, (**f**) After cycling at the current density of 100 mA g^{-1} for 100 cycles. (**g**) LED bulb powered by the battery with CuNi-pSi anodes.

To further analyze the effects of Cu and Ni doping on lithium-ion reaction kinetics and material conductivity, DFT calculations were performed. Four different models were constructed, including pure Si, single Cu-doped Si (Cu-pSi), single Ni-doped Si (Ni-pSi), and Cu/Ni co-doped Si (CuNi-pSi). The models of four kinds of materials after adsorbing lithium-ions are shown in Figure 6a. The calculation formula of adsorption energy is [61]:

$$E_{ad} = E_{SiLi} - E_{Si} - E_{Li} \quad (4)$$

The adsorption energies of pSi, Cu-pSi, Ni-pSi, and CuNi-pSi are −1.58, −1.89, −1.93, and −2.16 eV, respectively (Figure 6b). The lower adsorption energy after doping indicates that the addition of Cu and/or Ni can promote the adsorption of lithium atoms. The doping effect of Ni toward Si is slightly higher than that of Cu. Moreover, when Cu and Ni are co-doped into Si, the lowest adsorption energy and best absorption efficiency can be obtained. Figure 6c shows the projected density of state (PDOS) of pSi, Cu-pSi, Ni-pSi, and CuNi-pSi; the PDOS diagram of pure Si shows a clear bandgap (0.3 eV), which is a classic semiconductor characteristic with low conductivity. In contrast, the bandgap of the doped PDOS pattern disappears, indicating an improvement in material conductivity. Near the Fermi levels, the contribution of Ni is higher than that of Cu, which is consistent with the adsorption energy, confirming that the doping of copper and nickel can regulate the electronic structure and further promote the storage of Li^+. To better understand the influence of Cu and Ni doping on the electrochemical performance, the differential charge densities of pSi, Cu-pSi, Ni-pSi, and CuNi-pSi are calculated, as shown in Figure 6d. The figure clearly shows the charge transfer of Si materials after the adsorption of lithium atoms. The yellow and cyan areas represent the increase and decrease in charge density, respectively. It can be seen from the figure that the addition of Cu and Ni makes a stronger charge transfer of lithium. In order to quantify the amount of electron transfer, Bader charge analysis (Table S1) was conducted. The results showed that the electron numbers transferred by lithium atoms in Si, Cu-pSi, Ni-pSi, and CuNi-pSi were 0.8653, 0.8696, 0.8748, and 0.8821 e, respectively. CuNi-pSi presents the highest charge transfer amount, and the charge transfer amount of Ni-pSi is higher than that of Cu-pSi, which corresponds to the adsorption energy, indicating that Ni doping plays a slightly greater contribution. From the above data, it can be reasonably inferred that the doping of copper and nickel enables Si-based anodes to achieve superior Li capture ability, enhance the surface reactivity and conductivity of the anode, and thus achieve excellent cycle life and rate performance. Moreover, CuNi-pSi contains a small amount of aluminum, which can shift the Fermi level guide band and valence band, increasing the conductivity of Si.

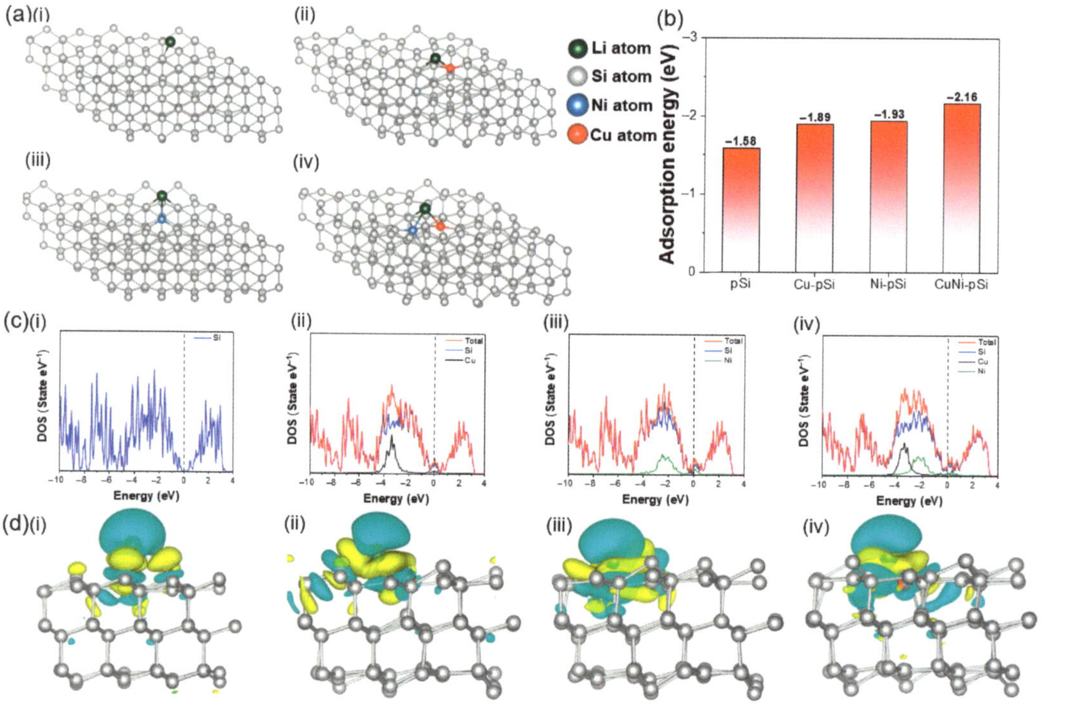

Figure 6. (**a**) optimized the adsorption of Li, (**b**)calculated the adsorption energy of Li on the surface, (**c**) projected DOS (PDOS) and (**d**) differential charge density of (**i**) pSi, (**ii**) Cu-pSi, (**iii**) Ni-pSi and (**iv**) CuNi-pSi.

In many dealloying reactions, the aluminum element is not completely removed. A small amount of its residue is generally considered not to have much effect on the electrochemical properties of LIBs. On the one hand, residual aluminum may cause capacity reduction, but the impact is limited. The distribution of these residual aluminum elements on the silicon ligament is relatively uniform, while not all aluminum is enriched on the surface of silicon. Only a small amount of aluminum is located on the surface of the ligament, causing limited capacity reduction by irreversible Al–Li alloys, while most aluminum is encased in silicon ligaments and does not react with lithium directly. On the other hand, trace amounts of aluminum may increase the electrical conductivity of silicon, thereby improving electrode performance. According to the above discussion, the positive and negative effects of aluminum on the material cancel each other out and ultimately do not have much impact on the electrochemical properties of the material.

Table 3 [48,56,59,62–68] shows the comparison of the electrochemical performance of CuNi-pSi and the previously reported Si-based anode. It can be found that CuNi-pSi displays excellent lithium storage performance in different studies. The main problem of the Si-based anode is that the continuous insertion and extraction of Li ions in the cycle process produce large mechanical stress in the Si material, resulting in material pulverization and electrode cracking. Moreover, a large number of volume fluctuations also make the fragile SEI unbearable, resulting in the unstable and continuous growth of the SEI layer. The CuNi-pSi presents good performance, which can be attributed to the following reasons. Firstly, CuNi-pSi has a porous nanowire network structure with rich porosity, which provides sufficient space for the expansion of the Si anode. Secondly, the addition of Cu and Ni can provide support to deal with the mechanical stress in the cycle process and maintain the integrity of the electrode structure. Thirdly, Cu and Ni are conductors, which can improve the conductivity of the Si anode. They not only improve the cycle stability of the battery but also improve the rate performance. In this paper, based on the structure adjustment strategy of a two-step dealloying process, Cu and Ni co-doping have been successfully implemented, which has a positive impact on the development of low-cost Si-based anodes.

Table 3. The comparison of electrochemical performance of Si-based anodes.

Anode Material	Current Density ($mA\ g^{-1}$)	Cycle Number	Reversible Capacity ($mAh\ g^{-1}$)	Reference
Si	100	100	1015	[62]
Si	100	100	980	[61]
Si	500	200	993	[56]
Si/C-P	200	200	883.4	[59]
Si/SiO_x	200	53	492	[64]
Si/N	200	150	690	[65]
Si/C	210	200	1110	[48]
Si/C	100	100	934	[66]
SiOC	500	300	501	[67]
$Mg\text{-}Si/SiO_x$	150	100	490	[68]
CuNi-pSi	100	100	1092.4	This work
	500	500	818.7	

Although the current preparation method uses acid and the processing time is long, which makes the process seem less green and time-consuming, there are many opportunities to optimize the existing process and reduce the use of acid and processing time. In addition, the technological route in this paper still has some advantages, as follows. Firstly, the samples were prepared by spray drying and dealloying methods, both of which are suitable for large-scale production. Secondly, the products fabricated by the related process present a good three-dimensional skeleton structure and rich hierarchical porous structure. The pores in the nanowire network structure and the pores on the nanowire are conducive to electrolyte permeation, allowing for more complete lithiation reactions and adapting to the significant volume changes in Si during cycling. Nanowire networks can establish rich three-dimensional channels, reduce ion transport distance, enhance reaction kinetics, and improve the electrochemical performance of Si anodes. As a result, the material shows promising electrochemical performance.

4. Conclusions

In summary, Cu and Ni co-doped porous Si nanowire networks were prepared by a two-step dealloying method and were used as Li-ion battery anodes, revealing an excellent electrochemical performance of 1092.4 mAh g^{-1} after 100 cycles at 100 mA g^{-1}. The reversible capacity of 818.7 mAh g^{-1} can be retained even after 500 cycles at 500 mA g^{-1}. DFT calculations show that the addition of copper and nickel can improve the adsorption capacity of the anode toward Li and can improve the conductivity and electrochemical reactivity of the Si material. The combination of a porous nanowire network structure and Cu/Ni co-doping contributes to good lithium storage performance, which is expected to promote the exploration of Si-based anode with low cost and high-performance.

Supplementary Materials: The following supporting information can be downloaded at: https://www.mdpi.com/article/10.3390/ma16216980/s1; Figure S1: Morphology and EDS of Al-Si-Cu-Ni particles; Figure S2: SEM images after the first step of dealloying; Figure S3: SEM images after the second step of dealloying; Figure S4: SEM images of pSi; Figure S5: EDS analysis of CuNi-pSi; Figure S6: EDS analysis of pSi; Figure S7: XPS spectra of CuNi-pSi; Figure S8: CV curves and Charge-discharge profiles at various current densities of pSi; Figure S9: Cycling performance of contrast materials. Table S1: Bader charge for pSi, Cu-pSi, Ni-pSi and CuNi-pSi.

Author Contributions: Conceptualization, Y.Z. and Z.W. (Zhifeng Wang); methodology, C.L., Y.Z. and Z.W. (Zhifeng Wang); investigation, C.M. and Z.W. (Zigang Wang), C.L. and S.Y.; data curation, C.M., C.L. and Z.W. (Zigang Wang); writing—original draft preparation, C.M. and C.L.; writing—review and editing, Y.Z. and Z.W. (Zhifeng Wang); visualization, C.M., Z.W. (Zigang Wang) and S.Y.; supervision, S.Y. and Y.Z.; project administration, Z.W. (Zhifeng Wang); funding acquisition, Z.W. (Zhifeng Wang). All authors have read and agreed to the published version of the manuscript.

Funding: The authors would like to acknowledge the financial support from the Science and Technology Project of Tianjin, China (S23TPD106), the Natural Science Foundation of Hebei Province, China (E2023202253), and the Hebei Higher Education Teaching Reform Research and Practice Project, China (2021GJJG050).

Informed Consent Statement: Not applicable.

Data Availability Statement: The data presented in this study are available upon request from the corresponding author.

Acknowledgments: The authors would like to thank HZWTECH for providing computation facilities and software.

Conflicts of Interest: The authors declare no conflict of interest.

References

1. Farahmandjou, M.; Zhao, S.; Lai, W.; Sun, B.; Notten, P.; Wang, G. Oxygen redox chemistry in lithium-rich cathode materials for Li-ion batteries: Understanding from atomic structure to nano-engineering. *Nano Mater. Sci.* **2022**, *4*, 322–338. [CrossRef]
2. Eshetu, G.G.; Zhang, H.; Judez, X.; Adenusi, H.; Armand, M.; Passerini, S.; Figgemeier, E. Production of high-energy li-ion batteries comprising silicon-containing anodes and insertion-type cathodes. *Nat. Commun.* **2021**, *12*, 5459. [CrossRef]
3. Wang, Z.; Yan, Y.; Zhang, Y.; Chen, Y.; Peng, X.; Wang, X.; Zhao, W.; Qin, C.; Liu, Q.; Liu, X.; et al. Single-atomic Co-B_2N_2 sites anchored on carbon nanotube arrays promote lithium polysulfide conversion in lithium-sulfur batteries. *Carbon Energy* **2023**, e306. [CrossRef]
4. Li, P.; Kim, H.; Myung, S.T.; Sun, Y.K. Diverting exploration of silicon anode into practical way: A review focused on silicon-graphite composite for lithium ion batteries. *Energy Storage Mater.* **2021**, *35*, 550–576. [CrossRef]
5. Wang, Z.; Wang, H.; Liu, X.; Chen, Y.; Zhao, Y.; Zhang, Y.; Han, Q.; Qin, C.; Bakenov, Z.; Wang, Y.; et al. Single Zn atoms anchored on hollow carbon nanofiber network for dendrite-free lithium metal anode of flexible Li–S full cell. *Rare Met.* **2023**, *42*, 3705–3717. [CrossRef]
6. Chae, S.; Choi, S.H.; Kim, N.; Sung, J.; Cho, J. Integration of graphite and silicon anodes for the commercialization of high-energy lithium-ion batteries. *Angew. Chem. Int. Ed. Engl.* **2020**, *59*, 110–135. [CrossRef]
7. Wang, N.; Liu, Y.; Shi, Z.; Yu, Z.; Duan, H.; Fang, S.; Yang, J.; Wang, X. Electrolytic silicon/graphite composite from SiO_2/graphite porous electrode in molten salts as a negative electrode material for lithium-ion batteries. *Rare Met.* **2022**, *41*, 438–447. [CrossRef]
8. Yan, Y.J.; Chen, Y.X.; Li, Y.Y.; Wu, X.Y.; Jin, C.; Wang, Z.F. Synthesis of Si/Fe_2O_3-anchored rGO frameworks as high-performance anodes for Li-ion batteries. *Int. J. Mol. Sci.* **2021**, *22*, 11041. [CrossRef]

9. Zhang, F.; Ma, Y.; Jiang, M.; Luo, W.; Yang, J. Boron heteroatom-doped silicon-carbon peanut-like composites enables long life lithium-ion batteries. *Rare Met.* **2022**, *41*, 1276–1283. [CrossRef]
10. McDowell, M.T.; Lee, S.W.; Nix, W.D.; Cui, Y. 25th anniversary article: Understanding the lithiation of silicon and other alloying anodes for iithium-ion batteries. *Adv. Mater.* **2013**, *25*, 4966–4984. [CrossRef]
11. Nguyen Thi, X.M.; Le, K.M.; Phung, Q.; Truong, D.Q.; Nguyen, H.V.; Nguyen, Q.N.; Huynh, T.T.K.; Pham, L.T.; Van, M.T.; Le, P.M.L. Improving the electrochemical performance of lithium-ion battery using silica/carbon anode through prelithiation techniques. *Battery Energy* **2023**, *2*, 20230003. [CrossRef]
12. Ashuri, M.; He, Q.; Liu, Y.; Shaw, L.L. Investigation towards scalable processing of silicon/graphite nanocomposite anodes with good cycle stability and specific capacity. *Nano Mater. Sci.* **2020**, *2*, 297–308. [CrossRef]
13. Casimir, A.; Zhang, H.G.; Ogoke, O.; Amine, J.C.; Lu, J.; Wu, G. Silicon-based anodes for lithium-ion batteries: Effectiveness of materials synthesis and electrode preparation. *Nano Energy* **2016**, *27*, 359–376. [CrossRef]
14. Roy, P.; Srivastava, S.K. Nanostructured anode materials for lithium ion batteries. *J. Mater. Chem. A* **2015**, *3*, 2454–2484. [CrossRef]
15. Gobena, H.T.; Lai, S.Y.; Koposov, A.Y.; Maehlen, J.P.; Ghamouss, F.; Lemordant, D. Cycling performance of silicon-carbon composite anodes enhanced through phosphate surface treatment. *Battery Energy* **2023**, *2*, 20220062. [CrossRef]
16. Kim, D.; Li, N.; Sheehan, C.J.; Yoo, J. Degradation of Si/Ge core/shell nanowire heterostructures during lithiation and delithiation at 0.8 and 20 A g^{-1}. *Nanoscale* **2018**, *10*, 7343–7351. [CrossRef] [PubMed]
17. Rashad, M.; Geaney, H. Vapor-solid-solid growth of silicon nanowires using magnesium seeds and their electrochemical performance in li-ion battery anodes. *Chem. Eng. J.* **2023**, *452*, 139397. [CrossRef]
18. Biserni, E.; Scarpellini, A.; Li Bassi, A.; Bruno, P.; Zhou, Y.; Xie, M. High-performance flexible nanoporous Si-carbon nanotube paper anodes for micro-battery applications. *Nanotechnology* **2016**, *27*, 245401. [CrossRef] [PubMed]
19. Rahman, M.A.; Song, G.S.; Bhatt, A.I.; Wong, Y.C.; Wen, C.E. Nanostructured silicon anodes for high-performance lithium-ion batteries. *Adv. Funct. Mater.* **2016**, *25*, 647–678. [CrossRef]
20. Kim, H.; Baek, J.; Son, D.K.; Raj, M.R.; Lee, G.B. Hollow porous N and Co dual-doped silicon@carbon nanocube derived by ZnCo-bimetallic metal-organic framework toward advanced lithium-ion battery anodes. *ACS Appl. Mater. Interfaces* **2022**, *14*, 45458–45475. [CrossRef]
21. Ashuri, M.; He, Q.R.; Liu, Y.Z.; Zhang, K.; Emani, S.; Sawicki, M.S.; Shamie, J.S.; Shaw, L.L. Hollow silicon nanospheres encapsulated with a thin carbon shell: An electrochemical study. *Electrochim. Acta* **2016**, *215*, 126–141. [CrossRef]
22. Wang, B.; Ryu, J.; Choi, S.; Zhang, X.; Pribat, D.; Li, X.; Zhi, L.; Park, S.; Ruoff, R. Ultrafast-charging silicon-based coral-like network anodes for lithium-ion batteries with high energy and power densities. *ACS Nano* **2019**, *13*, 2307–2315. [CrossRef]
23. Lee, J.H.; Yoon, C.S.; Hwang, J.Y.; Kim, S.J.; Maglia, F.; Lamp, P.; Myung, S.T.; Sun, Y.K. High-energy-density lithium-ion battery using a carbon-nanotube-Si composite anode and a compositionally graded $Li[Ni_{0.85}Co_{0.05}Mn_{0.10}]O_2$ cathode. *Energy Environ. Sci.* **2016**, *9*, 2152–2158. [CrossRef]
24. Prakash, S.; Zhang, C.F.; Park, J.D.; Razmjooei, F.; Yu, J.S. Silicon core-mesoporous shell carbon spheres as high stability lithium-ion battery anode. *J. Collid. Interface Sci.* **2018**, *534*, 47–54. [CrossRef]
25. Chaudhari, K.N.; Rajeev, K.K.; Kim, S.; Nersisyan, H.; Kirakosyan, A.; Jang, W.; Choi, J.; Lee, J.H.; Kim, T.H.; Kim, Y. Performance enhancement of carbon-coated Si nanoparticles for lithium-ion batteries through the generation of lithophilic sites by a simple oxidation process. *Appl. Surf. Sci.* **2022**, *602*, 154361. [CrossRef]
26. Hwa, Y.; Kim, W.S.; Hong, S.H.; Sohn, H.J. High capacity and rate capability of core-shell structured nano-Si/C anode for Li-ion batteries. *Electrochim. Acta* **2012**, *71*, 201–205. [CrossRef]
27. Sarang, K.T.; Li, X.Y.; Miranda, A.; Terlier, T.; Oh, E.S.; Verduzco, R.; Lutkenhaus, J.L. Tannic acid as a small-molecule binder for silicon anodes. *ACS Appl. Energy Mater.* **2020**, *3*, 6958–6994. [CrossRef]
28. Parekh, M.H.; Sediako, A.D.; Naseri, A.; Thomson, M.J.; Pol, V.G. In situ mechanistic elucidation of superior Si-C-graphite Li-ion battery anode formation with thermal safety aspects. *Adv. Energy Mater.* **2020**, *10*, 1902799. [CrossRef]
29. Wang, M.S.; Wang, G.L.; Wang, S.; Zhang, J.; Wang, J.; Zhong, W.; Tang, F.; Yang, Z.L.; Zheng, J.; Li, X. In situ catalytic growth 3D multi-layers graphene sheets coated nano-silicon anode for high performance lithium-ion batteries. *Chem. Eng. J.* **2019**, *356*, 895–903. [CrossRef]
30. Ke, C.; Liu, F.; Zheng, Z.; Zhang, H.; Cai, M.; Li, M.; Yan, Q.; Chen, H.; Zhang, Q. Boosting lithium storage performance of Si nanoparticles via thin carbon and nitrogen/phosphorus co-doped two-dimensional carbon sheet dual encapsulation. *Rare Met.* **2022**, *4*, 1347–1356. [CrossRef]
31. Cho, S.; Jung, W.; Jung, G.Y.; Eom, K. High-performance boron-doped silicon micron-rod anode fabricated using a mass-producible lithography method for a lithium ion battery. *J. Power Sources* **2020**, *454*, 227931. [CrossRef]
32. Rousselot, S.; Gauthier, M.; Mazouzi, D.; Lestriez, B.; Guyomard, D.; Roué, L. Synthesis of boron-doped Si particles by ball milling and application in Li-ion batteries. *J. Power Sources* **2012**, *202*, 262–268. [CrossRef]
33. Im, J.; Kwon, J.D.; Kim, D.H.; Yoon, S.; Cho, K.Y. P-doped SiO_x/Si/SiO_x sandwich anode for Li-ion batteries to achieve high initial coulombic efficiency and low capacity decay. *Small Methods* **2021**, *6*, 2101052. [CrossRef] [PubMed]
34. Nulu, A.; Nulu, V.; Sohn, K.Y. Influence of transition metal doping on nano silicon anodes for Li-ion energy storage applications. *J. Alloys Compd.* **2022**, *911*, 164976. [CrossRef]

35. Wang, Y.; Zhang, L.; Yin, K.; Zhang, J.; Gao, H.; Liu, N.; Peng, Z.; Zhang, Z. Nanoporous iridium-based alloy nanowires as highly efficient electrocatalysts toward acidic oxygen evolution reaction. *ACS Appl. Mater. Interfaces* **2019**, *43*, 39728–39736. [CrossRef] [PubMed]
36. Dong, C.; Kou, T.; Gao, H.; Peng, Z.; Zhang, Z. Eutectic-derived mesoporous Ni-Fe-O nanowire network catalyzing oxygen evolution and overall water splitting. *Adv. Energy Mater.* **2018**, *8*, 1701347. [CrossRef]
37. Wang, Z.; Fei, P.; Xiong, H.; Qin, C.; Zhao, W.; Liu, X. $CoFe_2O_4$ nanoplates synthesized by dealloying method as high performance Li-ion battery anodes. *Electrochim. Acta* **2017**, *252*, 295–305. [CrossRef]
38. Wang, Z.; Zhang, X.; Yan, Y.; Zhang, Y.; Wang, Y.; Qin, C.; Bakenov, Z. Nanoporous $GeO_2/Cu/Cu_2O$ network synthesized by dealloying method for stable Li-ion storage. *Electrochim. Acta* **2019**, *300*, 363–372. [CrossRef]
39. Kim, S.Y.; Kim, C.H.; Yang, C.M. Binder-free silicon anodes wrapped in multiple graphene shells for high-performance lithium-ion batteries. *J. Power Sources* **2021**, *486*, 229350. [CrossRef]
40. Chae, S.; Ko, M.; Park, S.; Kim, N.; Ma, J.; Cho, J. Micron-sized Fe-Cu-Si ternary composite anodes for high energy Li-ion batteries. *Energy Environ. Sci.* **2016**, *9*, 1251–1257. [CrossRef]
41. Men, X.; Wang, T.; Xu, B.H.; Kong, Z.; Liu, X.H.; Fu, A.P.; Li, Y.H.; Guo, P.Z.; Guo, Y.G.; Li, H.L. Hierarchically structured microspheres consisting of carbon coated silicon nanocomposites with controlled porosity as superior anode material for lithium-ion batteries. *Electrochim. Acta* **2019**, *324*, 134850. [CrossRef]
42. Hongzhiwei Technology, Device Studio, Version 2022A, China. 2022. Available online: https://iresearch.net.cn/cloudSoftware (accessed on 7 November 2022).
43. Blöchl, P.E. Projector augmented-wave method. *Phys. Rev. B* **1994**, *50*, 17953. [CrossRef]
44. Sohn, M.; Lee, D.G.; Park, H.I.; Park, C.; Choi, J.H.; Kim, H. Microstructure controlled porous silicon particles as a high capacity lithium storage material via dual step pore engineering. *Adv. Funct. Mater.* **2018**, *28*, 1800855. [CrossRef]
45. Sohn, M.; Kim, D.S.; Park, H.I.; Kim, J.H.; Kim, H. Porous silicon-carbon composite materials engineered by simultaneous alkaline etching for high-capacity lithium storage. *Electrochim. Acta* **2016**, *196*, 197–205. [CrossRef]
46. Wang, Z.; Zhang, X.; Liu, X.; Zhang, W.; Zhang, Y.; Li, Y.; Qin, C.; Zhao, W.; Bakenov, Z. Dual-network nanoporous $NiFe_2O_4/NiO$ composites for high performance Li-ion battery anodes. *Chem. Eng. J.* **2020**, *388*, 124207. [CrossRef]
47. Kim, N.; Park, H.; Yoon, N.; Lee, J.K. Zeolite-templated mesoporous silicon particles for advanced lithium-ion battery anodes. *ACS Nano* **2018**, *12*, 3853–3864. [CrossRef] [PubMed]
48. Sun, J.; Shi, J.; Ban, B.; Li, J.; Wei, M.; Wang, Q.; Chen, J. Porous Si/C anode materials by Al-Si dealloying method with PEA surfactant assisted cross-linked carbon coating for lithium-ion battery applications. *Electrochim. Acta* **2019**, *327*, 134995. [CrossRef]
49. Mulaudzi, I.; Zhang, Y.; Ndlovu, G.F.; Wu, Y.P.; Legodi, M.A.; van Ree, T. Copper doped Li_3VO_4 as anode material for lithium-ion batteries. *Electroanalysis* **2020**, *32*, 2635–2641. [CrossRef]
50. Fatile, B.O.; Pugh, M.; Medraj, M. Nickel-doped $Nb_{18}W_{16}O_{93}$ nanowires with improved electrochemical properties for lithium-ion battery anodes. *Mater. Chem. Phys.* **2023**, *307*, 128179. [CrossRef]
51. Tian, H.; Tan, X.; Xin, F.; Wang, C.; Han, W. Micro-sized nano-porous Si/C anodes for lithium ion batteries. *Nano Energy* **2015**, *11*, 490–499. [CrossRef]
52. Dong, L.; Wang, Z.; Li, Y.; Jin, C.; Dong, F.; Zhao, W.; Qin, C.; Wang, Z. Spinel-structured, multi-component transition metal oxide $(Ni,Co,Mn)Fe_2O_{4-x}$ as long-life lithium-ion battery anode material. *Batteries* **2023**, *9*, 54. [CrossRef]
53. Park, C.M.; Choi, W.; Hwa, Y.; Kim, J.H.; Jeong, G.; Sohn, H.J. Characterizations and electrochemical behaviors of disproportionated SiO and its composite for rechargeable Li-ion batteries. *J. Mater. Chem.* **2010**, *20*, 4854–4860. [CrossRef]
54. Zeng, Y.; Huang, Y.; Liu, N.; Wang, X.; Zhang, Y.; Guo, Y.; Wu, H.-H.; Chen, H.; Tang, X.; Zhang, Q. N-doped porous carbon nanofibers sheathed pumpkin-like Si/C composites as free-standing anodes for lithium-ion batteries. *J. Energy Chem.* **2021**, *54*, 727–735. [CrossRef]
55. Ashuri, M.; He, Q.R.; Shaw, L.L. Silicon as a potential anode material for Li-ion batteries: Where size, geometry and structure matter. *Nanoscale* **2016**, *8*, 74–103. [CrossRef] [PubMed]
56. Cao, W.; Chen, M.; Liu, Y.; Han, K.; Chen, X.; Ye, H.; Sang, S. $C_2H_2O_4$ etching of AlSi alloy powder:an efficient and mild preparation approach for high performance micro Si anode. *Electrochim. Acta* **2019**, *320*, 134615. [CrossRef]
57. Parekh, M.H.; Palanisamy, M.; Pol, V.G. Reserve lithium-ion batteries: Deciphering in situ lithiation of lithium-ion free vanadium pentoxide cathode with graphitic anode. *Carbon* **2023**, *203*, 561–570. [CrossRef]
58. Ahad, S.A.; Bhattacharya, S.; Kilian, S.; Ottaviani, M.; Ryan, K.M.; Kennedy, T.; Thompson, D.; Geaney, H. Lithiophilic Nanowire Guided Li Deposition in Li Metal Batteries. *Small* **2022**, *19*, 2205142. [CrossRef]
59. Nulu, A.; Hwang, Y.G.; Nulu, V.; Sohn, K.Y. Metal (Cu/Fe/Mn)-doped silicon/graphite composite as a cost-effective anode for Li-ion batteries. *Nanomaterials* **2022**, *12*, 3004. [CrossRef] [PubMed]
60. Zuo, X.; Xia, Y.; Ji, Q.; Gao, X.; Yin, S.; Wang, M.; Wang, X.; Qiu, B.; Wei, A.; Sun, Z.; et al. Self-templating construction of 3D hierarchical macro-/mesoporous silicon from 0D silica nanoparticles. *ACS Nano* **2017**, *11*, 889–899. [CrossRef]
61. Zou, H.; Zhang, C.; Xue, H.; Wu, Z.; Wang, Z.L. Boosting the solar cell efficiency by flexo-photovoltaic effect? *ACS Nano* **2019**, *13*, 12259–12267. [CrossRef]
62. Yu, H.; Gu, F.; Chen, S.; Du, Y.; Wang, L.; Song, Y. A N-doped carbon nanocages@silicon nanoparticles microcapsules for high-performance Li-storage. *Compos. Part. B Eng.* **2022**, *247*, 110334. [CrossRef]

63. Ruan, H.; Guo, S.; Zhang, L.; Liu, Y.; Li, L.; Huang, Y.; Gao, S.; Tian, Y. Boosting lithium storage performance of diatomite derived Si/SiO_x micronplates via rationally regulating the composition, morphology and crystalline structure. *Ceram. Int.* **2022**, *48*, 17510–17517. [CrossRef]
64. Kalidas, N.; Shen, X.; Yuan, M.; Zhao, X.; Lehto, V.P. Controlled surface oxidation of mesoporous silicon microparticles to achieve a stable Si/SiO_x anode for lithium-ion batteries. *Microporous Mesoporous Mater.* **2022**, *344*, 112243. [CrossRef]
65. Park, S.W.; Shim, H.W.; Kim, J.C.; Kim, D.W. Uniform Si nanoparticle-embedded nitrogen-doped carbon nanofiber electrodes for lithium ion batteries. *J. Alloys Compd.* **2017**, *728*, 490–496. [CrossRef]
66. Guan, P.; Li, J.; Lu, T.; Guan, T.; Ma, Z.; Peng, Z.; Zhu, X.; Zhang, L. Facile and scalable approach to fabricate granadilla-like porous-structured silicon-based anode for lithium ion batteries. *ACS Appl. Mater. Interfaces* **2018**, *10*, 34283–34290. [CrossRef] [PubMed]
67. Zhu, G.; Zhang, F.; Li, X.; Luo, W.; Li, L.; Zhang, H.; Wang, L.; Wang, Y.; Jiang, W.; Liu, H.K.; et al. Engineering the distribution of carbon in silicon oxide nanospheres at the atomic level for highly stable anodes. *Angew. Chem. Int. Ed. Engl.* **2019**, *58*, 6669–6673. [CrossRef] [PubMed]
68. Park, Y.K.; Boyer, M.; Hwang, G.S.; Lee, J.W. Synthesis of Si/SiO_x from talc and its characteristics as an anode for lithium-ion batteries. *J. Electroanal. Chem.* **2019**, *833*, 552–559. [CrossRef]

materials

MDPI

Article

Experimental and Computational Study of Optimized Gas Diffusion Layer for Polymer Electrolyte Membrane Electrolyzer

Javid Hussain [1,2,†], Dae-Kyeom Kim [2,†], Sangmin Park [1,2], Muhammad Waqas Khalid [1,2], Sayed-Sajid Hussain [3], Ammad Ali [1,2], Bin Lee [1,2,‡], Myungsuk Song [2,*] and Taek-Soo Kim [1,2,*]

1 Industrial Technology, University of Science and Technology, Daejeon 34113, Republic of Korea; javidmohsin77@kitech.re.kr (J.H.); jhsm8920@kitech.re.kr (S.P.); waqas@kitech.re.kr (M.W.K.); ammad5125@kitech.re.kr (A.A.); leebin@khu.ac.kr (B.L.)

2 Korea Institute for Rare Metals, Korea Institute of Industrial Technology, Incheon 21999, Republic of Korea; kyeom@kitech.re.kr

3 Chemical Engineering and Applied Chemistry, Chungnam National University, Daejeon 34134, Republic of Korea; sayedsajidh506@gmail.com

* Correspondence: mssong@kitech.re.kr (M.S.); tskim@kitech.re.kr (T.-S.K.)

† These authors contributed equally to this work.

‡ Current address: Department of Advanced Materials Engineering for Information and Electronics, Kyung Hee University, Yongin 17104, Republic of Korea.

Abstract: Polymer electrolyte membrane fuel cells (PEMFCs) and PEM electrolyzer are emerging technologies that produce energy with zero carbon emissions. However, the commercial feasibility of these technologies mostly relies on their efficiency, which is determined by individual parts, including the gas diffusion layer (GDL). GDL transfers fluid and charges while protecting other components form flooding and corrosion. As there is a very limited attention toward the simulation work, in this work, a novel approach was utilized that combines simulation and experimental techniques to optimize the sintering temperature of GDL. Ti_{64} GDL was produced through tape casting, a commercial method famous for producing precise thickness, uniform, and high-quality films and parameters such as slurry composition and rheology, casting parameters, drying, and debinding were optimized. The porosity and mechanical properties of the samples were tested experimentally at various sintering temperatures. The experimental results were compared with the simulated results achieved from the GeoDict simulation tool, showing around 96% accuracy, indicating that employing GeoDict to optimize the properties of Ti_{64} GDL produced via tape casting is a critical step towards the commercial feasibility of PEMFCs and electrolyzer. These findings significantly contribute to the development of sustainable energy solutions.

Keywords: PEMFC; GDL; tape casting; GeoDict; optimization; physical properties

Citation: Hussain, J.; Kim, D.-K.; Park, S.; Khalid, M.W.; Hussain, S.-S.; Ali, A.; Lee, B.; Song, M.; Kim, T.-S. Experimental and Computational Study of Optimized Gas Diffusion Layer for Polymer Electrolyte Membrane Electrolyzer. *Materials* **2023**, *16*, 4554. https://doi.org/10.3390/ma16134554

Academic Editors: Zhifeng Wang, Weiqing Zhang and Yichao Wang

Received: 12 May 2023
Revised: 8 June 2023
Accepted: 14 June 2023
Published: 23 June 2023

1. Introduction

Global warming, the high emission of carbon gases, and the energy crisis compel the world to move toward green energy. The fuel cell is one of the crucial technologies to solve this problem. The fuel cell is the best alternative for storing energy rather than batteries as it has an issue of self-discharge. There are many applications of fuel cells such as portable electronic devices, vehicles, power generation, transportation, backup power system, aerospace, military, and marine [1]. The basic five types of fuel cells on the basis of electrolytes are (PEM) fuel cells, solid oxide fuel cells, alkaline fuel cells, molten carbonate fuel cells, and phosphoric acid fuel cells. Among them, the PEM fuel cell received more attention due to the quick start and shut down, high efficiency, low operating temperature, and simple design. The PEM fuel cell uses hydrogen as the intake fuel and converts it into electrical energy [2,3]. The high combustion value, purity, and stability of hydrogen make it one of the most attractive sources of renewable energy [4]. To sufficiently provide hydrogen to the PEM fuel cell, enough hydrogen production is needed [5].

There are various techniques through which hydrogen can be produced. The most advanced technique is the alkaline electrolyzer, but its drawbacks include a corrosive electrolyte, a low maximum current density, and low operating pressure. In addition, the solid oxide electrolyzer has excellent efficiency and quick reaction time, but the high-temperature operation is the main limitation. Hence, the production of hydrogen by water electrolysis using a PEM electrolyzer can be a prominent solution. The PEM electrolyzer is at the door of commercialization. This electrolyzer allows various advantages over other electrolyzers such as the simple design, high productivity, fast response system, compactness, being safe at high current densities, easy maintenance, and environmental friendliness [6–9].

PEM electrolyzer convert electrical energy into chemical energy with oxygen as a by-product. It performs electrolysis of water into oxygen and hydrogen. The proton travels through the proton exchange membrane which is a solid electrolyte, while the electron travels through the external circuit, reacts again at the cathode side and releases hydrogen as the final product. The excess water and oxygen are the by-product in the whole process, which is very important for the control of the greenhouse effect [10,11].

The cathode and anode reactions of the electrolyzer are given as

$$\text{Anode: } 2H_2O \leftrightarrow O_2 + 4H^+ + 4e$$

$$\text{Cathode: } 4H^+ + 4e^- \leftrightarrow 2H$$

The overall reaction is as [7]

$$2H_2O + \text{electrical energy} \leftrightarrow 2H_2 + O_2$$

The electrolyzer is composed of the PEM membrane, anode, cathode, GDL, bipolar plate, and end plate. The major challenge in the electrolyzer is the cost of its individual parts. Since the output of one single cell is very limited, multiple cells are connected in series to increase the final production of hydrogen. The efficiency of the whole cell is based on the performance of each component including GDL [8].

1.1. GDL

The gas diffusion layer (GDL) is one of the essential parts of PEMFC and electrolyzer. It greatly affects the performance of PEMFC and facilitates fluid movement to and from the electrodes. GDL also shields other components from corrosion and erosion during the process. GDL, which exists in single and double layer configurations and is comprised of porous material, has a number of industrial, biomedical, and power generation uses. GDL must be designed with high thermal and electrical conductivity, strong mechanical properties, high corrosion resistance, and high permeability. Good thermal and electrical properties are important for efficient ion transport and fluid management. The high mechanical strength along with the high porosity is required to withstand the fluid pressure at high temperatures and ensure the high permeability of the fluid [12]. In addition, the hydrophilic and hydrophobic properties must also be considered in the design. The hydrophobic nature of certain materials can help manage water in and out of fuel cells, and coatings may be necessary for this purpose. To ensure all these properties, proper design and open pores are necessary for accurate handling of capillary pressure, and uniform transportation in order to avoid transport and flooding problems. The porosity and pore morphology of a GDL significantly affect its performance and, consequently, the efficiency of a PEM fuel cell. To achieve all these properties, it is important to select the appropriate material for the GDL [13,14].

Conventionally, carbon and steel were used as GDL materials, but their commercial feasibility was limited due to their mechanical and chemical degradation. Steel has high mechanical strength, but corrosion limits the commercialization of steel. The strong corrosion resistance of GDL is essential in the corrosive environment of the fuel cell and electrolyzer due to the presence of high reactive ions of H_2 and O_2 [15]. To solve this problem, titanium and its alloy can be a good candidate due to it having the same strength as steel with almost half density and high corrosion resistance. The high strength and light weight of titanium are also important for the smart manufacturing and weight control of the fuel cell and electrolyzer stack. Minh Young et al. reported that titanium is ten times more resistant to degradation compared to carbon [16].

This makes titanium a potentially attractive material for use in GDL and other applications where degradation resistance is important. The platinum coating makes this material very much feasible for commercial use with satisfactory corrosion resistance and mechanical strength [6,17]. After the selection of appropriate material, the fabrication and optimization of various properties of GDL is very important and challenging. Experimental optimization is very difficult due to the very low thickness of GDL and its porous nature. There is very limited attention given to the simulation work; therefore, in this study, GeoDict was employed for the modeling, simulation, and optimization process.

1.2. GeoDict

Currently, there is limited simulated research on titanium GDL manufactured by powder metallurgy. However, software prediction is expected to play a crucial role in the future for GDL design and property optimization. Computational fluid dynamics (CFD) provides insights into macroscopic fluid behavior, while finite element analysis (FEA) focuses only on mechanical aspects. Molecular dynamics (MD) simulations has specialty in offering atomistic-level insights. GeoDict possesses distinctive functionalities that enable the generation of realistic 3D micro-sized geometries, simulation of fluid flow and transport phenomena, and analysis of the mechanical behavior of GDL materials. These exclusive capabilities make GeoDict an ideal tool for investigating complex transport phenomena and optimizing the performance of GDLs in PEMFCs.

The GeoDict simulation tool (Release version: service pack 4) is a very adaptable and flexible software that needs several inputs based on real calculations, such as powder size and prototype dimensions, the volumetric porosity, distribution of the powder, volumetric shrinkage, etc. This also includes some manual addition of physical characteristics of the material. For instance, in this article, Ti_{64} was used and various modules were employed for the simulation with the manual inputs of the basic properties. The young modulus was measured and a comparison was then made with the experimental data. Moreover, several predictions were made to analyze and optimize other properties of GDL at various sintering temperatures (porosity).

Two studies, one by Gervais et al. and the other by Sudhakar et al., showed that GeoDict is highly accurate. Gervais et al. were able to achieve 97% accuracy, while Sudhakar et al. found that there was only a 0.07 deviation (0.3%) when comparing simulated water permeability with experimental results. This shows the verification and the authentication of GeoDict. Overall, the GeoDict simulation tool is widely recognized as a reliable and precise tool for simulating and predicting various characteristics of porous materials [18,19]. In a study conducted by Dennis Hoch et al., a comparison was made between GeoDict and StarCCM+, using the experimental results. The findings revealed that while StarCCM+ exhibited almost 1% higher accuracy, its simulation time was significantly longer and less convenient compared to GeoDict [20]. According to M. Amin et al. [21], who compared the results of GEODICT and COMSOL with the experimental results, GeoDict demonstrated a high level of accuracy in its predictions. H. Bai et al. utilized FilterDict to simulate the effective diffusion coefficient and filtration performance in relation to particle diameter. The results showed a deviation of approximately 0.6% from the experimental data, thus confirming the remarkable accuracy of GeoDict [22].

In conclusion, this paper aimed to investigate and develop a single-layer Ti_{64} GDL via tape casting. The tape was cast and sintered at different temperatures that ranged from 800 to 1400 °C. Then, the characterization was carried out and nanoindentation was used to find the mechanical properties and few readings were taken of each sample and the average was then carried out to minimize the error. For the validation of the GeoDict, first, a comparison was made of mechanical properties between the experimental and simulated data. A very decent agreement was seen. On the basis of these simulations, a further prediction and optimization of sintering w.r.t porosity was performed to successfully fabricate and develop a Ti_{64} GDL with high efficiency.

2. Experiment and Modeling

2.1. Material and Fabrication

Spherical APATM Ti-6Al-4V grade 23 powder of (10–60) µm size manufactured by AP& C a GE Additive company was taken as the starting material, as shown in Figure 1b. The powder size was measured with the particle analyzer. An aqueous-based slurry was preferred to avoid the hazardous effect. The slurry consisted of the binder polyvinyl alcohol (363073-500G, Sigma-Aldrich, USA) and the dispersant ammonium polymethacrylate (DARVAN® C-N, R.T. Vanderbilt, USA). To further enhance the rheological properties, the plasticizer polyethylene glycol (300, Sigma-Aldrich, Germany) was added to the slurry. Additionally, the surfactant Tergitol TMTMN-6 was utilized to improve the surface properties of the powder. The composition of each additive was carefully optimized through the use of a viscometer. The optimized composition of each component w.r.t to the powder is shown in Table 1.

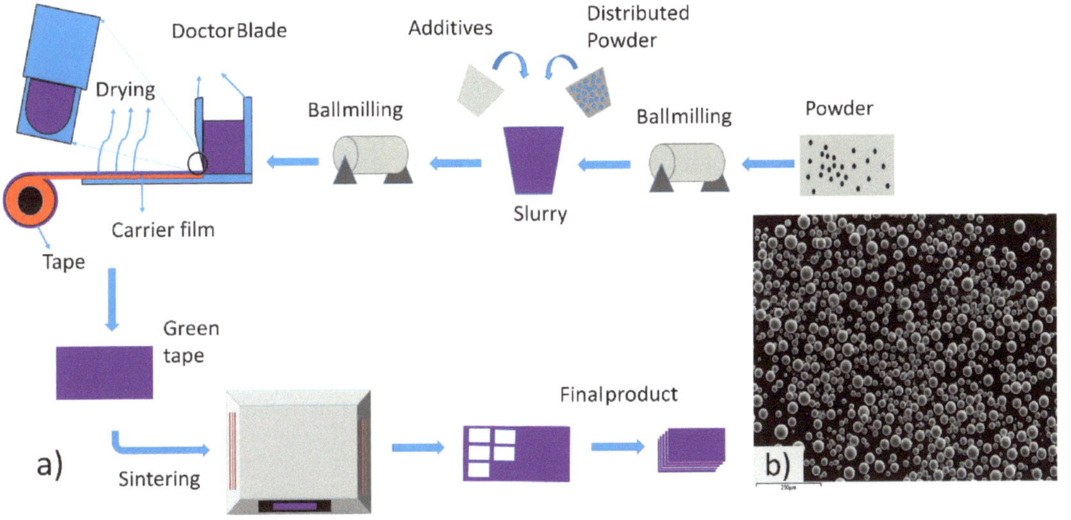

Figure 1. (**a**) Complete process of tape casting, and (**b**) SEM of Ti_{64} powder.

Table 1. Composition of the additive w.r.t Ti_{64} powder.

Additive	Material	(Wt. %)
Water	Demi water	31.8
Dispersant	Ammonium Polymethacrylate	2
Binder	Polyvinyl alcohol	6.46
Plasticizer	Polymethacrylate	6.46
Surfactant	Tergitol TMN 6	2

The mixing process was performed using a centrifugal mixer, with a one-step mixing technique at 1000 rpm for 10 min. Subsequently, de-foaming was carried out for 40 min at 600 rpm to remove any trapped air bubbles. The slurry was poured into a doctor blade machine, where the thickness was precisely controlled using the doctor blade. The initial thickness was set at 500 μm and it was almost 300–350 μm after drying. The tape was cast on the doctor blade machine at 1.5 mm/s speed and 18.5 °C temperature. The slow speed and low temperature were maintained to avoid cracking of the tape. After drying for 24 h in room temperature, the green tape was cut into a 3 mm × 3 mm pieces and debinding was carried out at 1 °C/min. Finally, the sintering was conducted at a heating rate of 5 °C/min, with a holding time of 2 h at a temperature range of 800 °C to 1400 °C. The whole process of tape casting is shown in Figure 1a.

The sintering and debinding were performed in a high temperature alumina furnace with high purity Ar atmosphere to prevent oxidation and other contamination. The porosity of each sample were measured by the image analyzer software version 1.0. This sophisticated instrument employs powerful algorithms to analyze the intricate microstructures and extract crucial porosity information. By employing high-resolution imaging and precise image segmentation techniques, we gained a comprehensive understanding of the sample's porosity. This meticulous approach ensured precise and reliable porosity measurements, enabling us to delve deeper into the material's properties and gain valuable insights. The samples were analyzed using scanning electron microscopy (SEM) and X-ray diffraction (XRD) techniques. Additionally, nano-indentation was employed to measure the shear modulus. The pressure and time for nano-indentation were set at 100 mN and 60 s, respectively. Several points were selected at different position and then average were taken to limit the error. For the modeling and simulation, GeoDict software tool was employed.

2.2. *Modeling*

Establishment of Grain Models

GeoDict is a user-friendly and integrated software, developed by Math2Market for multi-scale 3D image processing, material modeling, visualization, and property analysis [23]. The grain geo sub-model was used for the 3D modeling of spherical particles that were used for the fabrication of GDL in tape casting. GDL modeling was started with the dimension of 300 × 300 × 300 microns. The iteration and random seed were set to 1 and 45, respectively. During the modeling process, the air was selected as a fluid in the pores. The density was given w.r.t each sintering temperature that was measured with the image analyzer. The Gaussian powder distribution was selected with the size range of 5–65 microns, as measured by the powder size analyzer. The pile grain model was employed to construct a new model based on the real experimental inputs. After that, a distribution process was carried out and 140 iterations were selected for a more uniform distribution. The higher iteration resulted in more uniform distribution. The shift distance/voxel of 10 were used with a random seed of 1. The sintering process was performed based on the porosity of each sample. The sintering procedure was carried out with precision, considering the unique porosity of each sample. The whole process of 3D modeling, including some 3D visualization, is shown stepwise in Figure 2. The 3D models before and after distribution is shown in Figure 2a,b.

After distribution, the sintering was carried out, which depended on the shrinkage percentage as shown in Figure 2c. To gain a deeper understanding of the process, visual representations of FlowDict, ElastoDict, PoroDict, and ConductoDict can be seen in Figure 2d–g. This illustration highlights the systematic and meticulous approach that was taken to ensure the highest degree of accuracy and quality in the results.

2.3. *Simulation and Prediction*

The modeling process included generating a 3D model, distributing it, and then, sintering it. However, simulation was performed using various models, as shown in Figure 2, to predict the results at various porosity, representing the sintered sample at each temperature.

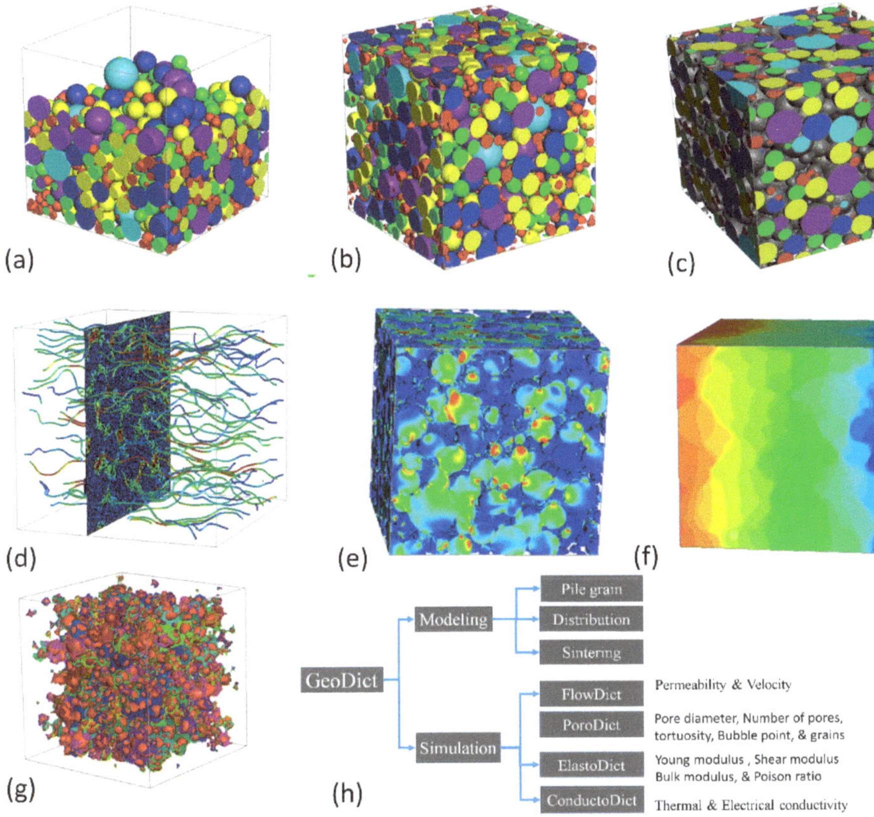

Figure 2. (**a**) 3D model before distribution, (**b**) distributed 3D model, (**c**) sintered 3D model, (**d**) fluid flow visual, (**e**) stress visual, (**f**) conductivity visuals, (**g**) pores visual, and (**h**) modeling and simulation of GDL by various module.

2.3.1. ElastoDict

The ElastoDict model was used to simulate the mechanical properties of the sample. Since some values of the titanium alloy (Ti_{64}) were not in the list of the material of GeoDict, the elastic modulus, shear modulus, and poison ratio of the titanium were given manually, which were 114 Gpa, 44 Gpa, and 0.33, respectively [24–26]. The pores and fluid were assigned the values 0 for each parameter. Based on these values, the mechanical properties at various porosity were simulated. These results were compared with the experimental values and further predictions were made to achieve optimal porosity and sintering conditions.

2.3.2. FlowDict

Permeability is a crucial aspect of GDL and plays a significant role in its performance. The FlowDict was used to simulate the effect of porosity on permeability and velocity in a porous material (GDL). The fluid flow in the porous material (GDL) usually takes place at low Reynolds number, which can be explained by the incompressible stokes equation and Darcy's law as given

$$\vec{u} = \frac{K}{\mu}\left(\nabla p - \vec{f}\right)$$

In this equation, $\vec{u}$ is the fluid velocity, ∇p is the pressure, K is the permeability, $\vec{f}$ is the force density and, μ is the fluid velocity. The Stokes–Brinkman equation was chosen due to the low velocity of the fluid in the porous material (GDL). The flow permeability tensor found from Darcy's law was used to find the permeability tensor

$$\vec{v}^{i} = -\mu^{-1} \begin{bmatrix} Kxx & Kxy & Kxz \\ Kyx & Kyy & Kyz \\ Kzx & Kzy & Kzz \end{bmatrix} \nabla p^{i}$$

where $\vec{v}$ is the average velocity, while i = 1, i = 2, and i = 3 represent the pressure drop in the X, Y, and Z direction, respectively. The three input parameters required for the flow experiment are the structure of porous material, the fluid, and the process parameters such as mass and flow rate, etc. [27,28]. The flow in the porous material GDL is very small; hence, the Stokes–Brinkman equation highly suits this phenomenon.

$$\nabla p = -\mu K^{-1} \vec{v} + \nabla \cdot \mu^{*} \left(\nabla \vec{v} + \nabla \vec{v}^{T} \right)$$

$$\nabla \cdot \vec{v} = 0$$

where p is the fluid pressure, $\vec{v}$ is the velocity, K is the permeability tensor, μ and μ^* are the fluid viscosity, and the fluid effective viscosity, respectively. For the prediction of permeability, water was taken as the fluid. Density and dynamic viscosity as an inputs were 998 kg/m^3 and 1.787×10^{-3} kg m^{-1} s^{-1}, respectively [29].

2.3.3. ConductoDict

The thermal properties are also a crucial feature of the GDL, which keeps the cell cool and also helps in the removal of fluid from the cell and makes it easy in the vapor form. Additionally, the electrical conductivity affects ion transport; hence, a high electrical conductivity is desirable for the GDL. ConductoDict was employed for the simulation and prediction of these two properties. The thermal conductivity is based on Fourier's law

$$\dot{q} = -K \nabla T$$

where K is the thermal conductivity, $\dot{q}$ is the heat flux, and T is the temperature. As the thermal conductivity is the second-order tensor, hence

$$K = \begin{pmatrix} K11 & K12 & K13 \\ K21 & K22 & K23 \\ K31 & K32 & K33 \end{pmatrix}$$

The material is isotropic and, therefore, instead of K, the scalar constant can be applied. For electrical conductivity, Ohm's law is applied

$$j = -\sigma \nabla \varphi$$

where j is the current density and φ is the electric potential [30].

Similar to thermal conductivity, the electrical conductivity is also the same in all direction; hence, the alpha is treated as a scalar.

Since the GeoDict did not contain all of the information of titanium alloys, the thermal and electrical conductivity of Ti_{64} and water (flows in GDL) were manually added. The thermal and electrical conductivity of Ti_{64} were 561,798 S/m and 6.7 W/mk, respectively. The thermal conductivity of the fluid (GDL) moving through the porous material was 1.787×10^{-3} W/mk, while electrical conductivity was 0.0005 S/m [31,32].

2.3.4. PoroDict

The structure of the pores significantly impacts all of the porous material's properties. The PoroDict was used to thoroughly examine the pores in the GDL at each porosity. The differential pore distribution in the PoroDict is based on the equation below.

$$\text{Pore volume distribution} = V_{cum}\ (d_{i+1}) - V_{cum}\ (d_i)/(\ln\ (d_{i+1}) - \ln\ (d_i)).m$$

where d is the pore diameter, v is the volume fraction, V_{cum} is the cumulative volume fraction, and m is the mass. The mass is computed on the base of the densities of the material. In addition, the bubble point pc was computed on the bases of the young Laplace equation

$$r = \frac{2\sigma}{pc}\cos\alpha$$

where α is the wetting phase contact angle, σ (sigma) is the surface tension, and r is through pore radius [33].

3. Result and Discussion

3.1. Model Verification and Analysis

The SEM microstructure of the gas diffusion layer (GDL) sintered at different temperatures was analyzed to better understand the properties of the material. As shown in Figure 3, at 800 °C, the particles of the GDL were not connected, leading to a brittle material. However, as the temperature increased to 900 °C, some particle connections formed and the porosity decreased. At 1000 °C, necking started to appear and continued to increase as the temperature was raised to 1100 °C and 1200 °C. When the sample was sintered at 1300 °C, the particles came together more closely, filling in the gaps, some grain formation began, which indicates a decrease in porosity and at 1400 °C, a completely sintered body with limited porosity and complete grain formation was achieved, providing optimal mechanical stability to the GDL. The 2D models generated by GeoDict at each porosity (sintering temperature) showed the same trend in microstructure, as seen in Figure 4. The comparison between the GeoDict-generated 2D structure and the SEM analysis showed that the model was generated using real experimental inputs and was a reliable representation of the material's properties. Furthermore, the XRD data were analyzed to gain a comprehensive understanding of the sample composition. The XRD pattern of GDL samples sintered at various temperatures (800 °C, 900 °C, 1000 °C, 1200 °C, and 1400 °C) is shown in Figure 5.

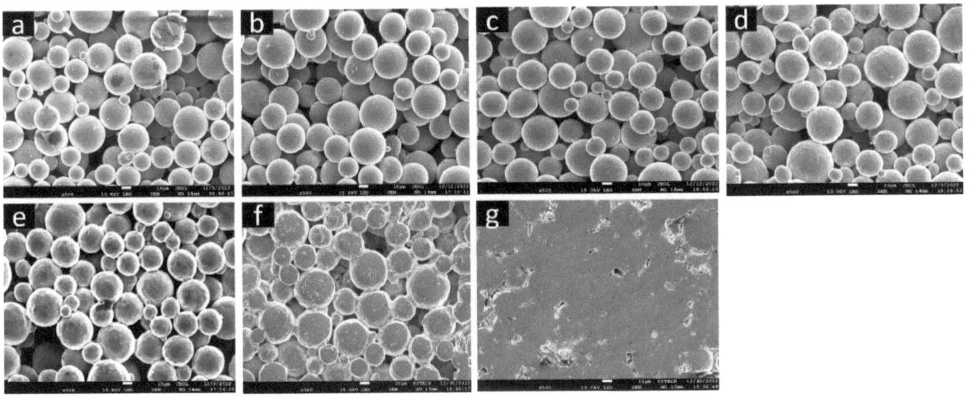

Figure 3. Microstructure of the GDL at various sintering temperature, (**a**) 800 °C, (**b**) 900 °C, (**c**) 1000 °C, (**d**) 1100 °C, (**e**) 1200 °C, (**f**) 1300 °C, and (**g**) 1400 °C.

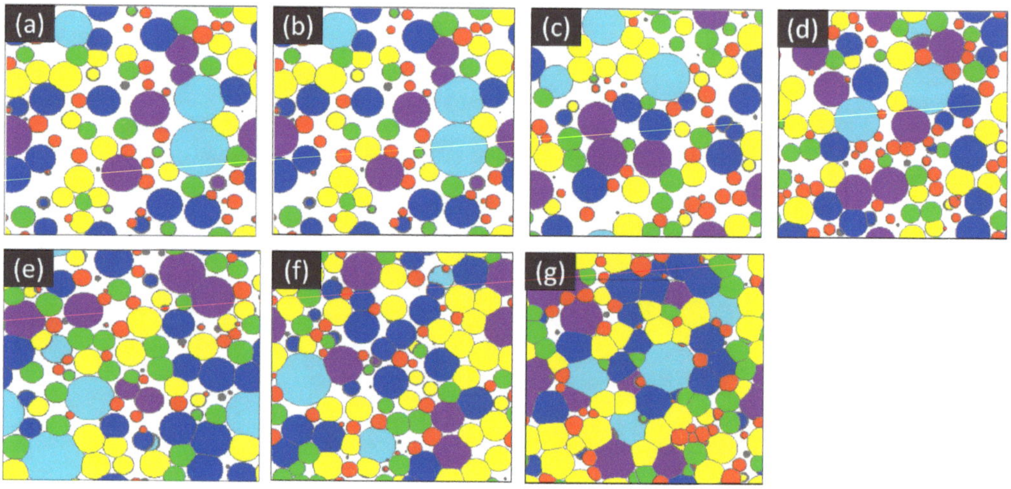

Figure 4. Microstructure of the GDL at various porosity (%), **(a)** 47, **(b)** 41, **(c)** 36, **(d)** 29, **(e)** 21, **(f)** 13, and **(g)** 5.

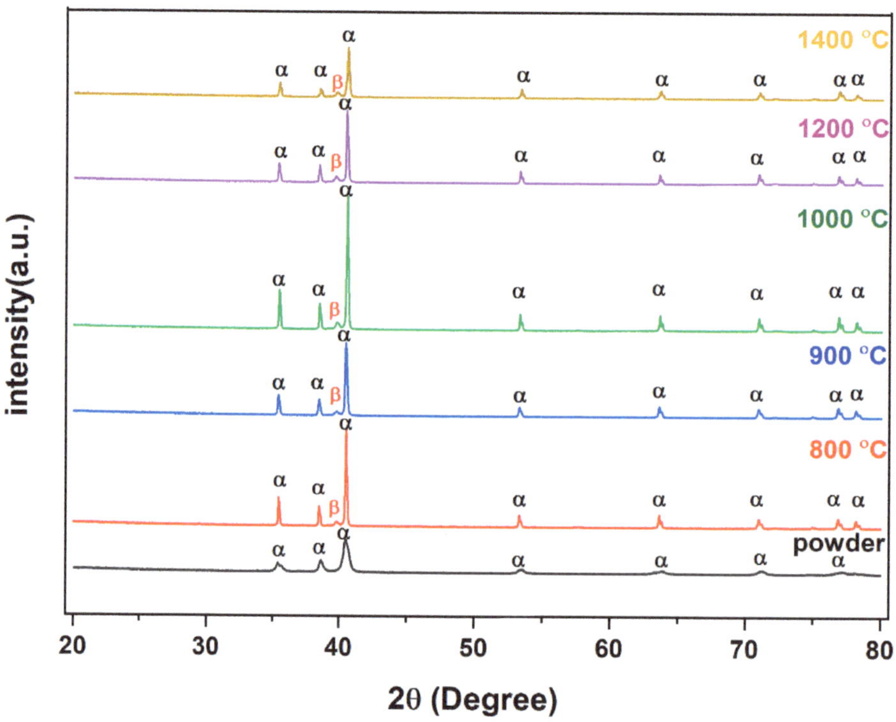

Figure 5. XRD analysis of powder and sintered Ti_{64}.

As there were no specific PDF cards for the Ti_{64} alloy, two cards, 00-044-1294 and 00-009-0098, were matched to identify the phase and structure of the XRD peaks. After the analysis of peaks at various 2θ, it was revealed that the majority of the peaks corresponded to the alpha (hcp) phase of titanium [34,35]. However, a small amount of the beta (bcc) phase was also detected at a specific 2θ value of 40. This suggests that the material had a dual-phase microstructure, with a mixture of hcp and bcc phases coexisting. It is worth mentioning that depending on the thermomechanical processing of the material, the proportion of each phase can be altered; hence, this information is important to understand the microstructure–property relationship of the material. Additionally, the XRD pattern showed that binder and other organic compounds were completely removed.

After the microstructural analysis, the porosity at each temperature was determined, as shown in Figure 6a. The porosity was around 47% at the start of 800 °C and it decreased to 5% at 1400 °C. The lower porosity was preferred for the higher mechanical, thermal, and electrical properties, but as a consequence, the permeability and other flow properties decreased. For this reason, a deep study of these properties at each porosity is very important for better optimization of porosity and, indirectly, the sintering temperature.

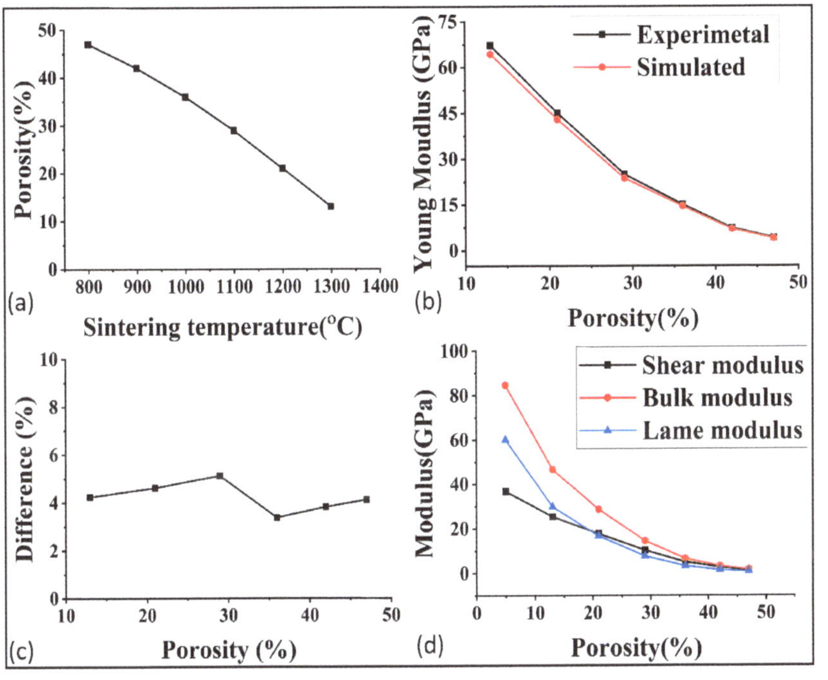

Figure 6. (**a**) Porosity of GDL at each sintering temperature, (**b**) experimental and simulated comparison of young modulus, (**c**) difference between the experimental and simulated result, and (**d**) relations of modulus with the porosity.

3.2. *Prediction of Various Properties via GeoDict*

The mechanical properties of porous materials are critical to withstand the fluid pressure and to ensure their performance in applications such as fuel cells. The validation of the simulation software GeoDict was demonstrated by comparing the simulated and experimental young modulus, as shown in Figure 6b. The comparison showed good agreement, with an average error of around 4% as demonstrated in Figure 6c. In order to further study the effect of porosity on the mechanical properties, the shear modulus, bulk modulus, and lame modulus were predicted at each porosity, as represented in Figure 6d. The poison ratio was also analyzed in relation to porosity, as shown in Figure 7a. The results showed that the mechanical properties were highly sensitive to even small changes in porosity. The optimization of mechanical properties was crucial for ensuring the high performance of the gas diffusion layer (GDL) in fuel cells.

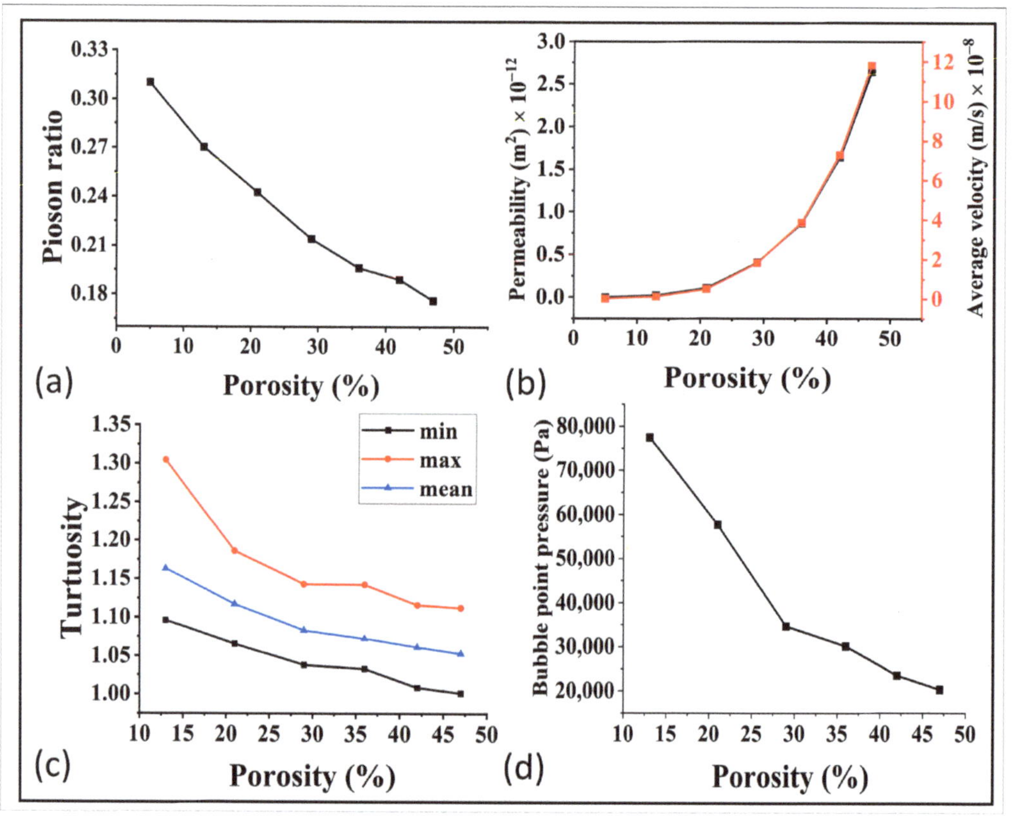

Figure 7. (**a**) Relation of porosity with the poison ratio. (**b**) Relation of Permeability and average velocity with porosity, (**c**) relation of tortuosity with porosity, and (**d**) porosity and bubble point relationship.

For further optimization, the relationship between porosity and permeability was simulated via GeoDict and the results are shown in Figure 7b. The permeability was in the range of 10^{-12} m^2 up to 1000 °C, and it then decreased to 10^{-13} m^2 and it became extremely less at higher temperature. This indicates that high sintering temperatures had a negative impact on the permeability. This trend was further confirmed by analyzing the average velocity and tortuosity, as shown in Figure 7b,c. The decrease in fluid velocity with decreasing porosity was a result of reduced available flow pathways, increased resistance to flow, enhanced fluid–solid interactions, and higher tortuosity within the porous material. These findings demonstrate the need for a balanced approach in optimizing the sintering temperature, taking into account both mechanical properties and flow properties to achieve optimal performance in the gas diffusion layer (GDL) of fuel cells. Furthermore, the relation of bubble point pressure and the porosity can be observed in Figure 7d. These results showed an easy bubble formation at higher porosity, which need to be further studied for the best control and optimization.

In addition to the mechanical and flow properties, the thermal and electrical properties of the GDL were also investigated. These properties play a crucial role in the transmission of charge and heat, which highly affect the overall efficiency of the cell. The relation of the porosity with the thermal and electrical conductivity can be seen in Figure 8a,b. With an increase in porosity, the electrical and thermal properties exhibited a discernible decline, highlighting a crucial relationship between these properties and the porosity of the material.

This phenomenon can be attributed to several underlying reasons. Firstly, as porosity increased, the material's effective cross-sectional area decreased. This reduction in the available pathway for electrical and thermal conduction led to a decrease in conductivity and thermal diffusivity. The presence of voids and air-filled spaces within the material interrupted the flow of electrons and heat, impeding their efficient transfer. Secondly, the interconnected voids and pores within the porous structure created additional interfaces, introducing resistance to the movement of charge carriers and thermal energy. These interfaces acted as barriers, hindering the smooth flow of electrons and heat through the material. Consequently, electrical and thermal conductivities were reduced. It is worth noting that the extent of the decrease in electrical and thermal properties depended on various factors, including the porosity volume fraction, pore size, and distribution. Materials with higher porosity and larger pore sizes tended to exhibit more pronounced decreases in these properties.

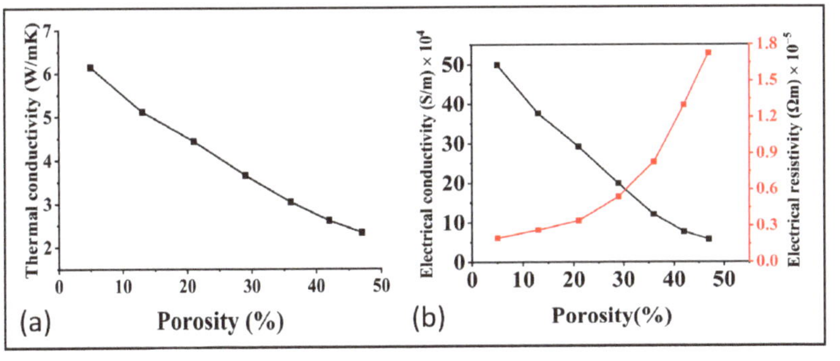

Figure 8. (**a**) Relation of porosity with thermal conductivity, and (**b**) relation of porosity and with electrical conductivity and electrical resistivity.

The morphology of the pores and grains within the GDL plays a crucial role in determining its overall properties. For example, the size, shape, and distribution of the pores decides the overall porosity and the flow properties of the material. Similarly, the shape, size, and orientation of the grains can have a significant effect on its mechanical, thermal and electrical properties. In summary, the porosity and grain morphology are essential microstructural features that significantly influence the macroscopic properties of the GDL. Some of the result related to the grains and pores at each sintering temperature (porosity) are as illustrated in Figure 9.

During the initial stages of the sintering process, the material exhibits a higher porosity, characterized by a low volume of grains and relatively large pore diameters, particularly open pores, as depicted in Figure 9a–d. As the temperature rises, the number of pores increased while their diameters decreased. Simultaneously, the number of grains decreased, and their volumes diminished with the escalating temperature. At higher porosity levels and larger pore diameters, a wider pathway was established for fluid flow, resulting in enhanced permeability and average velocity. These conditions facilitated a smoother and more defined movement of the fluid within the material On the other hand, at the high sintering temperature (lower porosity than 30%), the grain volume increased significantly while the pores volume declined. This shift from pores to grain ensured higher mechanical properties and lower flow properties. These understandings enable us to optimize the sintering temperature (porosity) to obtain the optimal mechanical, thermal, electrical, and flow properties.

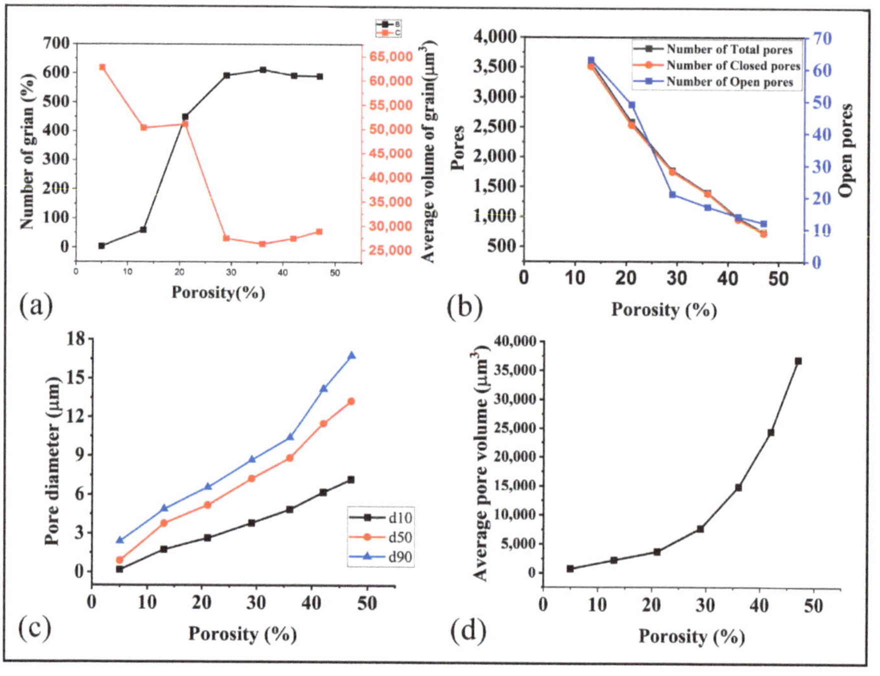

Figure 9. (**a**) Relation of porosity with number of grain and volume of grains, (**b**) porosity and pores, (**c**) porosity and pore diameter, and (**d**) porosity and average pore volume.

3.3. Optimization of Sintering Temperature w.r.t Various Properties

In a world where we demand sustainable energy solutions, the optimization of PEM fuel cell and electrolyzer materials is crucial. We observed that mechanical, thermal, and electrical properties increased with the decrease in the porosity, but the permeability, tortuosity, and the average velocity decreased as the porosity declined. For this purpose, the higher and lower porosity was not suitable for the application of PEM fuel cell and electrolyzer.

The optimization of the sintering temperature for GDL fabrication in PEM electrolyzers is crucial, taking into account two key factors such as permeability and mechanical properties. Extensive studies showed that a permeability above 10^{-12} m^2 is desirable for achieving high efficiency in GDLs. Unfortunately, the sample sintered at 1100 °C and above results in lower permeability values that did not meet this threshold, rendering them unsuitable for efficient applications [36–41].

On the other hand, lower sintering temperatures such as 800 °C and 900 °C may yield higher permeability, but they lack the necessary mechanical strength required for handling and testing in real stacks. These samples tend to be fragile and prone to breakage, making them impractical for implementation.

By selecting a sintering temperature of 1000 °C, a delicate balance was achieved between mechanical strength and permeability. Samples sintered at this temperature exhibited sufficient strength to withstand handling and fluid pressure, while still maintaining a relatively high permeability compared to high temperature (above 1000 °C) sintered samples. Overall, the optimization of the sintering temperature at 1000 °C ensures that the GDL exhibits both desirable mechanical strength and adequate permeability for efficient operation in PEM electrolyzers.

To achieve better results, there is a need for some special changes in the process to make, align, and control pores and grains. The pores and grains mainly depend on the sintering temperature, the heating and the cooling rate, the additive used in the slurry, the size and composition of powder, etc. In the bigger picture, these conditions control the overall properties of the whole material. However, this paper aimed to explore the optimization of only sintering temperature, and found that 1000 °C was the best optimal condition for the better performance. This study successfully optimized the sintering temperature and demonstrated that 1000 °C was the best sintering condition for the better performance of PEM fuel cell and electrolyzer materials.

4. Conclusions and Future Recommendation

In a world that is increasingly focused on sustainable energy solutions, the optimization of materials used in PEM fuel cells and electrolyzers is of utmost importance. The gas diffusion layer (GDL) is a critical component of PEM fuel cells, and its efficiency greatly impacts the overall performance of the system. This study demonstrated that tape casting is a highly effective method for producing Ti_{64} GDL. In addition, we observed that the flow properties increased with increasing po rosity, while mechanical properties, electrical, and thermal conductivity decreased. On the other hand, the sample sintered at higher temperatures had a reverse trend. The higher and lower temperature cannot be the ideal because at higher sintering temperature (porosity), the flow properties were very small and at lower sintering temperature, the material did not appreciate w.r.t to mechanical and conductive properties. Therefore, the optimized sintering temperature is 1000 °C and at this temperature, the mechanical, permeability, and the other properties showed an optimum value. Hence, we can say that the casted tape sintered at 1000 °C having porosity 36% is the optimized condition for this process. Moreover, the use of simulation tools such as GeoDict proved highly effective for optimizing the properties of Ti_{64} GDL produced via tape casting, and provided valuable insights for predicting and optimizing various properties.

The future of GDL fabrication lies in further increasing the porosity to enhance permeability without compromising mechanical properties. This can be achieved through the use of pore agents and high-temperature sintering techniques. Additionally, the application of fiber in the tape casting process can improve the properties of the Ti_{64} GDL. The control of pore structure is a crucial aspect of future research in GDL fabrication, and GeoDict simulation tool can provide a powerful information for the development of GDL with versatile properties. The coating of Ti_{64} for improved properties is an exciting avenue for future research. Overall, the optimization of GDL materials is a critical step towards the commercial feasibility of PEMFCs.

Author Contributions: Conceptualization, J.H., M.S. and T.-S.K.; Methodology, J.H., D.-K.K., S.P., M.W.K., B.L., M.S. and T.-S.K.; Software, J.H.; Validation, D.-K.K., S.P., A.A., B.L., M.S. and T.-S.K.; Investigation, D.-K.K., S.P., S.-S.H. and M.S.; Writing—original draft, J.H.; Writing—review & editing, M.S.; Supervision, M.S. and T.-S.K. All authors have read and agreed to the published version of the manuscript.

Funding: This research was funded by the Construction of the Foundation for Advanced Powder Material Development Purification funded by the Ministry of Trade, Industry & Energy (MOTIE) (No. P0018009) This work was supported by the Korea Institute of Industrial Technology as "International cooperation on the development of metallic compounds separation/recovery and byproducts materialization for total recycling during eco-friendly process" project (No. kitech JE-230013)" and "Based on the utilization of waste resources, the high-added value and development of rare earth metal recovery technology through pyrometallurgical extraction for ecosystem conversion" project (No. kitech UR230006), as well as by the UST Young Scientist+ Research Program 2022 through the University of Science and Technology (No. 2022YS28).

Institutional Review Board Statement: Not applicable.

Informed Consent Statement: Not applicable.

Data Availability Statement: Not applicable.

Acknowledgments: This work was supported by the Construction of the Foundation for Advanced Powder Material Development Purification funded by the Ministry of Trade, Industry & Energy (MOTIE) (No. P0018009) This work was supported by the Korea Institute of Industrial Technology as " International cooperation on the development of metallic compounds separation/recovery and byproducts materialization for total recycling during eco-friendly process" project (No. kitech JE-230013)" and "Based on the utilization of waste resources, the high-added value and development of rare earth metal recovery technology through pyrometallurgical extraction for ecosystem conversion" project (No. kitech UR230006), as well as by the UST Young Scientist+ Research Program 2022 through the University of Science and Technology (No. 2022YS28).

Conflicts of Interest: The authors declare no conflict of interest.

Abbreviations

PEMFC	Proton exchange membrane fuel cell
GDL	Gas diffusion layer
2D	Two dimensional
3D	Three dimensional
SEM	Scanning electron microscope
XRD	X-ray diffraction
BCC	Body center cubic
HCP	Hexagonal crystalline phase
Ti_{64}	Titanium-6Aluminum-4Vanadium

References

1. Garche, J.; Jürissen, L. Application of Fuel Cell Technology: Status and Perspectives. *Electrochem. Soc.* **2015**, *24*, 39. [CrossRef]
2. Wang, Y.; Chen, K.S.; Mishler, J.; Cho, S.C.; Adroher, X.C. A review of polymer electrolyte membrane fuel cells: Technology, applications, and needs on fundamental research. *Appl. Energy* **2011**, *88*, 981–1007. [CrossRef]
3. Vaghari, H.; Jafarizadeh-Malmiri, H.; Berenjian, A.; Anarjan, N. Recent advances in application of chitosan in fuel cells. *Sustain. Chem. Process.* **2013**, *1*, 16. [CrossRef]
4. Yang, Z.; Lin, J.; Zhang, H.; Lin, B.; Lin, G. A new direct coupling method for photovoltaic module-PEM electrolyzer stack for hydrogen production. *Fuel Cells* **2018**, *18*, 543–550. [CrossRef]
5. Ozden, A.; Shahgaldi, S.; Li, X.; Hamdullahpur, F. A review of gas diffusion layers for proton exchange membrane fuel cells—With a focus on characteristics, characterization techniques, materials and designs. *Prog. Energy Combust. Sci.* **2019**, *74*, 50–102. [CrossRef]
6. Maamouri, R.; Guilbert, D.; Zasadzinski, M.; Rafaralahy, H. Proton exchange membrane water electrolysis: Modeling for hydrogen flow rate control. *Int. J. Hydrog. Energy* **2021**, *46*, 7676–7700. [CrossRef]
7. IRENA. *Green Hydrogen Cost Reduction: Scaling Up Electrolysers to Meet the 1.5 °C Climate Goal*; International Renewable Energy Agency: Abu Dhabi, United Arab Emirates, 2020; 105p.
8. Selamet, F.; Becerikli, F.; Mat, M.D.; Kaplan, Y. Development and testing of a highly efficient proton exchange membrane (PEM) electrolyzer stack. *Int. J. Hydrog. Energy* **2011**, *36*, 11480–11487. [CrossRef]
9. Grigoriev, S.; Porembsky, V.; Fateev, V. Pure hydrogen production by PEM electrolysis for hydrogen energy. *Int. J. Hydrog. Energy* **2006**, *31*, 171–175. [CrossRef]
10. Marshall, A.; Børresen, B.; Hagen, G.; Tsypkin, M.; Tunold, R. Hydrogen production by advanced proton exchange membrane (PEM) water electrolysers—Reduced energy consumption by improved electrocatalysis. *Energy* **2007**, *32*, 431–436. [CrossRef]
11. Falcão, D.; Pinto, A. A review on PEM electrolyzer modelling: Guidelines for beginners. *J. Clean. Prod.* **2020**, *261*, 121184. [CrossRef]
12. Forner-Cuenca, A.; Biesdorf, J.; Gubler, L.; Kristiansen, P.M.; Schmidt, T.J.; Boillat, P. Engineered water highways in fuel cells: Radiation grafting of gas diffusion layers. *Adv. Mater.* **2015**, *27*, 6317–6322. [CrossRef] [PubMed]
13. Kang, Z.; Mo, J.; Yang, G.; Retterer, S.T.; Cullen, D.A.; Toops, T.J.; Green, J.B., Jr.; Mench, M.M.; Zhang, F.-Y. Investigation of thin/well-tunable liquid/gas diffusion layers exhibiting superior multifunctional performance in low-temperature electrolytic water splitting. *Energy Environ. Sci.* **2017**, *10*, 166–175. [CrossRef]
14. Hernandez-Aldave, S.; Andreoli, E. Fundamentals of gas diffusion electrodes and electrolysers for carbon dioxide utilisation: Challenges and opportunities. *Catalysts* **2020**, *10*, 713. [CrossRef]
15. Mo, J.; Steen, S.M.; Retterer, S.; Cullen, D.A.; Terekhov, A.; Zhang, F.-Y. Mask-patterned wet etching of thin titanium liquid/gas diffusion layers for a PEMEC. *ECS Trans.* **2015**, *66*, 3. [CrossRef]
16. Aşık, E.E.; Bor, Ş. Fatigue behavior of Ti–6Al–4V foams processed by magnesium space holder technique. *Mater. Sci. Eng. A* **2015**, *621*, 157–165. [CrossRef]

17. Wei, J.; Ning, F.; Bai, C.; Zhang, T.; Lu, G.; Wang, H.; Li, Y.; Shen, Y.; Fu, X.; Li, Q.; et al. An ultra-thin, flexible, low-cost and scalable gas diffusion layer composed of carbon nanotubes for high-performance fuel cells. *J. Mater. Chem. A* **2020**, *8*, 5986–5994. [CrossRef]
18. Gervais, P.-C.; Bourrous, S.; Dany, F.; Bouilloux, L.; Ricciardi, L. Simulations of filter media performances from microtomography-based computational domain. Experimental and analytical comparison. *Comput. Fluids* **2015**, *116*, 118–128. [CrossRef]
19. Jaganathan, S.; Tafreshi, H.V.; Pourdeyhimi, B. A case study of realistic two-scale modeling of water permeability in fibrous media. *Sep. Sci. Technol.* **2008**, *43*, 1901–1916. [CrossRef]
20. Hoch, D.; Azimian, M.; Baumann, A.; Behringer, J.; Niessner, J. Comparison of Voxel-Based and Mesh-Based CFD Models for Aerosol Deposition on Complex Fibrous Filters (vol 43, pg 2538, 2020). *Chem. Eng. Technol.* **2021**, *43*, 1153. [CrossRef]
21. Zarandi, M.A.F.; Arroyo, S.; Pillai, K.M. Longitudinal and transverse flows in fiber tows: Evaluation of theoretical permeability models through numerical predictions and experimental measurements. *Compos. Part A Appl. Sci. Manuf.* **2019**, *119*, 73–87. [CrossRef]
22. Bai, H.; Qian, X.; Fan, J.; Shi, Y.; Duo, Y.; Guo, C. Probing the effective diffusion coefficient and filtration performance of micro/nanofibrous composite layered filters. *Ind. Eng. Chem. Res.* **2021**, *60*, 7301–7310. [CrossRef]
23. Math 2 Market. 2023. Available online: https://www.math2market.de/ (accessed on 1 July 2022).
24. Welsch, G.; Boyer, R.; Collings, E. *Materials Properties Handbook: Titanium Alloys*; ASM International: Almere, Netherlands, 1993.
25. ASM International. *Metals Handbook Vol. 2: Properties and Selection: Nonferrous Alloys and Special-Purpose Materials*; ASM: Almere, Netherlands, 1990.
26. *Structural alloys handbook*; John, M.; Technical, Y.H. (Eds.) CINDAS/Purdue University: West Lafayette, IN, USA, 1996.
27. Ebadi, M.; Armstrong, R.T.; Mostaghimi, P.; Da Wang, Y.; Alqahtani, N.; Amirian, T.; James, L.A.; Parmar, A.; Zahra, D.; Hamze, H.; et al. Predictive Soft Computing Methods for Building Digital Rock Models Verified by Positron Emission Tomography Experiments. *Water Resour. Res.* **2022**, *58*, e2021WR031814. [CrossRef]
28. FlowDict User Guide GeoDict. 2022. Available online: https://www.geodict.com/geodict-software/geodict-base-modules/simulation/flowdict.html (accessed on 11 July 2022).
29. Krotkiewski, M.; Ligaarden, I.S.; Lie, K.-A.; Schmid, D.W. On the importance of the stokes-brinkman equations for computing effective permeability in karst reservoirs. *Commun. Comput. Phys.* **2011**, *10*, 1315–1332. [CrossRef]
30. Conductodict User Guide Geodict. Available online: https://www.geodict.com/geodict-software/geodict-base-modules/simulation/conductodict.html (accessed on 27 August 2021).
31. Csoklich, C.; Sabharwal, M.; Schmidt, T.J.; Büchi, F.N. Does the thermal conductivity of gas diffusion layer matter in polymer electrolyte fuel cells? *J. Power Sources* **2022**, *540*, 231539. [CrossRef]
32. Bolzoni, L.; ERuiz-Navas, M.; Gordo, E. Flexural properties, thermal conductivity and electrical resistivity of prealloyed and master alloy addition powder metallurgy Ti–6Al–4V. *Mater. Des. (1980–2015)* **2013**, *52*, 888–895. [CrossRef]
33. Porodict Guide to User GeoDict. 2022. Available online: https://www.geodict.com/geodict-software/geodict-base-modules/analysis/porodict.html (accessed on 21 July 2022).
34. Panin, A.; Kazachenok, M.; Kolmakov, A.; Chizhik, S.; Heifetz, M.; Chugui, Y. Microstructure and mechanical behaviour of additive manufactured Ti–6Al–4V parts under tension. *EPJ Web Conf.* **2019**, *221*, 01037. [CrossRef]
35. Wysocki, B.; Maj, P.; Sitek, R.; Buhagiar, J.; Kurzydłowski, K.J.; Święszkowski, W. Laser and electron beam additive manufacturing methods of fabricating titanium bone implants. *Appl. Sci.* **2017**, *7*, 657. [CrossRef]
36. Steen, S.M.; Mo, J.; Kang, Z.; Yang, G.; Zhang, F.-Y. Investigation of titanium liquid/gas diffusion layers in proton exchange membrane electrolyzer cells. *Int. J. Green Energy* **2017**, *14*, 162–170. [CrossRef]
37. Feser, J.P.; Prasad, A.K.; Advani, S.G. Experimental characterization of in-plane permeability of gas diffusion layer. *J. Power Sources* **2006**, *162*, 1226–1231. [CrossRef]
38. Tan, Z.; Jia, L.; Zhang, Z. Water vapor transport with condensation in gas diffusion Layer of a Proton Exchange Membrane Fuel Cell. *Heat Transf. Res.* **2012**, *43*, 139–150. [CrossRef]
39. Xia, L.; Ni, M.; He, Q.; Xu, Q.; Cheng, C. Optimization of gas diffusion layer in high temperature with the focus on thickness and porosity. *Appl. Energy* **2021**, *300*, 117357. [CrossRef]
40. Grigoriev, S.A.; Millet, P.; Volobuev, S.A.; Fateev, V.N. Optimization of porous current collector for PEM water electrolyzers. *Int. J. Hydrog. Energy* **2009**, *34*, 4968–4973. [CrossRef]
41. Gostick, J.T.; Fowler, M.W.; Pritzker, M.D.; Ioannidis, M.A.; Behra, L.M. In-plane and through-plane gas permeability of carbon fiber electrode backing layers. *J. Power Source* **2006**, *162*, 228–238. [CrossRef]

materials MDPI

Article

Towards High Surface Area α-Al_2O_3–Mn-Assisted Low Temperature Transformation

Tim Jähnichen [1,*], Simon Carstens [1], Maximilian Franz [1], Otto Laufer [1], Marianne Wenzel [2], Jörg Matysik [2] and Dirk Enke [1]

[1] Institute of Chemical Technology, Leipzig University, Linnéstr. 3, D-04103 Leipzig, Germany
[2] Institute of Analytical Chemistry, Leipzig University, Linnéstr. 3, D-04103 Leipzig, Germany
* Correspondence: tim.jaehnichen@uni-leipzig.com

Abstract: When impregnated with manganiferous precursors, γ-Al_2O_3 may be converted into α-Al_2O_3 under relatively mild and energy-saving conditions. In this work, a manganese assisted conversion to corundum at temperatures as low as 800 °C is investigated. To observe the alumina phase transition, XRD and solid-state ^{27}Al-MAS-NMR are applied. By post-synthetical treatment in concentrated HCl, residual manganese is removed up to 3 wt.-%. Thereby, α-Al_2O_3 with a high specific surface area of 56 m^2 g^{-1} is obtained after complete conversion. Just as for transition alumina, thermal stability is an important issue for corundum. Long-term stability tests were performed at 750 °C for 7 days. Although highly porous corundum was synthesized, the porosity decreased with time at common process temperatures.

Keywords: manganese-assisted synthesis; low-temperature α-Al_2O_3; high specific surface area of 56 m^2 g^{-1}

Citation: Jähnichen, T.; Carstens, S.; Franz, M.; Laufer, O.; Wenzel, M.; Matysik, J.; Enke, D. Towards High Surface Area α-Al_2O_3–Mn-Assisted Low Temperature Transformation. *Materials* **2023**, *16*, 3047. https://doi.org/10.3390/ma16083047

Academic Editors: Zhifeng Wang, Weiqing Zhang and Yichao Wang

Received: 20 March 2023
Revised: 6 April 2023
Accepted: 8 April 2023
Published: 12 April 2023

1. Introduction

Aluminum oxide is widely employed for catalytic applications, most commonly in its γ-form. γ-Al_2O_3 is readily available, easily moldable, e.g., for wash coating, and its porosity can be precisely controlled. In addition, the material has large specific surface areas in the range of 150 to 300 m^2 g^{-1} and attainable pore sizes in the mesopore and macropore range [1]. For certain applications, catalyst supports need to endure high temperatures above 1000 °C [2]. This includes, among others, petroleum refining processes like steam reforming [3,4], combined heat and power units (CHP) [5], and wood-burning stoves [6]. While γ-alumina is a suitable material for most catalytic applications, it is subjected to an irreversible phase transition to thermodynamically stable α-modification at temperatures above 1000 °C [7,8].

This irreversible transition from γ- to α-Al_2O_3 entails several problems. As the underlying crystal structure fundamentally changes from cubic to hexagonal, a complete rearrangement of the entire alumina microstructure occurs. An unwanted but inevitable side effect thereof is the loss of porosity, and hence of internal surface area [9]. This could possibly be circumvented by substituting γ-Al_2O_3 with the thermodynamically stable α-modification. However, the application of α-Al_2O_3 is difficult because the material usually has a very low specific surface area, which diminishes catalyst loadings and dispersion, greatly impairing its application as catalyst support. Many attempts have been undertaken to introduce porosity into intrinsically non-porous corundum [10]. One approach to preserving γ-Al_2O_3 porosity consists of lowering the α-transition temperature, which is also less costly and saves energy. This may be achieved, for instance, by impregnating γ-Al_2O_3 catalyst supports with a manganiferous precursor, as reported by Tsyrulnikov et al. [11]. Upon subsequent drying, Mn(II)-ions are dispersed on the surface of the transition alumina. Calcination at 500–800 °C leads to solid solutions of the now oxidized Mn(III)-ions in the

Al_2O_3 structure. These Mn-doped corundum-type domains act as seeding crystals, resulting in the complete conversion of the transition alumina into α-Al_2O_3 at 900 °C. We recently applied this procedure to sol-gel samples [12,13]. While pure alumina samples require a temperature of 1200 °C for complete α-transition, this was attained at temperatures as low as 900 °C after Mn-impregnation. Successive extraction of manganiferous species by dissolution in aqua regia yields α-Al_2O_3 with an increased A_{BET} of 23 $m^2\ g^{-1}$. This is a remarkably high surface area for an α-Al_2O_3 without any traceable transition alumina modifications.

Most often, the phase composition, or more precisely the complete transition to corundum, is monitored by XRD. In this work, in addition to XRD, solid-state ^{27}Al-MAS-NMR is used as a complementary method to observe phase transitions. The ^{27}Al isotope is a highly sensitive NMR nucleus yielding broad lines over a wide chemical shift range [14]. As a spin 5/2 nucleus, it is quadrupolar. Hence, the line width is dependent on the structure, where highly symmetric environments lead to narrower lines. An issue regarding the ^{27}Al-NMR spectra is whether these spectra show intensities properly, representing the proportional amounts of different aluminum species in the samples. This is particularly true for ^{27}Al-NMR signals associated with aluminum sites with very low symmetry, such as θ- and γ-alumina [15]. In general, ^{27}Al-MAS-NMR is used to detect the presence of aluminum and measure its relaxation time to elucidate binding relations. However, there are also studies regarding different aluminum structures and their conversions [15–18]. In 1997, Fitzgerald et al. [15] used ^{27}Al-MAS-NMR for dehydration studies of high surface area alumina. They published spectra of γ- and α-Al_2O_3 alongside various aluminum oxide materials and were able to determine a difference between 6-fold (9 ppm), 5-fold (37 ppm), and 4-fold (69 ppm) coordinated Al sites. In a later work, O'Dell et al. [18] investigated the conversion from boehmite to corundum from RT to 1200 °C. Thereby, they were able to show that different γ-, δ- and θ-alumina phases could be detected by NMR with increasing synthesis temperature, while the material was X-ray amorphous. The group assumed that the previous alumina phases were X-ray amorphous, due to small nanocrystal sizes between 10 to 20 nm, before grain growth occurs during the transition to corundum [19,20]. Hence, they recognized ^{27}Al-MAS-NMR as an important method in studying phase transitions in alumina.

In this publication, we investigated the Mn-assisted α-transition of sol-gel alumina materials, using ^{27}Al-MAS-NMR and XRD as complementary methods. Thereby, we were able to show that the temperature threshold for a complete formation of α-Al_2O_3 is as low as 800 °C and results in a specific surface area A_{BET} of 56 $m^2\ g^{-1}$, which is an unusually high value for corundum. Moreover, we investigated the long-term thermal stability of the thus obtained α-Al_2O_3. Here, we observed a degradation in specific surface area with prolonged exposure at 750 °C, similar to observations for diaspore-derived corundum [21–23].

2. Materials and Methods

2.1. Preparation of Alumina Starting Material

To ensure all samples originated from the exact same batch, a large batch of γ-Al_2O_3 was prepared, similar to the previously reported method [13]. To this end, 156 g of $AlCl_3 \cdot 6H_2O$ (*Fluka*) were dissolved in 137 mL of distilled water and 200 mL ethanol (*Bioenergie Icking GmbH*). After complete dissolution, 10.2 g of oxalic acid (dihydrate, *Merck*) were added and dissolved. Without delay, the reaction mixture was placed in a flask surrounded by an ice bath and cooled down to 4 °C, then 140 mL of propylene oxide (*Acros Organics*) were added, all at once, under vigorous stirring. The ice bath was removed after 6.5 min, and the reaction mixture was stirred at room temperature for another 6.5 min. The obtained sol was poured into a beaker, sealed and placed in an oven at 40 °C for gelation. To ensure a uniform heat transfer, the beaker resided in a water bath. After 24 h, solvent exchange to pure ethanol was carried out under static conditions for 3 days. The opaque white gel was then transferred to a drying cabinet pre-heated at 70 °C, with the lid removed. Drying was terminated after 7 days. Ultimately, the obtained xerogel was pre-calcined at 650 °C for 6 h, with a heating rate of 3 K min^{-1}. Characterization of the starting material was carried out as described in Section 3.1.

2.2. Mn-Assisted Conversion into α-Al_2O_3

The pre-calcined starting material was impregnated with a 1 M aqueous solution of $Mn(NO_3)_2 \cdot 4H_2O$ (*Alfa Aesar*). Samples were completely immersed in the manganese precursor solution and outgassed at 80 mbar for 1 h. When no more air emerged from the alumina pores, excess solution was removed via filtration and the impregnated samples were then dried in a drying cabinet at 120 °C for 18 h. Finally, $Mn(NO_3)_2$ was thermally converted into manganese oxide at 200 °C for 6 h.

Conversion into α-Al_2O_3 of both Mn-impregnated and pristine alumina samples was attempted at different temperatures ranging from 750 to 1200 °C. The heating rate was set to 3 K min^{-1}, while dwell times varied from 6 to 48 h. After calcination, excess Mn species from the impregnated samples were removed by stirring in concentrated HCl (37 wt.-%) for 4 h at room temperature.

Impregnated samples are designated "Mn" and reference samples as "Ref", followed by the calcination conditions. For instance, sample Mn-800-12 is a Mn-impregnated sample calcined at 800 °C for 12 h and leached with HCl. Samples that are investigated before leaching are marked with * (e.g., Mn-800-12*).

To verify thermal stability, a Mn-800-12 sample and a reference sample Ref-650-6 were exposed to a temperature of 750 °C for another 7 days. These are designated as Mn-800-12@750-172 and Ref-650-6@750-172, respectively.

2.3. Characterization Techniques

Nitrogen sorption (*Autosorb iQ*, Quantachrome) was used to determine the specific surface area A_{BET} of reference and corundum samples. Prior to analysis, the samples were dried and activated at 250 °C for 10 h under ultra-high vacuum. The determination of specific surface areas was conducted using the linearized form of the BET equation in the range of $0.05 \leq p\,p_0^{-1} \leq 0.30$.

For mercury intrusion, Pascal 140 and Pascal 440 (Porotec) were used. In Pascal 140, the samples were evacuated to 0.2 mbar and then filled with mercury. In Pascal 440, the intrusion measurement up to 400 MPa was performed at room temperature. The contact angle of mercury was set to 140° and the surface tension to 0.48 N m^{-1}.

For X-ray powder diffraction analysis (XRD), the samples were finely ground before measuring. The measurement was performed on a STADIP instrument from STOE & Cie GmbH with a Mythen1K detector. Cu-K_α-radiation was used at 40 kV and 40 mA. Phase identification was carried out in Match! Software (Crystal Impact).

Magic angle spinning (MAS) ^{27}Al-NMR spectra were recorded on an Avance-III 400-MHz WB NMR spectrometer (Bruker BioSpin, Rheinstetten, Germany) equipped with a 4 mm MAS BB/^{1}H probe at a Larmor frequency of 104.26 MHz. All spectra were collected at a spinning frequency of 10 kHz at 20 °C and referenced externally on the Xsi-scale to water at 4.7 ppm. For magic angle spinning (MAS) experiments, 1024 scans were accumulated with a recycle delay of 0.1 s. All samples were ground and subsequently transferred to the MAS rotor before measuring.

The manganese content of selected synthesized materials was determined using inductively coupled plasma optical emission spectrometry (ICP-OES) with a Perkin ElmerOptimal 8000.

Scanning electron microscopy (SEM) images were obtained using a Leo Gemini 1530 by Zeiss with an Everhart-Thornley detector for collecting secondary electrons. EDX measurements were also carried out on said device, using a SUTW-Sapphire detector. Samples were fixated on a carbon foil and vapor coated with a gold film. The accelerating voltage was 20 kV for both devices.

3. Results and Discussion

3.1. Properties of the Starting Material

Alumina starting materials for all tests were taken from a single sol-gel batch, which was prepared as described above and previously reported [13]. Reproducibility issues should thus be circumvented. The alumina product was characterized after calcination at 650 °C for 6 h. XRD and ^{27}Al-MAS-NMR indicates no long-range order, as is characteristic for transition alumina such as γ-Al_2O_3 [10]. A specific surface area A_{BET} of 276 m^2 g^{-1} was determined by nitrogen sorption. A modal mesopore diameter of 11 nm with a corresponding mesopore volume of 0.55 cm^3 g^{-1} was established via mercury intrusion. In the SEM images shown in Figure 1, interconnected mesopores embedded in a macropore structure can be identified.

Figure 1. SEM images of the alumina starting material Ref-650-6 depicting its overall morphology (**right**) and in more detail (**left**).

3.2. Phase Identification of α-Al_2O_3 by XRD and NMR

The pre-calcined alumina material was subjected to calcination at different temperatures with varying dwell times. The temperature range spanned from 750 to 1150 °C, depending on the nature of the samples (pristine or impregnated with manganese precursor solution). Based on the applied calcination temperature, different aluminum phases can be synthesized. While the conversion of γ-Al_2O_3 to α-Al_2O_3 in general requires temperatures above 1100 °C [24–26], the necessary transition temperature is reduced by Mn-impregnation. As a result, α-phase transition can be achieved under much milder conditions [11]. Afterward, Mn-impregnated samples were leached for Mn removal. Conversion of transitional alumina into corundum was monitored by XRD. An error of up to 5 wt.-% for undetectable alumina modifications of lesser crystallinity, e.g., γ, θ, η or κ-alumina is implied. Emerging α-Al_2O_3 crystallites only contribute to the diffraction intensity once they surpass the threshold of $\approx$30 Å, marking the detection limit of XRD. This reservation being stated, complete transition to α-Al_2O_3 was observed at 850 °C for a dwell time of 6 h, as the diffraction patterns shown in Figure 2 reveal.

In the reference samples, similar crystallinities can only be obtained by a calcination above 1100 °C at equal calcination times of 6 h (Figure 3). For manganese impregnated samples, conversion attempts at lower temperatures remained unsuccessful up to 750 °C, as the diffraction pattern in Figure 2 demonstrates. Increasing the calcination temperature to 800 °C results in corundum reflexes with very low intensity. However, a substantial growth of both the α-Al_2O_3 fraction and its individual crystallites takes place during the following 6 h. Extending the dwell time to 12 h at 800 °C leads to the complete conversion into corundum. To the best of our knowledge, these are the mildest calcination conditions reported to date in obtaining α-Al_2O_3 by excluding diaspore-derived corundum [21–23]. For comparison, the calcination of non-impregnated alumina at the same conditions (Ref-800-12, Figure 3) leads to a diffraction pattern of an amorphous material, with the major reflexes of γ-Al_2O_3 emerging as extremely broad and of very low intensity.

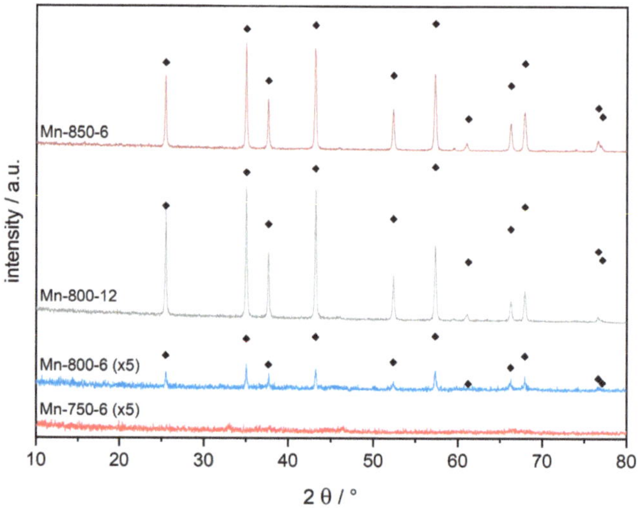

Figure 2. XRD patterns of Mn-impregnated samples after calcination at 750–850 °C for 6 or 12 h. Patterns of samples Mn-750-6 and Mn-800-6 are expanded (×5) for better visibility. Black rhombs mark corundum reflexes.

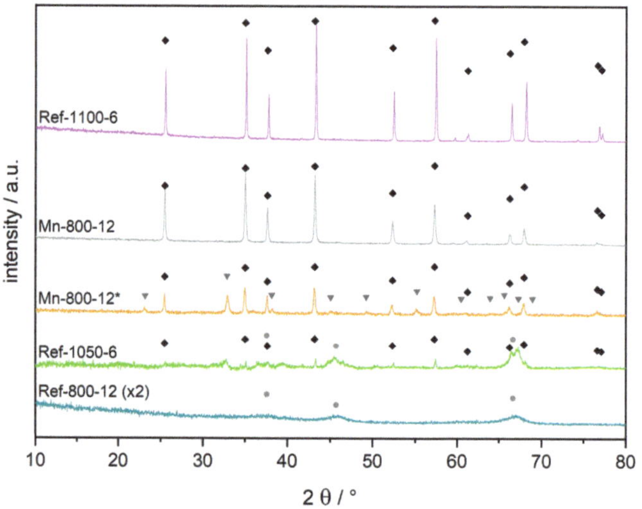

Figure 3. XRD patterns of sample Mn-800-12, calcined at 800 °C for 12 h, before (yellow, marked *) and after (grey) removal of manganese species. For comparison, Mn-free reference samples are given at the top and bottom positions, calcined at 800 °C and 1100 °C, respectively. The pattern of reference sample Ref-800-12 is expanded (×2) for better visibility. Black rhombs mark corundum reflexes, gray triangles show bixbyite (Mn_2O_3), and light gray circles indicate the major reflexes of γ-Al_2O_3.

Hence, by the Mn-impregnation, the α-transition threshold temperature can be lowered to 800 °C which is even lower than the 900 °C reported earlier by our group [10]. The reason for this further reduction in the α-transition temperature may be found in a more thoroughly performed impregnation step, and therefore in higher amounts of manganese oxide species on the converted samples.

To further investigate the Mn impregnation, XRD, ICP-OES and SEM-EDX were performed before and after acidic leaching. The different analytical methods were used to determine the successful impregnation and removal of Mn and the subsequent purity of the product. In Figure 3, the XRD pattern of sample Mn-800-12* before leaching depicts well-crystallized bixbyite Mn_2O_3 and corundum. The bixbyite pattern vanishes after leaching for 4 h in concentrated HCl (Mn-800-12), implying a complete removal of Mn_2O_3. However, ICP-OES analysis (Table 1) reveals that 4.0 wt.-% of the initial 17.6 wt.-% of manganese remain in the leached Mn-800-12. Those findings are reinforced by SEM-EDX whereby 5.5 wt.-% of manganese is found on the materials surface after leaching. The discrepancy in manganese content between the ICP-OES and SEM-EDX is attributable to the different analysis methods. While the former method (ICP-OES) is a bulk analysis, SEM-EDX can only penetrate the surface ($\approx$1 μm) of the sample. As Mn is applied solely by impregnation, it will necessarily be present in higher concentrations near the surface rather than in the bulk (as shown for Mn-800-12*, Table 1).

Table 1. Mn weight-contents of selected samples as determined by ICP-OES and SEM-EDX.

Sample	Mn Weight-Content Determined by …	
	ICP-OES/wt.%	SEM-EDX/wt.%
Mn-800-12	4.0	5.5
Mn-800-12*	17.6	50.7
Mn-900-6	3.0	3.6

Although XRD depicts no crystalline manganese phases after leaching (Figure 3), ICP-OES and SEM-EDX show that a Mn content of 3 to 4 wt.-% is always retained. The value cannot be undercut, regardless of the harshness of the leaching conditions, even when boiling aqua regia is used [10]. This observation is very important regarding two aspects, Firstly, it seems to fortify the mechanism postulated by Tsyrulnikov et al. [11], hypothesizing the permanent integration of Mn^{3+}-ions into the hexagonal α-Al_2O_3 structure. Secondly, the integration of a certain percentage of Mn^{3+}-ions into the corundum structure appears to be unavoidable, as previous studies have already suggested [27].

Besides XRD, ^{27}Al-MAS-NMR spectroscopy was used to elucidate the structure of the present samples. As depicted in Figure 4, non-impregnated materials synthesized below 1100 °C (Ref-800-12, Ref-1050-6) exhibit two isolated main peaks, with high intensities at 9 and 69 ppm. In addition to the main peaks, symmetric spinning sidebands can be observed at an interval of ~100 ppm (due to the set spinning frequency of 10 kHz). The most intense peak at 9 ppm is assigned to octahedrally coordinated aluminum (θ-Al_2O_3), while the less intense peak at 69 ppm can be assigned to either γ- or δ-Al_2O_3. Based on the phase transformation of γ- to δ-Al_2O_3 around 650 °C, we assume that primarily the δ-phase is present in the samples. Compared to the XRD findings in Figure 3, where XRD amorphous patterns (Ref-800-12) or reflexes with very low intensities (Ref-1050-6) were depicted, ^{27}Al-MAS-NMR spectroscopy provided much more detailed information on the present alumina phases, as suggested by O'Dell et al. [18]. At higher temperatures of 1100 °C (Ref-1100-6), a single main peak at 13 ppm and the associated spinning side bands can be observed, which is characteristic for ordered α-Al_2O_3 [15]. The sample Mn-800-12 shows a similar periodic pattern, with an even higher intensity. Hence, the synthesis of highly ordered corundum at 800 °C by Mn-impregnation can also be confirmed by ^{27}Al-MAS-NMR spectroscopy. In addition, it was shown that no other aluminum species is present, and thus a complete conversion to corundum at 800 °C has taken place.

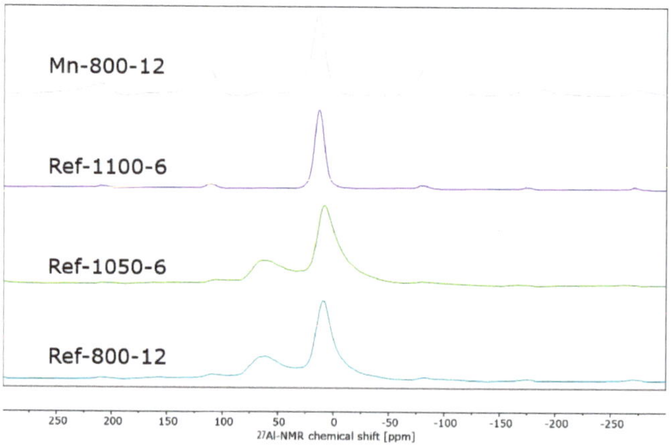

Figure 4. ^{27}Al-MAS-NMR spectrum of sample Mn-800-12, calcined at 800 °C for 12 h after acid treatment to remove Mn-species. For comparison, spectra of Mn-free reference samples are given, which were calcined at 800 °C for 12 h, 1050 °C for 6 h and 1100 °C for 6 h.

3.3. Porosity and Long-Term Stability of Al_2O_3

The pre-calcined γ-alumina reference Ref-650-6 exhibits a specific surface area A_{BET} of 276 $m^2\ g^{-1}$. It decreases to 15 $m^2\ g^{-1}$ after calcination at 1100 °C (Ref-1100-6), and even to 8 $m^2\ g^{-1}$ after calcination at 1150 °C without impregnation. In comparison, a large surface area of 56 $m^2\ g^{-1}$ can be maintained for the Mn-impregnated sample Mn-800-12. This is an unprecedented value for α-Al_2O_3, surpassing all previously reported examples [10]. In other publications, sol-gel synthesized α-Al_2O_3 has values around 10 to 20 $m^2\ g^{-1}$, much lower than the newly synthesized material [12,28,29]. However, with increasing calcination durations, the specific surface area decreases, as Figure 5 illustrates. During the first 12 h, the mesopore diameter increases from 11 to 20 nm and the mesopore volume decreases from 0.55 $cm^3\ g^{-1}$ to approximately 0.1 $cm^3\ g^{-1}$. Correspondingly, the specific surface area decreases, due to the transformation of γ-Al_2O_3 into α-Al_2O_3. (cf. Figures 3 and 4 in Section 3.2). With prolonged calcination times (>12 h), the mesopore diameter increases further, with no significant change in the pore volume, resulting in lower specific surface areas. The associated nitrogen sorption isotherms are given in Figure S1.

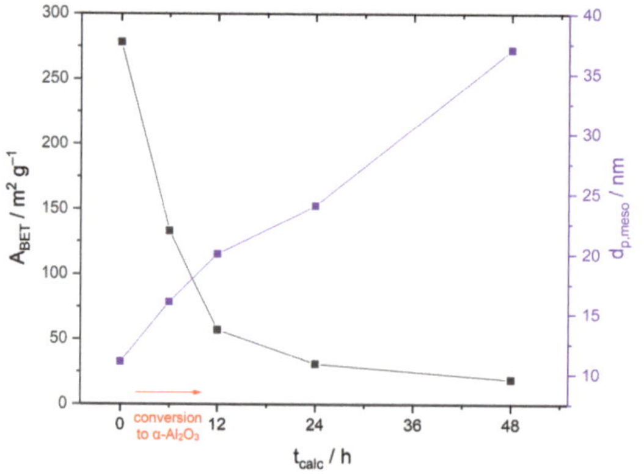

Figure 5. Specific surface areas A_{BET} (grey) and mesopore diameter $d_{p,meso}$ (purple) of Mn-impregnated samples calcined at 800 °C, plotted as a function of the calcination duration. The sharp decrease of A_{BET} within the first few hours of the calcination is due to the transformation of γ-Al_2O_3 into α-Al_2O_3, as indicated by the red arrow.

As suggested by those findings, thermal stability seems to be a significant issue of the synthesized α-Al_2O_3. This is already known for diaspore-derived corundum and transition alumina. To further investigate the long-term thermal stability of the synthesized sample Mn-800-12, the material is heated for 7 days at 750 °C. As shown in Table 2 and Figure S2, the thermal treatment resulted in a loss of specific surface area. The A_{BET} decreased by ~48% to 29 $m^2\ g^{-1}$. Interestingly, this decrease in specific surface area was far less pronounced for the starting material that was treated at the same conditions (Ref-650-6@750-172, Figure S3). Here A_{BET} decreased by just ~35% to 180 $m^2\ g^{-1}$. This means that, although porous corundum was synthesized at low temperatures, its porosity is not stable over time at common process temperatures. SEM images given in Figure 6 reveal a densification of the microstructure for Mn-800-12 during the thermal treatment, while primary and secondary particles retain approximately the same size. We assume that this effect contributes to the reduction in specific surface area as well.

Table 2. Cumulative pore volume, mesopore volume and mesopore diameter obtained by mercury intrusion and specific surface area (A_{BET}) obtained by nitrogen sorption.

Sample	$A_{BET}/m^2\ g^{-1}$	Pore Diameter/nm	Mesopore Volume/$cm^3\ g^{-1}$	Cumulative Pore Volume/$cm^3\ g^{-1}$
Ref-650-6	276	12	0.51	2.87
Ref-650-6@750-172	180	13	0.51	2.79
Mn-800-6	122	16	0.11	0.61
Mn-800-12	56	20	0.07	0.47
Mn-800-12@750-172	29	31	0.10	0.54
Mn-800-24	31	24	0.10	0.52
Mn-800-48	18	37	0.08	0.72

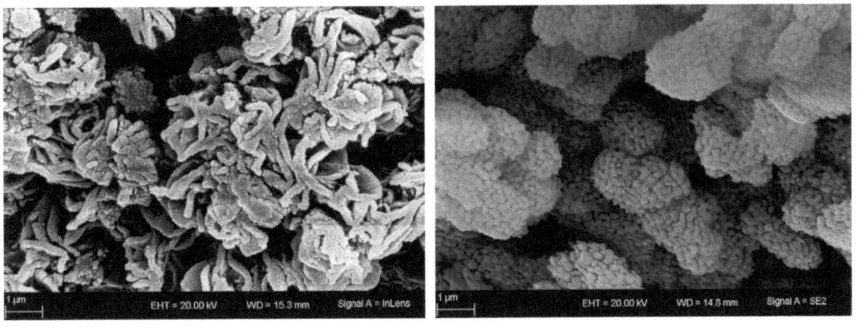

Figure 6. SEM images of samples Mn-800-12 (**left**) and Mn-800-12@750-172 (**right**).

4. Conclusions

Sol-gel derived γ-Al_2O_3 samples were impregnated with a manganese precursor. This led to a considerably lower conversion temperature of 800 °C, compared to 1100–1200 °C, which is usually required to obtain α-Al_2O_3. After complete conversion into corundum, a high specific surface area of 56 $m^2\ g^{-1}$ was obtained. Treatment in concentrated HCl led to a removal of Mn, while about 3 wt.-% remained in the sample. Along with the well-established XRD, solid-state ^{27}Al-MAS-NMR was introduced as a reliable complementary method to determine the different alumina phases.

We also investigated the long-term thermal stability of the synthesized corundum samples, which is an important issue for any application. After having been exposed to a temperature of 750 °C for 7 days, the specific surface areas decreased by ~48% to 29 $m^2\ g^{-1}$. In conclusion, although porous corundum was synthesized by impregnation with manganese precursors, its porosity does not appear to be stable over time at common process temperatures.

Supplementary Materials: The following supporting information can be downloaded at: https://www.mdpi.com/article/10.3390/ma16083047/s1. Nitrogen sorption and pore size distribution data.

Author Contributions: Conceptualization, S.C., J.M. and D.E.; Formal analysis, T.J., S.C., M.F. and M.W.; Investigation, S.C., M.F. and O.L.; Supervision, S.C., M.W., J.M. and D.E.; Validation, S.C. and M.F.; Visualization, T.J. and S.C.; Writing—original draft, T.J. and S.C.; Writing—review & editing, T.J., S.C., M.F. and D.E. All authors have read and agreed to the published version of the manuscript.

Funding: Funded by the Open Access Publishing Fund of Leipzig University, which is supported by the German Research Foundation within the program Open Access Publication Funding.

Data Availability Statement: The data presented in this study are available on request from the corresponding author.

Acknowledgments: We want to thank H. Rudzik for the measurement of ICP-OES. In addition, we want to thank P. Seidel for help with the revisions.

Conflicts of Interest: The authors declare no conflict of interest.

References

1. Bartholomew, C.H.; Farrauto, R.J. (Eds.) *Fundamentals of Industrial Catalytic Processes*, 2nd ed.; Wiley & Sons: New York, NY, USA, 2006.
2. Heck, R.M.; Farrauto, R.J.; Gulati, S.T. (Eds.) *Catalytic Air Pollution Control: Commercial Technology*, 3rd ed.; Wiley: Hoboken, NJ, USA, 2009.
3. Faure, R.; Rossignol, F.; Chartier, T.; Bonhomme, C.; Maître, A.; Etchegoyen, G.; Del Gallo, P.; Gary, D. Alumina foam catalyst supports for industrial steam reforming processes. *J. Eur. Ceram. Soc.* **2011**, *31*, 303–312. [CrossRef]
4. Murrell, L.L.; Grenoble, D.C.; DeLuca, J.P. Process for Preparing Ultra-Stable, High Surface Area Alpha-Alumina. U.S. Patent 4,169,883, 2 October 1979.
5. Hartmann, I.; Billig, E.; Bindig, R.; Carstens, S.; Liebetrau, J. *Möglichkeiten, Limitierungen und Entwicklungsbedarf zur katalytischen Emissionsminderung*; DBFZ: Frankfurt/Main, Germany, 2014.
6. Dvoracek, D. Festkörper-, Metathese- und Thermit-Reaktion. Neue Wege zu porösen Metall-bzw. Mischmetalloxid Monolithen. Ph.D. Dissertation, Universität Leipzig, Leipzig, Germany, 2015.
7. Zhang, K.; Fu, Z.; Nakayama, T.; Niihara, K. Structural evolution of hierarchically macro/mesoporous Al2O3 monoliths under heat-treatment. *Microporous Mesoporous Mater.* **2012**, *153*, 41–46. [CrossRef]
8. Wefers, K.; Misra, C. (Eds.) *Oxides and Hydroxides of Aluminum: Alcoa Technical Paper No. 19, Revised;* Aluminum Company of America: Pittsburgh, PA, USA, 1987.
9. Schaper, H.; Doesburg, E.; Dekorte, P.; Vanreijen, L. Thermal stabilization of high surface area alumina. *Solid State Ion.* **1985**, *16*, 261–265. [CrossRef]
10. Carstens, S.; Meyer, R.; Enke, D. Towards Macroporous α-Al_2O_3-Routes, Possibilities and Limitations. *Materials* **2020**, *13*, 1787. [CrossRef]
11. Tsyrulnikov, P.G.; Tsybulya, S.V.; Kryukova, G.N.; Boronin, A.I.; Koscheev, S.V.; Starostina, T.G.; Bubnov, A.V.; Kudrya, E.N. Phase transformations in the thermoactivated MnOx–Al_2O_3 catalytic system. *J. Mol. Catal. A Chem.* **2002**, *179*, 213–220. [CrossRef]
12. Carstens, S.; Enke, D. Investigation of the formation process of highly porous α-Al_2O_3 via citric acid-assisted sol-gel synthesis. *J. Eur. Ceram. Soc.* **2019**, *39*, 2493–2502. [CrossRef]
13. Carstens, S.; Splith, C.; Enke, D. Sol-gel synthesis of α-Al_2O_3 with enhanced porosity via dicarboxylic acid templating. *Sci. Rep.* **2019**, *9*, 19982. [CrossRef]
14. Meija, J.; Coplen, T.B.; Berglund, M.; Brand, W.A.; de Bièvre, P.; Gröning, M.; Holden, N.E.; Irrgeher, J.; Loss, R.D.; Walczyk, T.; et al. Isotopic compositions of the elements 2013 (IUPAC Technical Report). *Pure Appl. Chem.* **2016**, *88*, 293–306. [CrossRef]
15. Fitzgerald, J.J.; Piedra, G.; Dec, S.F.; Seger, M.; Maciel, G.E. Dehydration Studies of a High-Surface-Area Alumina (Pseudo-boehmite) Using Solid-State 1 H and 27 Al NMR. *J. Am. Chem. Soc.* **1997**, *119*, 7832–7842. [CrossRef]
16. Nýblová, D.; Senna, M.; Düvel, A.; Heitjans, P.; Billik, P.; Filo, J.; Šepelák, V. NMR study on reaction processes from aluminum chloride hydroxides to alpha alumina powders. *J. Am. Ceram. Soc.* **2019**, *102*, 2871–2881. [CrossRef]
17. Ferreira, A.; Viana-Gomes, J.; Bludov, Y.V.; Pereira, V.; Peres, N.M.R.; Castro Neto, A.H. Faraday effect in graphene enclosed in an optical cavity and the equation of motion method for the study of magneto-optical transport in solids. *Phys. Rev. B* **2011**, *84*, 235410. [CrossRef]
18. O'Dell, L.A.; Savin, S.L.P.; Chadwick, A.V.; Smith, M.E. A (27)Al MAS NMR study of a sol-gel produced alumina: Identification of the NMR parameters of the theta-Al(2)O(3) transition alumina phase. *Solid State Nucl. Magn. Reson.* **2007**, *31*, 169–173. [CrossRef]
19. Wen, H.-L.; Yen, F.-S. Growth characteristics of boehmite-derived ultrafine theta and alpha-alumina particles during phase transformation. *J. Cryst. Growth* **2000**, *208*, 696–708. [CrossRef]
20. Legros, C.; Carry, C.; Bowen, P.; Hofmann, H. Sintering of a transition alumina: Effects of phase transformation, powder characteristics and thermal cycle. *J. Eur. Ceram. Soc.* **1999**, *19*, 1967–1978. [CrossRef]

21. Tsuchida, T. Preparation of high surface area α-Al_2O_3 and its surface properties. *Appl. Catal. A Gen.* **1993**, *105*, L141–L146. [CrossRef]
22. Wefers, K. Über die thermische Umwandlung des Diaspors. *Z. Erzbergbau Met.* **1962**, *15*, 339–343.
23. Mao, C.-F.; Vannice, M.A.s. High surface area a-alumina. I. *Appl. Catal. A Gen.* **1994**, *111*, 151–173. [CrossRef]
24. Zaki, T.; Kabel, K.I.; Hassan, H. Using modified Pechini method to synthesize α-Al_2O_3 nanoparticles of high surface area. *Ceram. Int.* **2012**, *38*, 4861–4866. [CrossRef]
25. Martín-Ruiz, M.M.; Pérez-Maqueda, L.A.; Cordero, T.; Balek, V.; Subrt, J.; Murafa, N.; Pascual-Cosp, J. High surface area α-alumina preparation by using urban waste. *Ceram. Int.* **2009**, *35*, 2111–2117. [CrossRef]
26. Kim, S.-M.; Lee, Y.-J.; Jun, K.-W.; Park, J.-Y.; Potdar, H.S. Synthesis of thermo-stable high surface area alumina powder from sol–gel derived boehmite. *Mater. Chem. Phys.* **2007**, *104*, 56–61. [CrossRef]
27. Dekker, E.H.L.J.; Rieck, G.D. Revised phase diagram and X-ray data of The Mn3O4-Al2O3 System in air. *Z. Anorg. Allg. Chem.* **1975**, *415*, 69–80. [CrossRef]
28. Roque-Ruiz, J.H.; Medellín-Castillo, N.A.; Reyes-López, S.Y. Fabrication of α-alumina fibers by sol-gel and electrospinning of aluminum nitrate precursor solutions. *Results Phys.* **2019**, *12*, 193–204. [CrossRef]
29. Xu, H.; Zhang, C.; Cai, J.; Wang, J.; Liu, K.; Cheng, X. Synthesis and characterization of activated alumina with high thermal stability by a low-heat solid-phase precursor method. *Microporous Mesoporous Mater.* **2022**, *337*, 111921. [CrossRef]

materials

Article

Synthesis of Coal-Fly-Ash-Based Ordered Mesoporous Materials and Their Adsorption Application

Miaomiao Tan [1], Dahai Pan [2,*], Shuwei Chen [2], Xiaoliang Yan [1], Lina Han [3], Ruifeng Li [1] and Jiancheng Wang [4]

1 College of Chemical Engineering and Technology, Taiyuan University of Technology, Taiyuan 030024, China
2 College of Chemistry, Taiyuan University of Technology, Taiyuan 030024, China
3 College of Materials Science and Engineering, Taiyuan University of Technology, Taiyuan 030024, China
4 State Key Laboratory of Clean and Efficient Coal Utilization, Taiyuan University of Technology, Taiyuan 030024, China; wangjiancheng@tyut.edu.cn
* Correspondence: pandahai@tyut.edu.cn

Abstract: A feasible approach was developed for the synthesis of ordered mesoporous SBA-15-type materials using coal fly ash (CFA) as raw material. In the proposed approach, CFA was, firstly, activated by subcritical water with the addition of NaOH, which allowed an efficient extraction of silicon species from CFA under strong acidic conditions at near room temperature. Subsequently, in the synthesis system, using silicon extraction solution as the silicon precursor, the introduction of anhydrous ethanol as a co-solvent effectively inhibited the polymerization of silanol species and promoted their collaborative self-assembly with surfactant molecules by enhancing the hydrogen bond interactions. The resultant SBA-15 material had a high purity, high specific surface area (1014 m^2/g) and pore volume (1.08 cm^3/g), and a highly ordered mesostructure, and, therefore, exhibited an excellent removal efficiency (90.5%) and adsorption capacity (160.8 mg/g) for methylene blue (MB) from simulated wastewater. Additionally, the generation of surface acid sites from the homogenous incorporation of Al atoms onto the mesoporous walls of SBA-15 combined with the perfect retention of the ordered mesostructure endowed the obtained Al-SBA-15 material with a further boost in the removal performance of MB. The MB removal efficiency can reach ~100%, along with a maximum adsorption capacity of 190.1 mg/g.

Keywords: coal fly ash; ordered mesoporous materials; subcritical activation treatment; Al doping; adsorption

check for updates

Citation: Tan, M.; Pan, D.; Chen, S.; Yan, X.; Han, L.; Li, R.; Wang, J. Synthesis of Coal-Fly-Ash-Based Ordered Mesoporous Materials and Their Adsorption Application. *Materials* **2023**, *16*, 2868. https://doi.org/10.3390/ma16072868

Academic Editor: Miguel Ángel Sanjuán

Received: 16 February 2023
Revised: 18 March 2023
Accepted: 21 March 2023
Published: 4 April 2023

1. Introduction

Coal fly ash (CFA) generated from the combustion of coal in electric power plants is one of the main solid wastes in China, whose emissions are increasing exponentially due to the rapid demand for energy [1]. The efficient recycling of CFA has been a key issue, due to the accumulation, disposal, and under-utilization of massive amounts of CFA, posing serious hazards to the environment and human health [2]. Taking CFA containing high amounts of silica (40–60 wt.%) into consideration, more attention is focused on the synthesis of nanoporous materials using CFA as an extensive, cheap, and readily available silicon source, which provides an available option for recycling CFA to prepare high-value-added products.

With the fast development of industry, the mass discharge of pollutants triggers an increasing water contamination, which exerts a drastic threat to the ecological system. Taking the common organic dye methylene blue (MB) for example, its release into water sources can reduce light penetration, and, in turn, affects photosynthesis and induces damage to the ecosystem's balance [3]. In addition, prolonged exposure to MB may provoke many diseases, such as eye and skin irritation, digestive infections, and mental confusion, for both aquatic animals and humans [4]. The key issue in treating contaminated water is the development of high-performance adsorbents, since the adsorption technology is characterized by high efficiency, simple operation, and environmental friendliness, as compared with other wastewater treatment techniques (such as chemical precipitation, membrane filtration, degradation, and so on) [5]. Ordered mesoporous silica materials with uniform and adjustable channels, high specific surface areas and pore volumes, and large and tunable pore sizes provide important impetuses for the development of high-performance adsorbents [6] and are widely used for the efficient removal of heavy metals and toxic organic dyes [7–9]. Among various ordered mesoporous silica materials, SBA-15 type materials have received much attention, owing to their non-toxicity and high structural stability [10]. Unfortunately, when CFA was used as a raw material, typical recipes for the preparation of SBA-15 did not act effectively, which could be attributed to the low extraction efficiency of silicon species from CFA [11]. Consequently, costly silicon precursors (such as TEOS) should be added to improve the quantity of the product [12,13]. In addition, the strong acidic synthesis conditions make it is extremely difficult to directly incorporate metal atoms into the siliceous framework of SBA-15 because metals only exist in their aqueous cationic form rather than their corresponding oxo species [14]. As a result, SBA-15 with a neutral Si framework has a weak acidity and lacks active sites, which severely hinder its practical applications [15]. Therefore, it is still a great challenge to develop more economical procedures for the synthesis of metal-doping SBA-15 materials with excellent structural, textural, and surface properties using CFA as raw material.

Recently, we introduced a novel supercritical water treatment method to activate coal gangue for extracting Si species [16]. Compared to the extensively used high-temperature alkali fusion method (a high temperature (600–900 °C) and a high usage amount of alkali were required), our method involves a much lower activation temperature, a shorter activation time, and a smaller usage amount of alkali; thus, it can effectively reduce the energy consumption. More importantly, through our method, an Si extraction efficiency of 69.2% from the coal gangue can be obtained, which is much higher than the reported activation methods such as high-temperature alkali fusion [17], strong alkali dissolution [18], mechanical activation [19], and so on. Inspired from our previous work [16], herein, CFA was activated using the subcritical water with the addition of NaOH, and ordered mesoporous siliceous SBA-15 material, derived from CFA, was designed and successfully synthesized. In addition, to improve the surface acidity and adsorption performance of the obtained SBA-15, aluminum (Al) atoms were successfully incorporated into the silica framework by adjusting the pH hydrothermal grafting method. The resultant materials of SBA-15 and Al-SBA-15 possessed highly ordered two-dimensional (2D) hexagonal mesostructures, narrow mesoporous size distributions, and high specific surface areas and pore volumes, and they were evaluated as efficient adsorbents for the removal of methylene blue (MB) from the simulated wastewater. The subcritical activation mechanism of CFA, the crucial role of ethanol as a co-solvent in promoting the collaborative self-assembly between the inorganic silicon species and the organic surfactant molecules, and the promoting effect of Al doping on MB removal performance were investigated.

2. Materials and Methods

2.1. Raw Materials and Chemicals

Coal fly ash (CFA) was derived from a coal-fired power plant (Shanxi, China). Triblock copolymer P123 ($EO_{20}PO_{70}EO_{20}$) was obtained from Sigma-Aldrich Pty., Ltd., St. Louis, MI, USA. Sodium hydroxide (NaOH), hydrochloric acid (HCl), acetic acid (HAc), aluminum nitrate nonahydrate ($Al(NO_3)_3 \cdot 9H_2O$), and anhydrous ethanol were purchased from Sinopharm Group Chemical Reagent Co., Ltd., China. All commercially available chemical reagents were of an analytical grade and used without further purification.

2.2. Preparation of SBA-15-Type Mesoporous Materials

As shown in Figure 1, the synthesis of ordered mesoporous Al-SBA-15 was carried out in three stages: (1) silicon extraction from CFA, (2) preparation of siliceous SBA-15, and (3) grafting Al atoms into the mesoporous silica framework of SBA-15.

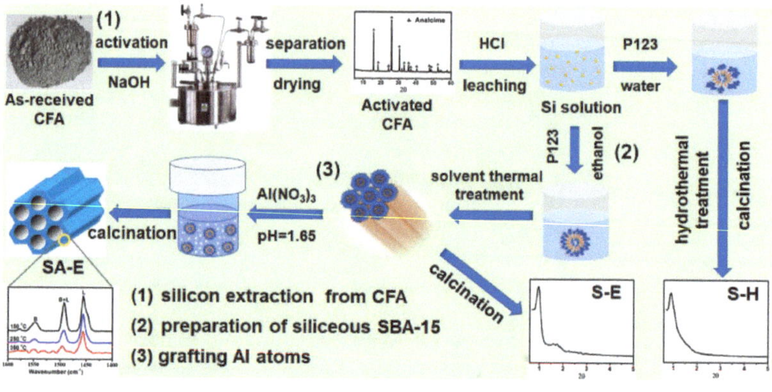

Figure 1. Schematic illustration for the synthesis of mesoporous materials of S-H, S-E, and SA-E.

Silicon extraction was performed through a subcritical hydrothermal method. The as-received CFA (10 g) was mixed with 120 mL of NaOH solution (0.5 mol/L) in an autoclave reactor, and the mixture was reacted at 250 °C with a pressure of 400 kPa for 2 h under stirring (500 rpm). The activated CFA was collected by filtration, washed, dried in air, and dispersed in 140 mL of 2.5 M HCl solution. After vigorously stirring at 45 °C for 1.5 h, the supernatant silicon-rich solution was obtained using a filtration and separation process, which was used as silicon precursor for the synthesis of mesoporous silica materials.

During the preparation of the as-synthesized siliceous SBA-15, 2.2 g of P123 and 5.5 g of HAc were dissolved in 18 mL of anhydrous ethanol, followed by the slow addition of silicon extraction solution prepared in the previous step under vigorous stirring at 38 °C. After reacting for 24 h, the mixture was transferred into an autoclave for solvothermal treatment at 100 °C for 24 h. The resulting precipitate was filtered and washed with deionized water.

During the third stage, the ordered mesoporous Al-SBA-15 was prepared using a post-synthesis grafting method. The as-synthesized SBA-15 was mixed with 60 g of deionized water containing 0.677 g of $Al(NO_3)_3{\cdot}9H_2O$. The pH value of the solution was adjusted to 1.65 using 0.5 M HCl. The mixture was transferred into an autoclave for hydrothermal Al-grafting treatment at 100 °C for 24 h. The final product was collected by filtration, washed, dried, calcined at 550 °C for 5 h to remove the organic template, and named SA-E. For comparison, siliceous SBA-15 material prepared during the second stage with a calcination treatment at 550 °C for 5 h was denoted as S-E.

In order to investigate the effect of co-solvent anhydrous ethanol on the synthesis of siliceous SBA-15, 18 mL of deionized water was used instead of anhydrous ethanol for the dissolution of template P123. The mesoporous silica material was synthesized following the same procedure as that for preparing S-E, and the final product was designated as S-H. In addition, a SBA-15 material was also prepared by the conventional method using the chemical TEOS as silicon precursor [20] and named S-O.

2.3. Adsorption Procedure

Adsorption of methylene blue (MB) dye was conducted using batch balance experiments. At 25 °C, a required amount (0.1, 0.3, and 0.5 g/L) of obtained mesoporous material (including S-O, S-H, S-E, and SA-E) was suspended in a 50 mL of 10–70 mg/L MB solution with various pH values under stirring at 200 rpm. The sampling was performed at different time intervals. The MB content in the solution before and after the adsorption was determined using a UV-vis spectrophotometer at $\lambda = 664$ nm. The removal efficiency (Y) and adsorption capacity (Q_e) were calculated using Equations (1) and (2), respectively.

$$Y = \frac{C_0 - C_e}{C_0} \times 100 \quad (1)$$

$$Q_e = \frac{(-C_e)V}{m} \quad (2)$$

where C_0 and C_e stand for the initial and equilibrium concentration of MB in solution, respectively (mg/L), V denotes the volume of MB solution (L), and m is the adsorbent weight (g).

2.4. Characterization

The chemical compositions of both the as-received CFA and subcritical activated CFA were analyzed using a Panalytical Epsilon1 X-ray fluorescence (XRF) spectrometer. The concentration of the silicon-rich solution was measured using a Thermo-Fisher ICAP 6300 inductively coupled plasma optical emission spectrometer (ICP-OES). Powder X-ray diffraction (XRD) patterns were recorded on a Rigaku MiniFlex 600 diffractometer using Ni-filtered Cu Kα radiation. Transmission electron microscopy (TEM) was carried out on a JEM 2100 microscope operated at 200 kV. Nitrogen adsorption–desorption isotherms were measured on an ASAP 2460 analyzer at 77 K. The BET method was used for the calculation of specific surface areas, while the BJH method was used for the determination of pore size distributions derived from the adsorption branches of isotherms. The total pore volumes were calculated according to the amount of nitrogen adsorbed at the relative pressure of 0.99. The in situ Fourier transform infrared (FTIR) spectra were obtained using a Bruker TENSOR II FTIR spectrometer. Before analysis, the samples were pressed into wafers with the same weight and area and then mounted onto a high-vacuum cell for dehydration treatment under a high vacuum (10^{-6} Pa) at 140 °C for 5 h. After the complete removal of physically adsorbed water, the concentration of silanol (Si-OH) groups existing within the mesoporous silica wall of samples was determined according to the integral area of the Si-OH absorbent band. The size of silanol species existing in the acid-leaching desilication solution with and without the introduction of anhydrous ethanol were measured on a Microtrac MN42X laser diffraction particle size analyzer. The thermogravimetric analysis (TGA) was performed using a NETZSCH STA 449F3 instrument. The samples were heated in air with the heating rate of 10 °C/min from 100 °C to 500 °C. Elemental mapping measurements were performed on a JEOL JSM-7900F scanning electron microscope instrument with an energy dispersive X-ray (EDX) spectrometer. The acidity of samples was measured by the temperature-programmed desorption of NH_3 (NH_3-TPD) using a Micromeritics AUTOCHEM II 2920 analyzer that was equipped with a thermal conductivity detector (TCD). Before measurement, 100 mg of the sample was pretreated at 550 °C for 1 h and then cooled to 100 °C in a pure He flow. A 10% NH_3-He flow (25 mL/min) was used to saturate the sample with ammonia for 30 min and then replaced with a pure He flow (25 mL/min) to remove the physically adsorbed NH_3 for 1 h. Subsequently, the NH_3-TPD was carried out from 100 °C to 550 °C at a rate of 10 °C/min in a pure He flow. The desorbed NH_3 was monitored using TCD. The adsorption of pyridine was conducted using a TENSOR II FTIR instrument, and the types of acid sites in the samples were evaluated. Before the adsorption of pure pyridine vapors at room temperature, the sample was pretreated at 500 °C for 4 h under a high vacuum to remove the impurities adsorbed in the sample. The FTIR spectra of the sample, evacuated at 150, 250 and 350 °C, were recorded using a Bruker TENSOR II FTIR spectrometer. The concentration of MB was determined using a Shanghai MC-721 spectrophotometer.

3. Results and Discussion

3.1. The Subcritical Activation of CFA

The chemical compositions of as-received and activated CFA and the elemental concentrations in the acid leaching desilication solution (silicon-rich solution) derived from the activated CFA are listed in Table 1.

Table 1. Chemical compositions of coal fly ash before and after the activation (wt.%) and of silicon-rich solution (mg/mL).

	SiO_2	Al_2O_3	Fe_2O_3	CaO	TiO_2
Original CFA	47.9	36.3	6.2	3.1	1.7
Activated CFA	52.9	28.7	6.6	3.6	1.8
	Si	Al	Fe	Ca	Ti
Silicon-rich solution	7.22	8.26	1.03	0.95	0.43

The principal components of the original CFA were SiO_2 (47.9 wt.%), Al_2O_3 (36.3 wt.%), and Fe_2O_3 (6.2 wt.%). After high-temperature subcritical water treatment, an evident increase in the content of SiO_2 (52.9 wt.%) was observed for the activated CFA. To illustrate the Si activation mechanism of CFA, the mineral composition of CFA before and after the activation treatment was characterized using XRD analysis. As shown in Figure 2, the major crystalline components of the original CFA were stable mullite and quartz along with other amorphous aluminosilicates. For activated CFA, the diffraction peaks of mullite and quartz disappeared; meanwhile, a new crystalline phase, square zeolite (Analcime, $Na(Si_2Al)O_6{\cdot}H_2O$; JCDPS 41-1478) [21], was detected through the diffraction peaks at the 2θ values of 15.76°, 18.22°, 25.92°, 30.50°, 33.20°, 35.78°, 40.46°, 47.72°, and 52.46°, which were assigned to the (211), (220), (400), (332), (431), (521), (611), (640), and (651) reflections of Analcime, respectively. This result indicates that the introduction of alkali can efficiently enhance the reactivity of SiO_2 and Al_2O_3 and broke the aluminosilicates framework of mullite to free the elements of Si and Al during the high-temperature subcritical water treatment process [22]. Under the promotion of Na^+, the secondary reaction between the dissolved SiO_3^{2-} and $Al(OH)_4^-$ accounted for the formation of square zeolite [23] and the increased content of SiO_2 for the activated CFA. The results from the elemental analysis confirmed that the square zeolite in activated CFA displayed a high solubility in the HCl solution. A silicon concentration of 0.258 mol/L was found in the acid leaching solution (Table 1). Based upon the calculations performed considering the amount of silica in the original CFA and the amount of silica actually dissolved in the extraction solution, the extraction efficiency of silica reached the value of 46.9%, which was relatively higher or comparable compared to the earlier reported Si extraction techniques (for example, alkali fusion [24], the ammonium fluoride microwave-assisted method [25], and the alkaline-acid leaching process [26]). Even so, it should be pointed out that the silicon concentration in the extraction solution was still far below the required concentration (about 0.33 mol/L) for the synthesis of SBA-15 material using the conventional synthesis method [27].

3.2. The Role of Ethanol as a Co-Solvent for the Synthesis of SBA-15-Type Materials from CFA

The supernatant acid leaching solution of silicon species extracted from CFA was used as the only silicon source to prepare mesoporous silica materials (Samples S-H and S-E). As a reference, the chemical TEOS was also used as the only silicon source for the preparation of ordered mesoporous SBA-15 material (Sample S-O). The small-angle XRD patterns of obtained samples of S-O, S-E, and S-H, prepared with different synthesis solutions, are shown in Figure 3a. Sample S-O, prepared using TEOS and water as silicon source and solvent, respectively, displayed three well-resolved diffraction peaks at 2θ = 0.97, 1.63, and 1.90°. According to the *d*-spacing reciprocal ratio, such three diffraction peaks can be indexed as the (100), (110), and (200) reflections of an ordered 2D hexagonal mesostructure (space group *p6mm*), respectively. Different from Sample S-O, Sample S-H, prepared using silicon species extracted from CFA as silicon precursor and water as the only solvent, only exhibited a strong peak at 0.90° in its XRD pattern. The absence of the other two weak diffraction peaks suggests that Sample S-H possesses a disordered worm-like mesoporous structure [28], corresponding to a less ordering in the mesoporous arrangement. This indicates that it is difficult to synthesize a highly ordered mesoporous SBA-15 material in an aqueous solution using silicon species extracted from CFA as a silicon source due to the relatively low silicon concentration (Table 1). Such a conclusion is in accordance with the previously reported results [29]. Interestingly, for the Sample S-E, prepared using the anhydrous ethanol as co-solvent, an intense diffraction peak with a narrower full width at half-maximum and two additional weak diffraction peaks were observed at 0.94°, 1.56°, and 1.78°, respectively. These three diffraction peaks could be assigned to the (100), (110), and (200) reflections, respectively [30]. Such a result revealed that the introduction of anhydrous ethanol in the synthesis solution was favorable for the synthesis of siliceous SBA-15 material with an ordered 2D hexagonal mesostructure (space group *p6mm*) when CFA was used as the raw material. According to the intense (100) diffraction peak, a d_{100} spacing of 9.39 nm was calculated for Sample S-E, which corresponded to a unit cell parameter (a) of 10.84 nm (Table 2). In addition, no diffraction peak was observed in the wide-angle XRD pattern of Sample S-E (Figure 3b), suggesting that Sample S-E was amorphous in nature. The chemical composition from ICP analysis showed that the content of Si in Sample S-E accounted for 100%. The absence of metal elements, such as Al, Fe, and Ti can be attributed to the use of strong acid synthesis solution, in which metals exist only in their cationic form and are unable to be introduced into the mesoporous siliceous skeleton of S-E.

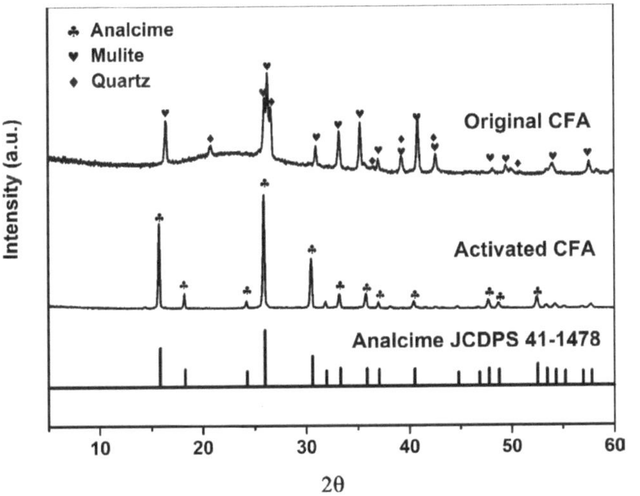

Figure 2. Wide-angle XRD patterns of CFA before and after the subcritical hydrothermal activation treatment.

TEM images of Sample S-E further provided a direct observation of highly ordered mesoporous arrangements. As seen in Figure 4a,b, honeycomb-like hexagonally packed mesopores and cylindrical parallel mesoporous channels along the (100) and (110) directions can be clearly discerned, respectively, indicating that Sample S-E has a typical 2D hexagonal ordered mesostructure with a uniform mesoporous size and wall thickness [31]. Based upon the dark and white contrast of TEM images, the distance between the two adjacent mesopores was ca. 10.52 nm, which is consistent with the value of the unit cell parameter calculated from the small-angle XRD analysis (Figure 3a). In addition, the results from SAED analysis (the inset in Figure 2b) further confirmed that Sample S-E is amorphous in the nature of mesoporous walls, which is well consistent with the observation from the wide-angle XRD analysis (Figure 3b).

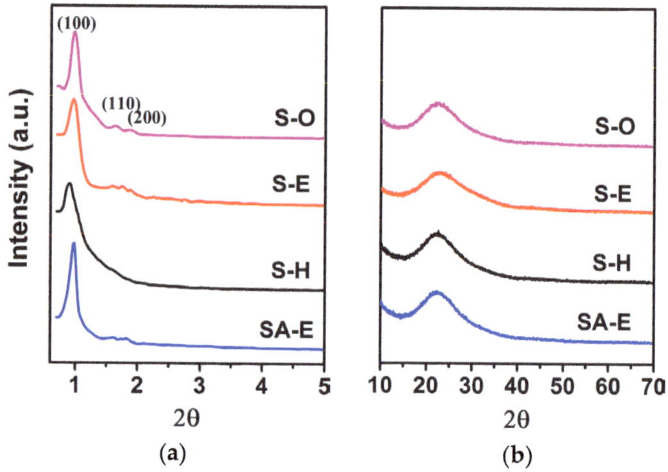

Figure 3. Small-angle (**a**) and wide-angle (**b**) XRD patterns of Samples S-O, S-H, S-E, and SA-E.

Table 2. Physicochemical properties of Samples S-H, S-E, and SA-E [a].

Sample	d (nm)	a (nm)	S_{BET} (m^2/g)	V_p (cm^3/g)	D_p (nm)
S-O	9.01	10.40	768	1.09	7.84
S-H	9.59	- [b]	855	0.90	8.62
S-E	9.39	10.84	1014	1.08	7.06
SA-E	9.03	10.42	871	1.14	8.34

[a]: d represents the d-spacing calculated from the first strong diffraction peak; a represents the unit-cell parameter; S_{BET} represents the BET specific surface area; V_P represents the total pore volume; D_p represents the average mesoporous diameter. [b]: The unit cell parameter of Sample S-H was not calculated due to its less mesostructural ordering.

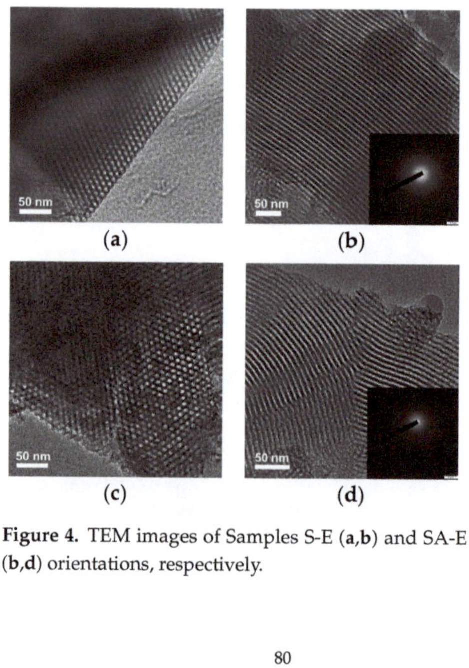

Figure 4. TEM images of Samples S-E (**a**,**b**) and SA-E (**c**,**d**) viewed along the [100] (**a**,**c**) and [110] (**b**,**d**) orientations, respectively.

To ascertain the effect of introducing anhydrous ethanol as the co-solvent on the textural properties of resultant Sample S-E, N_2 physisorption measurement was performed. Figure 5 shows the N_2 adsorption–desorption isotherms and the corresponding pore size distribution curves of Samples S-O, S-E and S-H, which were prepared with different synthesis solutions [32]. The detailed textural properties of samples are summarized in Table 2. From Figure 5a, it can be seen that all Samples S-O, S-E and S-H exhibited a typical Type IV isotherm with a Type H1 hysteresis loop, which is typical characteristics of mesoporous materials with large cylindrical mesopores. It is noteworthy that, compared with the Sample S-H, the Samples S-O and S-E demonstrated a steeper capillary condensation step occurring at a relative pressure (P/P_0) within the range of 0.65–0.75, which corresponded to a narrower mesopore size distribution (Figure 5b). Additionally, the porosities reflected from the isotherms of Samples S-O and S-E were much higher than that of Sample S-H. Based upon the calculations, the specific surface areas, pore volumes, and average mesoporous sizes of Samples S-O, S-H, and S-E were 768, 855, and 1014 m^2/g; 1.09, 0.90, and 1.08 cm^3/g; and 7.84, 8.62, and 7.06 nm, respectively (Table 2). The highly regular and ordered mesostructure can be responsible for the significantly improved textural properties for Sample S-E, as compared with those of Sample S-H. In addition, as shown in Figure 5b, the Samples S-E and S-H demonstrated higher microporous content than that of Sample S-O, which can account for the higher specific surface areas for Samples S-E and S-H than that of Sample S-O.

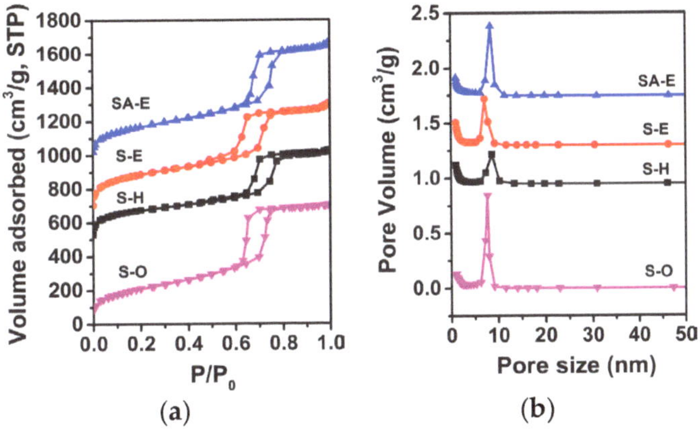

Figure 5. N_2 adsorption–desorption isotherms and the corresponding pore size distribution curves of samples S-O, S-E, S-H, and SA-E. For clarity, in (**a**), the isotherms of S-H, S-E, and SA-E are offset along the Y axis by 440, 610, and 930 cm^3/g, respectively. In (**b**), the Y axis values are 0.9, 1.3, and 1.7 cm^3/g higher for Samples S-H, S-E, and SA-E, respectively.

According to the characterization results from XRD, TEM, and N_2 physisorption analyses, it can be inferred that the introduction of anhydrous ethanol as a co-solvent played a crucial role in promoting the collaborative self-assembly between the surfactant P123 molecules and the silicon species extracted from CFA for the synthesis of ordered mesoporous SBA-15 material with excellent properties. To understand the role of introduced anhydrous ethanol, high-vacuum in situ FTIR analysis was used to semi-quantitatively compare the relative concentration of silanol existing within the mesoporous walls of Samples S-E and S-H [33]. The relative silanol concentration was estimated from the integral area of the Si-OH band after completely removing the physisorbed water (Figure 6). Taking Sample S-E as an example, as shown in Figure 6a, before the evacuation treatment under high vacuum, its FTIR spectrum showed one sharp adsorption peak at 1630 cm^{-1}, which can be ascribed to the bending mode of water molecules [34]. The broad adsorption band at 2800–3900 cm^{-1} can be attributed to the stretching hydroxyl (νO-H) from the contribution of both the silanol groups and the physisorbed water, provided no other impurity components were present in Sample S-E. After evacuation treatment under a high vacuum, the adsorption peak at 1630 cm^{-1} could not be detected in the FTIR spectrum of Sample S-E, confirming that the physisorbed water was completely removed. Therefore, the integral area of the νO-H band can be used to estimate the relative concentration of silanol groups.

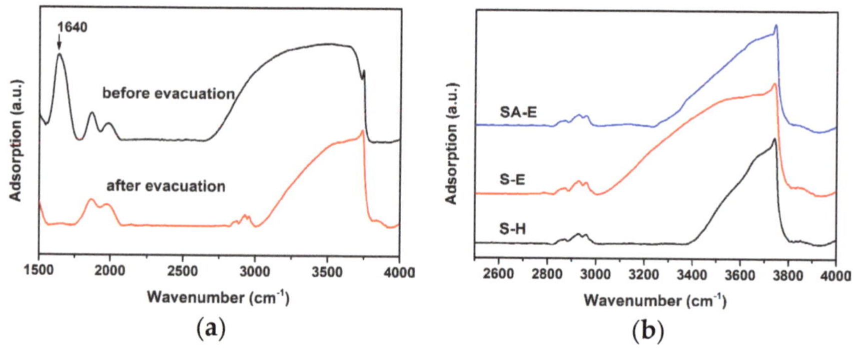

Figure 6. FTIR spectra of Sample S-E before and after the evacuation under vacuum (**a**); FTIR spectra of Samples S-H and S-E after evacuation under vacuum (**b**).

Figure 6b shows the FTIR spectra of Samples S-E and S-H after evacuation treatment. It can be seen that a larger integral area of the νO-H band was distinguished for Sample S-E, compared to Sample S-H, indicating that the silanol concentration of Sample S-E was much higher than that of Sample S-H. Such an observation strongly suggests that the introduction of anhydrous ethanol with weak polarity as a co-solvent combined with the coordination role of HAc can effectively inhibit the polymerization of silanol species in the acid-leaching solution extracted from CFA. As a result, a large number of silanol species with enough small sizes remained in the synthesis solution, which is vital in enhancing the hydrogen bond interactions between the silicon species and the hydrophilic segment (PEO) of surfactant P123 [35], thus driving their collaborative self-assembly to form an organic–inorganic hybrid composite with an ordered mesoporous structure. On the contrary, when only water was used as the solvent, the rapid hydrolysis and condensation of silicon species catalyzed by the hydrochloric acid led to the generation of polymeric silicon species and/or nano-silica particles with low silicon content and a prominently increased size, due to which the hydrogen bond interactions between the silicon species and the surfactant molecules were significantly weakened. Taking this fact into account, once the CFA extraction solution with a very low silicon concentration was used as the only silicon source, it was difficult to prepare the ordered mesoporous SBA-15 materials using the common synthesis method without the introduction of anhydrous ethanol.

To underpin the proposed deduction, a mixture solution of acid-leaching desilication solution and anhydrous ethanol was used to measure the polymerization extent of silanol species after aging 1 h, in which the volume ratio of two liquids was same as that of the synthetic solution for Sample S-E. As shown in Figure 7, an average size of 3.0 nm was detected for the silicon species, which was much smaller than that of silicon species existing in the acid-leaching desilication solution after the same aging time. Such a result sufficiently confirmed that the introduction of anhydrous ethanol with weak polarity could effectively inhibit the polymerization of silanol species. As a result, the silicon species existing in the synthetic solution possessed enough small sizes and rich surface hydroxyl contents. In addition, TG analysis was also used to compare the retained amount of P123 in different as-synthesized samples. As shown in Figure 8, one mass loss step was observed within the temperature range of 130–250 °C for the as-synthesized Samples S-E and S-H, which could be assigned to the decomposition of surfactant P123 that was retained within the mesopores [36]. Interestingly, compared to the as-synthesized Sample S-H, the as-synthesized Sample S-E exhibited a much steeper decomposition behavior of P123, which could be ascribed to the highly uniform dispersion of surfactant P123 in confined mesopores with an ordered arrangement (Figures 3a and 5b). It was also noticeable that the as-synthesized Sample S-E showed a higher mass loss and higher initial decomposition temperature of P123 than those of the as-synthesized Sample S-H, indicating that the increased surface hydroxyl content of silicon species (Figure 6b) could significantly enhance the interaction between silicon species with surfactant P123 due to the introduction of anhydrous ethanol in the synthesis solution. Consequently, a larger amount of surfactant P123 in the synthesis solution acted as a template for the formation of ordered mesostructures.

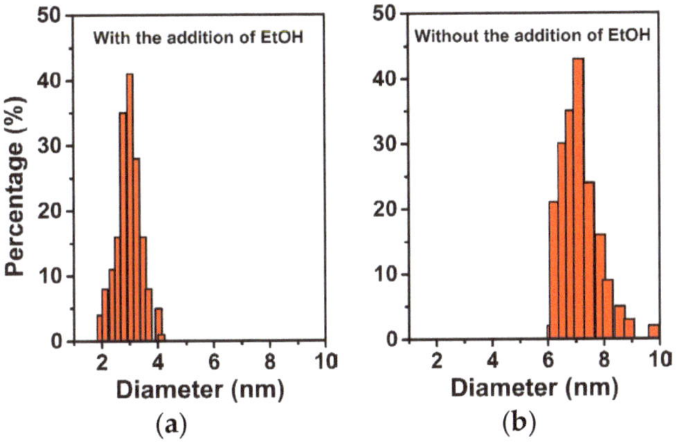

Figure 7. The size distribution of silicon species existing in the acid-leaching solution with (**a**) and without (**b**) the introduction of anhydrous ethanol.

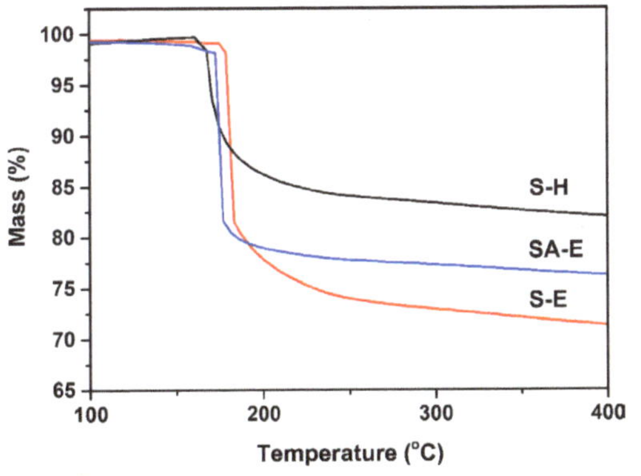

Figure 8. TG curves of as-synthesized Samples S-H, S-E, and SA-E.

It is reasonably accepted that the abundant silanol groups in mesoporous walls of SBA-15 were beneficial to the grafting reactions between Si-OH and Al-OH for the preparation of Al-containing SBA-15 material with strong acidity [37]. In this regard, Sample S-E was used as a silica matrix for the incorporation of Al using the pH-adjusting hydrothermal treatment method [22], resulting in the successful synthesis of ordered mesoporous Al-SBA-15 material. As shown in Figure 3a, the resultant Sample SA-E exhibited three well-resolved diffraction peaks indexed as the (100), (110), and (200) reflections, indicating that the highly ordered mesostructure of Sample S-E was well-preserved during the hydrothermal grafting

Al treatment process. A unit cell parameter (*a*) of 10.42 nm was calculated for the Sample SA-E (Table 2). The ordered mesostructure of SA-E was further reflected in the TEM images. The ordered 2D hexagonal arrangement of mesopores and the alignment of cylindrical mesopores were clearly observed along the [100] and [110] directions, respectively, as shown in Figure 4c,d. Sample SA-E displayed a typical Type IV isotherm along with a H1 hysteresis loop (Figure 5a) and a narrow mesopore size distribution (Figure 5b). As shown by the results presented in Table 2, the specific surface area, pore volume, and mesoporous size of SA-E were 871 m^2/g, 1.14 cm^3/g, and 8.34 nm, respectively. Further framework shrinkage of Sample S-E during the second hydrothermal treatment can be responsible for the decrease in the unit cell parameter and specific surface area for Sample SA-E, as compared with those of Sample S-E. In addition, compared to the matrix Sample S-E, the condensation between Si-OH and Al-OH and the formation of Si-O-Al bonding during the post-synthesis grafting Al process contribute to a remarkable decrease in the silanol concentration of Sample SA-E, reflected from its decreased integral area of the νO-H band (Figure 6b), which can be responsible for the lower mass loss for Sample SA-E (Figure 8). It is worthy of noting that no characteristic reflection peaks associated with crystalline alumina were detected in the wide-angle XRD pattern of Sample SA-E (Figure 3b), suggesting that aluminum species had been highly homogenously incorporated into the amorphous mesoporous walls of Sample SA-E (Figure 4d). Such expected distribution of Al and Si instead of the presence of alumina aggregations was further confirmed by the element mapping images of Sample SA-E. As shown in Figure 9, for a randomly scanned region, elements O, Si, and Al were highly uniformly distributed throughout the whole sample. Compared with the initial Si/Al ratio of 20 for the synthesis of Sample SA-E, a final Si/Al ratio of 29 was measured by the inductively coupled plasma (ICP) analysis, indicating that most of the Al species existing in the synthesis solution could be, ultimately, homogeneously incorporated into the mesoporous silica walls of S-E.

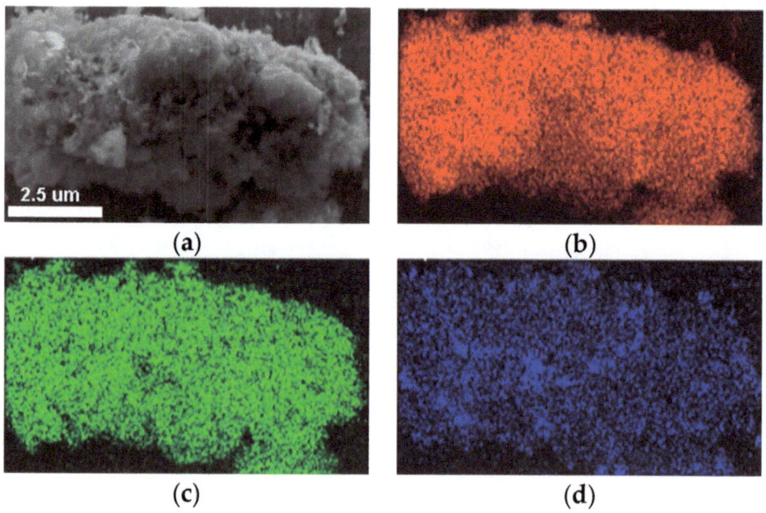

Figure 9. SEM image (**a**) and the elemental analysis mapping showing the distribution of O (**b**), Si (**c**), and Al (**d**) for the Sample SA-E.

Furthermore, NH_3-TPD measurement was used to evaluate the surface acidity of SA-E, and Samples S-E and S-H were used for comparison. It can be seen from Figure 10a that different from Samples S-E and S-H that demonstrated an inconspicuous desorption peak of NH_3, Sample SA-E exhibited a broad and asymmetric NH_3 desorption peak within the temperature range of 120–400 °C, which corresponded to a wide distribution of acid strength. In terms of the difference in the desorption temperature of NH_3, the acidic sites could be classified into three categories regarded as weak (100–250 °C), medium (250–350 °C), and strong (>350 °C) acid sites [22]. The Gaussian peak fitting method was used to deconvolute the NH_3-TPD profile of Sample SA-E for comparing the distribution of acidic sites. As shown in Figure 10a, three fitted desorption peaks with the maximum temperature at 205, 275, and 405 °C were observed, suggesting that the introduction of Al atoms favored the generation of acid sites with different strengths. Moreover, the calculated fitted peak areas were 3.29, 4.02, and 1.02 (a.u.), respectively, indicating that the amount of desorbed NH_3 mainly came from the acid sites with weak and medium acid strengths.

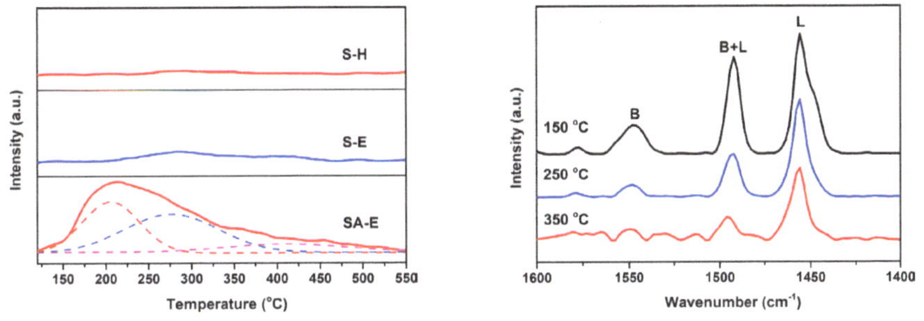

Figure 10. NH_3-TPD profiles of Samples S-H, S-E, and SA-E (**a**), and pyridine-FTIR spectra of Sample SA-E after degassing at different temperature (**b**).

To further ascertain the types of acid sites on the surface of Sample SA-E, pyridine adsorption spectroscopy measurement was performed. As shown in Figure 10b, after evacuation at 150 °C, three sharp adsorption bands arising from the C-C stretching vibrations of pyridine were clearly discerned within the wavenumber range of 1600–1400 cm^{-1}. It is widely accepted that the bands at 1547 and 1456 cm^{-1} can be attributed to the adsorption of pyridine molecules by surface B acid sites and L acid sites, respectively. Moreover, the band at 1492 cm^{-1} can be ascribed to the simultaneous adsorption of pyridine molecules by both surface B and L acid sites [22]. Evidently, Sample SA-E possessed both the B and L acid sites. The generation of B acid sites reveals that Al atoms had been successfully incorporated into the mesoporous silica framework of S-E to form bridging hydroxyl groups (Si-OH-Al; B acid sites) during the hydrothermal treatment process under weakly acidic conditions. Based upon the calculations, the total acid content was determined to be 0.871 mmol/g for Sample SA-E, in which the ratio of B acid sites to L acid sites was 0.33. The numbers of B and L acid sites with medium and strong acid strengths were further quantitatively estimated by degassing at 250 and 350 °C, respectively. From Figure 10b, it can be seen that the intensity of characteristic adsorption bands identifying B and L acid sites distinctly decreased with the increase in the evacuation temperature. This variation is more noticeable for the absorption band at 1547 cm^{-1}, suggesting that the surface of Sample SA-E was rich in weak and medium L acid sites. After degassing at 250 °C, the calculated acid content with medium and strong strengths was 0.546 mmol/g, in which the quantity of the L acid sites was 0.486 mol/L. Further degassing at 350 °C, the numbers of strong B acid sites and L acid sites were calculated to be 0.016 and 0.126 mmol/g, respectively.

3.3. Removal of Methylene Blue Using the Prepared Mesoporous Materials

Methylene blue (MB) is a common toxic organic dye, and its massive discharge seriously endangers the aquatic ecosystem and exerts a drastic threat to human health [4]. The potential of Samples S-H, S-E, and SA-E derived from CFA in the removal of MB dye from simulated wastewater was evaluated using adsorption experiments and compared with that of ordered mesoporous material SBA-15 (Sample S-O). For an adsorption process, contact time exerts considerable influence on the adsorption kinetics, and the shorter equilibrium time corresponds to the higher adsorption efficiency. In addition, the pH value is a vital parameter influencing the adsorption efficiency of adsorbents. Thus, to investigate the effects of contact time and pH on the MB adsorption, 15 mg of absorbent (including S-O, S-H, S-E, and SA-E) was added to 50 mL of 50 mg/L MB solution with different pH values at 25 °C. As shown in Figure 11, for four samples at different pH values, the amount of MB adsorbed increased with the duration of the contact time, and the process of MB adsorption seemed to have occurred in two steps, namely, the rapid initial kinetic adsorption process occurring within 10 min, followed by an assuasive adsorption process. After 40 min of contact time, no evident increase in the adsorbed amount of MB was observed over the four absorbents. The high specific surface area and large mesoporous size (Table 2) endowed mesoporous silica materials (Samples S-O, S-H, and S-E) with a large amount of accessible active adsorption sites and a fine performance for the mass transfer of MB, which can be responsible for their high removal efficiency of MB during the initial adsorption process. As the adsorption sites were rapidly occupied by MB molecules, the adsorption capacity tended to stabilize.

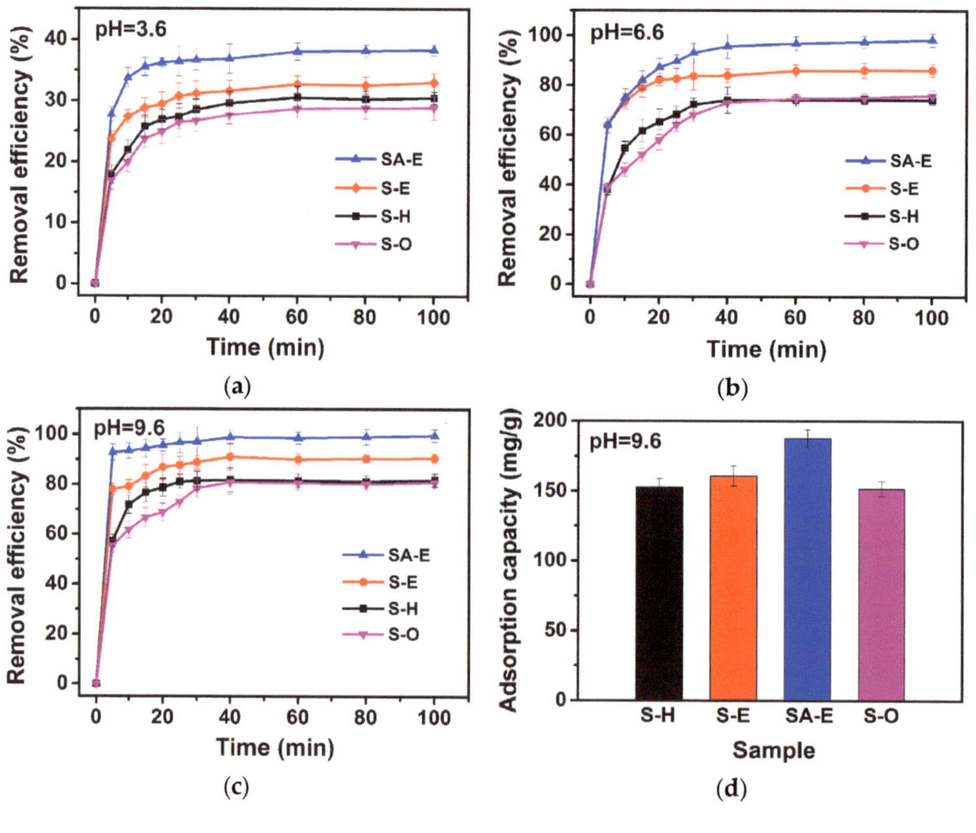

Figure 11. Effects of pH on the removal efficiency of methylene blue (**a**–**c**) and adsorption capacity of methylene blue (**d**) over Samples S-O, S-H, S-E, and SA-E.

In addition, it can be found from Figure 11 that under different pH value conditions, the Samples S-H and S-E (especially Sample S-E) exhibited a much faster MB removal rate, as compared to Sample S-O. Among all mesoporous silica adsorbents, Sample S-E demonstrated the highest MB removal efficiency. The significantly improved MB removal performance for S-E can be attributed to its higher specific surface area than those of Samples S-O and S-H (Table 2), which exerted considerable influences on increasing surface active adsorption sites. Interestingly, the initial pH value of MB solution significantly affects the removal performance of mesoporous silica adsorbents. It was observed that with the increase in the pH of MB solution, both the removal rate and the adsorption capacity distinctly increased for Sample S-O, S-H, and S-E. As reported by the previous literatures [38,39], for the silica adsorbents, the surface Si-OH groups were identified as the active sites for the adsorption of cationic MB dye by the hydrogen bond and electrostatic interactions. When the pH is higher than the isoelectric point of silicon (2.5–3.0), the surface Si-OH groups of silica absorbents easily dissociate to surface Si-O^- groups with a negative surface charge. With the increase in the pH of MB solution, the surface charge of Si-O^- groups become more negative, which is more favorable for enhancing the electrostatic attractions between the silica adsorbent with the cationic MB molecules [4]. When a pH of 9.6 was used, a maximum MB removal efficiency of 90.5% and an adsorption capacity of 160.8 mg/g were observed for the Sample S-E.

It is worth noticing that the highly homogeneous incorporation of Al onto the surface of Sample S-E provided a further boost in the removal performance of MB. The highest MB removal efficiency of 98.8% along with a MB adsorption capacity of 188.1 mg/g was obtained for Sample SA-E. Taking into consideration that Sample SA-E exhibited a relatively lower specific surface area than that of Sample S-E (Table 2), it is reasonable to believe that the introduction of Al atoms onto the surface of ordered mesoporous silica material played a positive role in the adsorption of MB. The presence of a large number of surface acid sites (Figure 10) from the doping of Al atoms can enhance the adsorption performance of MB through the coordination interaction between surface acid sites and MB molecules [40]. As a result, Sample SA-E possessed an optimal MB removal performance.

To determine the optimized amount of adsorbent, the dosage of Sample SA-E was varied in a range from 0.1 g/L to 0.5 g/L for the MB adsorption, at a pH of 9.6 and an initial MB concentration of 50 mg/L. As shown in Figure 12a, with the increase in the dosage of Sample SA-E, the removal rate and removal efficiency of MB increased remarkably, due to the increase in the number of active sites of adsorbent. When the dosage increases from 0.1 g/L to 0.3 g/L, the MB removal efficiency can increase from 39.0% to 92.8% within 5 min of contact time. Further increasing the dosage to 0.5 g/L, the adsorption equilibrium can be reached within 5 min, and a MB removal efficiency of 99.6% can be obtained, even though the MB adsorption capacity has decreased from 188.1 mg/g to 116.8 mg/g (Figure 12b). From these observations, an optimized dosage of 0.3 g/L can be determined.

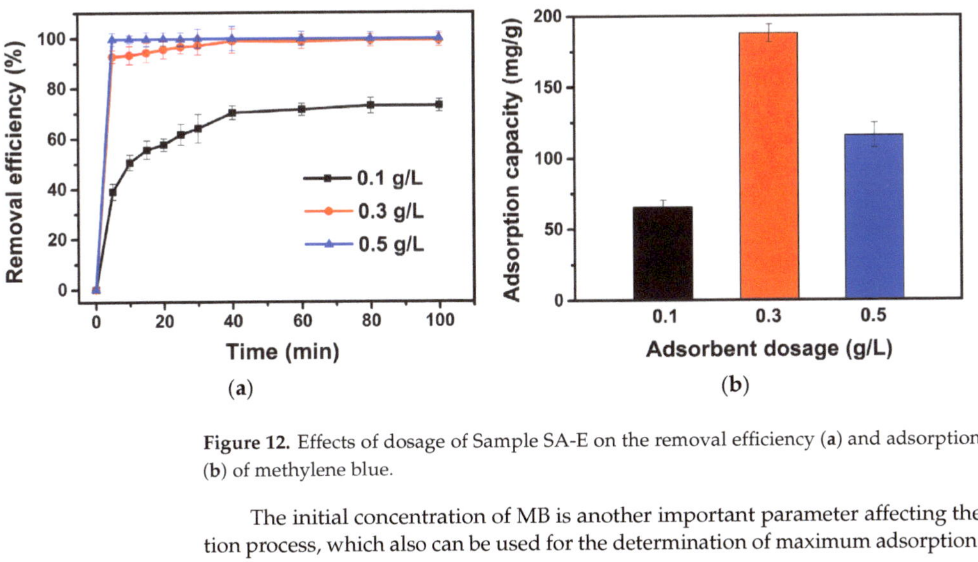

Figure 12. Effects of dosage of Sample SA-E on the removal efficiency (**a**) and adsorption capacity (**b**) of methylene blue.

The initial concentration of MB is another important parameter affecting the adsorption process, which also can be used for the determination of maximum adsorption capacity of adsorbents. The effect of MB initial concentration on adsorption capacity was studied by varying the MB concentration in a range of 10–70 mg/L at a pH of 9.6 and an optimized dosage of Sample SA-E. From Figure 13a, it can be seen that the removal efficiency decreases gradually from 99.9% for 10 mg/L to 86.7% for 70 mg/L with the increase in the MB concentration, during which the adsorption capacity increases gradually (Figure 13b). According to the effects of MB initial concentration on the removal efficiency and adsorption capacity, 50 mg/L was found to be the optimized concentration, and the maximum adsorption capacity of Sample SA-E was 190.1 mg/g.

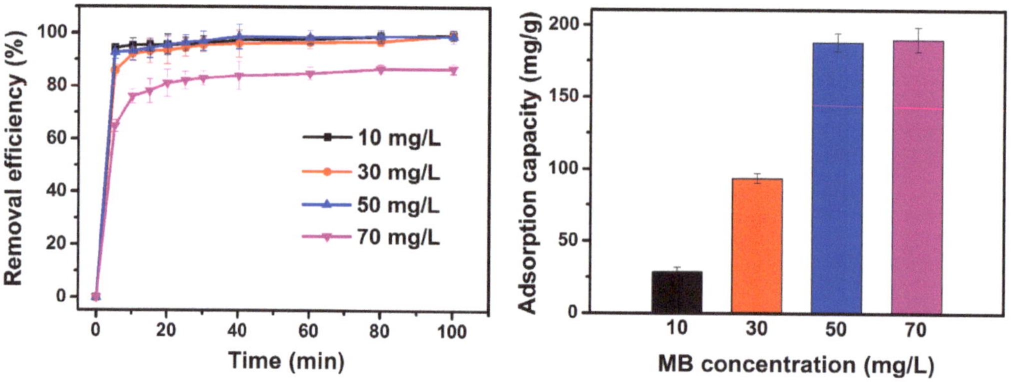

(a) (b)

Figure 13. Effects of initial concentration on the removal efficiency (**a**) and adsorption capacity (**b**) of methylene blue over Sample SA-E.

4. Conclusions

Ordered mesoporous SBA-15 and Al-SBA-15 materials have been successfully synthesized using the silicon species derived from industrial waste coal fly ash (CFA) and were used as adsorbents for the efficient removal of organic dye methylene blue (MB) from an aqueous solution. The resultant SBA-15 material (Sample S-E) exhibited a highly ordered 2D mesoporous structure with a high purity, high specific surface area (1014 m^2/g) and pore volume (1.08 cm^3/g) and, therefore, displayed an excellent removal efficiency (90.5%) and adsorption capacity (160.8 mg/g) for methylene blue (MB) from simulated wastewater. In addition, the generation of surface acid sites from the homogenous incorporation of Al atoms into the mesoporous silica framework combined with the perfect retention of ordered mesostructures endowed the obtained Al-SBA-15 material (Sample SA-E) with a further boost in MB removal performance. For Sample SA-E, the MB removal efficiency reached almost 100% along with a maximum adsorption capacity of 190.1 mg/g, with that too within a contact time of only 40 min. The results obtained in the current work provides a new contribution in designing and utilizing CFA for the large-scale synthesis of ordered mesoporous SBA-15 type materials with excellent properties, which can be widely applied as effective and low-cost materials for adsorption and catalysis.

Author Contributions: Investigation, Data curation, Writing—original draft, M.T.; Conceptualization, Writing—review and editing, Supervision, Funding acquisition, D.P.; Methodology, Data curation, Formal analysis, Funding acquisition, S.C.; Investigation, Formal analysis, Visualization, X.Y.; Investigation, Validation, Funding acquisition, L.H.; Methodology, Project administration, Resource, R.L.; Conceptualization, Writing—review and editing, Supervision, Resource, J.W. All authors have read and agreed to the published version of the manuscript.

Funding: This research was funded by the National Natural Science Foundation of China, grant numbers 21975174 and 21978195, the Foundation of State Key Laboratory of Coal Conversion, grant number J20-21-309, and the Central Guidance on Local Science and Technology Development Fund of Shanxi Province, grant number YDZJSX2021A014.

Institutional Review Board Statement: Not applicable.

Informed Consent Statement: Not applicable.

Data Availability Statement: Not applicable.

Conflicts of Interest: The authors declare no conflict of interest.

References

1. Lin, S.; Jiang, X.; Zhao, Y.; Yan, J. Zeolite greenly synthesized from fly ash and its resource utilization: A review. *Sci. Total Environ.* **2022**, *851*, 158182. [CrossRef] [PubMed]
2. Wang, N.N.; Sun, X.Y.; Zhao, Q.; Yang, Y.; Wang, P. Leachability and adverse effects of coal fly ash: A review. *J. Hazard. Mater.* **2020**, *396*, 122725. [CrossRef] [PubMed]
3. Ulu, A.; Alpaslan, M.; Gultek, A.; Ates, B. Eco-friendly chitosan/kappa-carrageenan membranes reinforced with activated bentonite for adsorption of methylene blue. *Mater. Chem. Phys.* **2022**, *278*, 125611. [CrossRef]
4. de Oliveira, T.F.; de Souza, C.P.; Lopes-Moriyama, A.L.; de Silva, M. In situ modification of MCM-41 using niobium and tantalum mixed oxide from columbite processing for methylene blue adsorption: Characterization, kinetic, isotherm, thermodynamic and mechanism study. *Mater. Chem. Phys.* **2023**, *294*, 127011. [CrossRef]
5. Sun, G.C.; Zhang, J.Y.; Hao, B.L.; Li, X.; Yan, M.W.; Liu, K.Q. Feasible synthesis of coal fly ash based porous composites with multiscale pore structure and its application in Congo red adsorption. *Chemosphere* **2022**, *298*, 134136. [CrossRef]
6. Wang, D.; Chen, X.; Feng, J.; Sun, M. Recent advances of ordered mesoporous silica materials for solid-phase extraction. *J. Chromatogr. A* **2022**, *1675*, 463157. [CrossRef]
7. Li, G.; Wang, B.D.; Sun, Q.; Xu, W.Q.; Han, Y.F. Adsorption of lead ion on amino-functionalized fly-ash-based SBA-15 mesoporous molecular sieves prepared via two-step hydrothermal method. *Micropor. Mesopor. Mat.* **2017**, *252*, 105–115. [CrossRef]
8. Diagboya, P.N.E.; Dikio, E.D. Silica-based mesoporous materials; emerging designer adsorbents for aqueous pollutants removal and water treatment. *Micropor. Mesopor. Mat.* **2018**, *266*, 252–267. [CrossRef]
9. Khan, A.J.; Song, J.X.; Ahmed, K.; Rahim, A.; Volpe, P.L.O.; Rehman, F. Mesoporous silica MCM-41, SBA-15 and derived bridged polysilsesquioxane SBA-PMDA for the selective removal of textile reactive dyes from wastewater. *J. Mol. Liq.* **2020**, *298*, 111957. [CrossRef]
10. Rathinavel, S.; Ekambaram, S.; Korrapati, P.S.; Sangeetha, D. Design and fabrication of electrospun SBA-15-incorporated PVA with curcumin: A biomimetic nanoscaffold for skin tissue engineering. *Biomed. Mater.* **2020**, *15*, 035009. [CrossRef]
11. Xu, Y.S.; Hu, E.Z.; Xu, D.Y.; Guo, Q.J. Activation of peroxymonosulfate by bimetallic CoMn oxides loaded on coal fly ash-derived SBA-15 for efficient degradation of Rhodamine B. *Sep. Purif. Technol.* **2021**, *274*, 119081. [CrossRef]
12. Chandrasekar, G.; Son, W.J.; Ahn, W.S. Synthesis of mesoporous materials SBA-15 and CMK-3 from fly ash and their application for CO_2 adsorption. *J. Porous. Mat.* **2009**, *16*, 545–551. [CrossRef]
13. Soongprasit, C.; Aht-Ong, D.; Sricharoenchaikul, V.; Vichaphund, S.; Atong, D. Hydrocarbon Production from Catalytic Pyrolysis-GC/MS of Sacha Inchi Residues Using SBA-15 Derived from Coal Fly Ash. *Catalysts* **2020**, *10*, 1031. [CrossRef]
14. Shakeri, M.; Shal, Z.K.; Van Der Voort, P. An Overview of the Challenges and Progress of Synthesis, Characterization and Applications of Plugged SBA-15 Materials for Heterogeneous Catalysis. *Materials* **2021**, *14*, 5082. [CrossRef]
15. Rakngam, I.; Osakoo, N.; Wittayakun, J.; Chanlek, N.; Pengsawang, A.; Sosa, N.; Butburee, T.; Faungnawakij, K.; Khemthong, P. Properties of mesoporous Al-SBA-15 from one-pot hydrothermal synthesis with different aluminium precursors and catalytic performances in xylose conversion to furfural. *Microporous. Mesoporous. Mat.* **2021**, *317*, 110999. [CrossRef]
16. Han, L.N.; Ren, W.G.; Wang, B.; He, X.X.; Ma, L.J.; Huo, Q.H.; Wang, J.C.; Bao, W.R.; Chang, L.P. Extraction of SiO_2 and Al_2O_3 from coal gangue activated by supercritical water. *Fuel* **2019**, *253*, 1184–1192. [CrossRef]
17. Li, D.; Jiang, X.X.; Wang, S.D.; Zhao, F.; Jiang, W.; Liu, W. Research on the alkali-digestion properties of alumina and silicon dioxide during phase transformation roasting process. *Fuel Process. Technol.* **2019**, *191*, 223–231. [CrossRef]
18. Li, C.C.; Qiao, X.C. A new approach to prepare mesoporous silica using coal fly ash. *Chem. Eng. J.* **2016**, *302*, 388–394. [CrossRef]
19. Terzic, A.; Pavlovic, L.; Obradovic, N.; Pavlovic, V.; Stojanovic, J.; Milicic, L.; Radojevic, Z.; Ristic, M.M. Synthesis and Sintering of High-temperature Composites Based on Mechanically Activated Fly Ash. *Sci. Sinter.* **2012**, *44*, 135–146. [CrossRef]
20. Zhao, D.Y.; Huo, Q.S.; Feng, J.L.; Chmelka, B.F.; Stucky, G.D. Nonionic triblock and star diblock copolymer and oligomeric surfactant syntheses of highly ordered, hydrothermally stable, mesoporous silica structures. *J. Am. Chem. Soc.* **1998**, *120*, 6024–6036. [CrossRef]
21. Hameed, A.M.; Alharbi, A.; Abdelrahman, E.A.; Mabrouk, E.M.; Hegazey, R.M.; Algethami, F.K.; Al-Ghamdi, Y.O.; Youssef, H.M. Facile Hydrothermal Fabrication of Analcime and Zeolite X for Efficient Removal of Cd(II) Ions From Aqueous Media and Polluted Water. *J. Inorg. Organomet. Polym. Mater.* **2020**, *30*, 4117–4128. [CrossRef]

22. Yu, H.B.; Jia, X.; Wang, P.; Xiong, Y.; Shan, W.J.; Zhang, C.G.; Song, Z.M. High efficiency Al modified mesoporous SBA-15 catalyst for biomass-derived gamma-valerolactone decarboxylation to butene. *Microporous. Mesoporous. Mat.* **2021**, *327*, 111432. [CrossRef]
23. Hui, K.S.; Chao, C.Y.; Kot, S.C. Removal of mixed heavy metal ions in wastewater by zeolite 4A and residual products from recycled coal fly ash. *J. Hazard. Mater.* **2005**, *127*, 89–101. [CrossRef] [PubMed]
24. Gong, Y.B.; Sun, J.M.; Zhang, Y.M.; Zhang, Y.F.; Zhang, T.A. Dependence on the distribution of valuable elements and chemical characterizations based on different particle sizes of high alumina fly ash. *Fuel* **2021**, *291*, 120225. [CrossRef]
25. Chen, W.T.; Song, G.Q.; Lin, Y.Y.; Qiao, J.T.; Wu, T.H.; Yi, X.Y.; Kawi, S. A green and efficient strategy for utilizing of coal fly ash to synthesize K-MER zeolite as catalyst for cyanoethylation and adsorbent of CO_2. *Microporous. Mesoporous. Mat.* **2021**, *326*, 111353. [CrossRef]
26. Bai, G.H.; Teng, W.; Wang, X.G.; Zhang, H.; Xu, P. Processing and kinetics studies on the alumina enrichment of coal fly ash by fractionating silicon dioxide as nano particles. *Fuel Process. Technol.* **2010**, *91*, 175–184. [CrossRef]
27. Kaur, P.; Chopra, H.K. SBA-15 Supported Benzimidazolium-based Ionic Liquids: Synthesis, Characterization, and Applications in the Fuel Desulfurization. *Fuel* **2022**, *328*, 125261. [CrossRef]
28. El-Fiqi, A.; Bakry, M. Facile and rapid ultrasound-mediated synthesis of spherical mesoporous silica submicron particles with high surface area and worm-like mesoporosity. *Mater. Lett.* **2020**, *281*, 128620. [CrossRef]
29. Shi, L.; Xu, Y.; Zhang, N.; Lin, S.; Li, X.P.; Guo, P.; Li, X.B. Direct synthesis of Al-SBA-15 containing aluminosilicate species plugs in an acid-free medium and structural adjustment by hydrothermal post-treatment. *J. Solid. State Chem.* **2013**, *203*, 281–290. [CrossRef]
30. Verma, P.; Kuwahara, Y.; Mori, K.; Raja, R.; Yamashita, H. Functionalized mesoporous SBA-15 silica: Recent trends and catalytic applications. *Nanoscale* **2020**, *12*, 11333–11363. [CrossRef]
31. Bukhari, S.N.; Chong, C.C.; Setiabudi, H.D.; Cheng, Y.W.; Teh, L.P.; Jalil, A.A. Ni/Fibrous type SBA-15: Highly active and coke resistant catalyst for CO_2 methanation. *Chem. Eng. Sci.* **2021**, *229*, 116141. [CrossRef]
32. Miao, H.M.; Li, M.S.; Wang, F.; Li, J.; Lin, Y.W.; Xu, J.K. Surface Functionalization of SBA-15 for Immobilization of Myoglobin. *Front. Bioeng. Biotechnol.* **2022**, *10*, 907855. [CrossRef]
33. Achilias, D.S.; Gerakis, K.; Giliopoulos, D.J.; Triantafyllidis, K.S.; Bikiaris, D.N. Effect of high surface area mesoporous silica fillers (MCF and SBA-15) on solid state polymerization of PET. *Eur. Polym. J.* **2016**, *81*, 347–364. [CrossRef]
34. Pan, D.H.; Yuan, P.; Zhao, L.Z.; Liu, N.A.; Zhou, L.; Wei, G.F.; Zhang, J.; Ling, Y.C.; Fan, Y.; Wei, B.Y.; et al. New Understanding and Simple Approach to Synthesize Highly Hydrothermally Stable and Ordered Mesoporous Materials. *Chem. Mat.* **2009**, *21*, 5413–5425. [CrossRef]
35. Wu, X.K.; Liu, J.; Gu, Y.W.; Chen, Y.; Shan, X.G.; Tang, B.Y. Inorganic salt assisted high temperature hydrothermal synthesis of ordered mesoporous silicas with enhanced stability and ultralow dielectric constants. *J. Porous Mat.* **2022**, *29*, 9–17. [CrossRef]
36. Aumond, T.; Pinard, L.; Batiot-Dupeyrat, C.; Sachse, A. Non-thermal plasma: A fast and efficient template removal approach allowing for new insights to the SBA-15 structure. *Microporous. Mesoporous. Mat.* **2020**, *296*, 110015. [CrossRef]
37. Pan, D.H.; Tan, L.; Qian, K.; Zhou, L.; Fan, Y.; Yu, C.Z.; Bao, X.J. Synthesis of highly ordered and hydrothermally stable mesoporous materials using sodium silicate as a precursor. *Mater. Lett.* **2010**, *64*, 1543–1545. [CrossRef]
38. Mahmoudi, F.; Amini, M.M. Confined crystallization of microporous metal-organic framework within mesoporous silica with enhanced hydrostability: Ultrafast removal of organic dyes from aqueous solutions by MIL-68(Al)@SBA-15 composite. *J. Water Process. Eng.* **2020**, *35*, 101227. [CrossRef]
39. Sareen, S.; Kaur, S.; Mutreja, V.; Sharma, A.; Kansal, S.K.; Mehta, S.K. Coral-Reef Shaped Mesoporous Silica Obtained from Coal Fly Ash with High Adsorption Capacity. *Top. Catal.* **2022**, *65*, 1791–1810. [CrossRef]
40. Wu, Y.H.; Ma, Y.L.; Sun, Y.G.; Ji, W.X.; Lin, F.; Yang, Y.P.; Ma, L.J.; Zhu, C.H.; Xu, Y.J.; Miao, Q. Effects of acid ionization on the formation mechanism of bimodal mesoporous Al-MCM-41s from coal gasification fine residue and evaluation of adsorption capabilities. *J. Hazard. Mater.* **2021**, *417*, 126052. [CrossRef]

materials

Article

Spherical Attapulgite/Silica Aerogels Fabricated via Different Drying Methods with Excellent Adsorption Performance

Zhixiang Zhu [1,2], Shengyuan Wang [1,2], Ya Zhong [1,2,3,*], Qi You [1,2], Jun Gao [1,2], Sheng Cui [1] and Xiaodong Shen [1]

1 State Key Laboratory of Materials-Oriented Chemical Engineering, College of Materials Science and Engineering, Nanjing Tech University, Nanjing 210009, China
2 Jiangsu Collaborative Innovation Center for Advanced Inorganic Function Composites, Nanjing Tech University, Nanjing 210009, China
3 Suqian Advanced Materials Industry Technology Innovation Center, Nanjing Tech University, Suqian 223800, China
* Correspondence: yzhong@njtech.edu.cn

Abstract: Dye wastewater has caused great harm to the environment, which is an urgent problem to be solved. As typical three-dimensional porous materials, aerogels have attracted great interest in dye wastewater treatment. In this work, spherical attapulgite/silica (ATP/SiO_2) gels were initially prepared by easily scalable sol-gel dripping methods and then dried to aerogels with three drying techniques, namely, supercritical CO_2 drying (SCD), freeze-drying (FD), and ambient pressure drying (APD). The effect of the drying techniques and heat-treated temperature on the physical characteristic, morphological properties, microstructure, and chemical structure of the spherical ATP/SiO_2 aerogels were investigated. The macroscopic morphology of the spherical ATP/SiO_2 aerogels was homogeneous and integrated without local cracking. The average pore diameter and specific surface area of the spherical ATP/SiO_2 aerogels prepared by the three drying techniques were in the range of 6.8–8.6 nm and 218.5–267.4 m^2/g, respectively. The heat treatment temperature had a significant effect on the pore structure and the wetting properties of the aerogels. The 600 °C heat-treated aerogels were subjected to adsorption tests in methylene blue (MB) solution (60 mg/g, 100 mL), which exhibited a great adsorption capacity of 102.50 mg/g. Therefore, the resulting spherical ATP/SiO_2 aerogels possessed multipath preparation and exhibited an efficient adsorption performance, with the potential to be applied as an adsorbent for dye wastewater.

Keywords: spherical ATP/SiO_2 aerogels; drying techniques; heat treatment; adsorption performance

Citation: Zhu, Z.; Wang, S.; Zhong, Y.; You, Q.; Gao, J.; Cui, S.; Shen, X. Spherical Attapulgite/Silica Aerogels Fabricated via Different Drying Methods with Excellent Adsorption Performance. *Materials* **2023**, *16*, 2292. https://doi.org/10.3390/ma16062292

Academic Editors: Zhifeng Wang, Weiqing Zhang and Yichao Wang

Received: 26 January 2023
Revised: 9 March 2023
Accepted: 9 March 2023
Published: 13 March 2023

1. Introduction

Highly toxic, carcinogenic, and teratogenic dye wastewater has caused serious damage to the ecological environment [1,2]. Dyes in wastewater can destroy the self-purifying function of water bodies, since they can reduce water clarity, consume oxygen in the water, and hinder photosynthesis in plants [3]. Remarkably, most organic dyes are difficult to degrade naturally and can persist in the natural environment [4]. Methyl blue (MB) is a hydrophilic organic compound with free solubility in water, representing typical cationic dyes [5,6]. Adsorption has been recognized as one of the most efficient methods for removing dyes from wastewater due to the fact of its widespread use, effectiveness, and reusability. As typical three-dimensional porous materials, SiO_2 aerogels have attracted great interest in adsorption fields [7]. SiO_2 aerogels can be employed as dye adsorbents owing to the following two advantages: a high specific surface area that enables numerous adsorption sites and properties that can be easily tuned by surface modification to adapt to different application environments [8]. Yao et al. [8] used water as a solvent for preparing MTMS/tetraethoxysilane silica aerogels with a high porosity of 96.36%. Jiang et al. [9] synthesized amine-grafted silica aerogels by sodium silicate to adsorb CO_2.

A SiO_2 aerogel adsorbent can be obtained by solvent substitution and surface modification via the APD method. Regrettably, the hydrophobicity is unfavorable to the dispersion of aerogel in water and limits the adsorption of soluble dye. Previous research has suggested that heat treatment is a facile and high-efficiency method for converting hydrophobic SiO_2 aerogels into hydrophilic ones. For instance, Yi et al. [10] compared the adsorption performance of hydrophobic and hydrophilic SiO_2 aerogels on MB. The adsorption capacity of the hydrophilic SiO_2 aerogel obtained by heat treatment at 500 °C was 9.53 mg/g, significantly higher than that of the hydrophobic sample (3.05 mg/g). Wei et al. [11] investigated the adsorption property of hydrophilic SiO_2 aerogels obtained by calcinating hydrophobic aerogels at 600 °C for 3 h, which exhibited good adsorption performance for MB (51.16 mg/g). Some researchers also proved that metal ion solution modification is beneficial for increasing the adsorption capacity of aerogels on MB. Yang et al. [12] provided SiO_2 aerogels modified with Mg^{2+} solutions, which displayed an adsorption capacity of 40.40 mg/g for MB. However, SiO_2 aerogel materials are known for their poor mechanical behavior, which limits the practical applications of aerogels and is attributed to their fragility and brittleness. Hence, enhancing the mechanical properties without sacrificing its adsorption performance is a factor worthy of attention. One way is to optimize the micro/macro morphology of the aerogel through preparation [13]. Another method is to add materials to form an enhanced composite structure [14]. Additionally, it has been demonstrated that spherical aerogels have good fluidity, uniform stress, and strain, which can improve the elastic–plastic deformability and mechanical properties [15,16]. Fiber reinforcement is one of the most effective strategies for improving the mechanical property of aerogel materials. Tian et al. [17] synthetized glass fiber enhanced SiO_2 aerogel, which exhibited a high flexural modulus (2.65 MPa). Slosarczyk [18] provided a synthesis method for a carbon fiber-SiO_2 aerogel possessing an excellent elastic–plastic deformation ability. However, an obvious defect is that the structures of the reinforcement material and the aerogel are usually different, which will lead to huge stress in the composite material during the preparation and drying process [19]. To maintain the three-dimensional skeleton structure of an aerogel, plentiful and nanoscale natural clay with a unique structure and function is usually used to strengthen silicon-based aerogels, such as attapulgite [20], montmorillonite [21], and sepiolite [22].

Attapulgite (ATP) is a naturally available 1D hydrophilic clay, which displays the hydrated magnesium aluminum silicate structure with a theoretical composition of $Mg(Al)_5 Si_8O_{20}(OH)_2(OH_2)_4 \bullet 4H_2O$ [23]. The diameter of ATP is approximately 30 nm, and the length is between 0.5 and 1 μm, which is very close to the pore size of SiO_2 aerogels (5–100 nm) [24]. In addition, ATP exhibits a 2:1 type chain laminar structure with plentiful -OH groups, which has good chemical compatibility with the chemical groups of SiO_2 aerogels. Because of the porous internal channels, large specific surface area, and certain ion exchange capability, club-shaped ATP is also extensively applied as an adsorbent material [25,26]. Chen et al. [27] studied the removal rate and adsorption mechanism of calcined ATP for MB. With the increasing temperature, the adsorption capacity of ATP reached a maximum of 700 °C (78.11 mg/g). All of the above characteristics of ATP are beneficial to the preparation of high-performance ATP/SiO_2 composite aerogel adsorbent. Zhang et al. [28] synthesized ATP/SiO_2 aerogels, which displayed an adsorption quantity of 5.00 g/g for total petroleum hydrocarbon. Meanwhile, the structures and properties of ATP/SiO_2 composite aerogels dried under air conditions were also investigated. Notably, drying technology is an important process for the synthesis of aerogels with a specific structure. The microstructures and properties of the final aerogels mainly depend on the drying technique. Hence, it is necessary to study the influence of different drying processes on the structure of ATP/SiO_2 composite aerogels.

Currently, aerogels are usually obtained by three drying techniques, such as supercritical CO_2 drying (SCD), freeze-drying (FD), and ambient pressure drying (APD) [29]. The SCD technique is an alternative drying process that avoids capillary pressure by replacing the solvent with supercritical fluid CO_2. It is an ideal method to obtain aerogels with an intact structure and low density. The main disadvantages of the SCD method are the required energy, maintenance cost, and high-pressure conditions [30]. FD is the process of removing water or other solvents from frozen samples through sublimation in a vacuum environment. The freezing process can evade the shrinkage of the aerogel skeleton, but the crystallization of the solvent will enlarge the pore size and destroy the original pore structure [31]. The principle of APD is to elevate the temperature of the solvent above the boiling point and transform it into gas, which is the most convenient technique for removing solvents from porous materials. The convenient evaporation of the APD method with its low cost is very suitable for mass production and wide application [19,32]. However, the capillary pressure during the drying process exceeds the elastic limit of the solid structure, which may result in the shrinkage and collapse of the porous aerogel materials [33]. To reduce the effect of capillary forces and control the irreversible shrinkage of SiO_2 aerogels during the APD process, surface hydrophobic modification by silylation is of great importance. Selay Sert et al. [34,35] investigated the effects of different silylation agents (including MTMS, trimethylchlorosilane, and methyltriethoxysilane) on the microstructure and chemical properties of SiO_2 aerogels under APD. The results show that the MTMS-modified SiO_2 aerogel had a highly developed three-dimensional network structure and high specific surface area. Therefore, MTMS was used as a silane modifier in this work.

To synthesize an excellent dye adsorbent, ecofriendly spherical ATP/SiO_2 aerogels were successfully prepared via three drying methods (SCD, FD, and APD). Hydrophobization of the spherical ATP/SiO_2 aerogels was achieved by MTMS modification. The structure of the as-prepared spherical ATP/SiO_2 aerogels with three drying techniques was analyzed and compared. The effects of the heat treatment temperatures on the physicochemical properties of the resulting aerogels were studied. Furthermore, the adsorption performance of the spherical ATP/SiO_2 aerogels dried by APD and heat treated at 600 °C was also investigated. The spherical ATP/SiO_2 aerogels have excellent adsorption properties and can be prepared in multiple ways. This suggests that the prepared aerogels have promise as an adsorbent for dye wastewater.

2. Experimental Section

2.1. Materials

Attapulgite (ATP) was purchased from Jiangsu Jiuchuan Nanotechnology Co., Ltd (Xuyi, China). Alkaline silica sols (30 wt%) were obtained from Shanghai Yuanye Biological Technology Co., Ltd. (Shanghai, China). Dimethicone oil (1000 ± 80 mPa.s), acetic acid, and methylene blue (MB) were purchased from Shanghai Macklin Biochemical Co., Ltd. (Shanghai, China). Trimethoxymethylsilane (MTMS) and tert-butanol were supplied by Shanghai Aladdin Biochemical Technology Co., Ltd. (Shanghai, China). N-hexane was provided by Shanghai Lingfeng Chemical Reagent Co., Ltd. (Shanghai, China). Ethanol and deionized water (H_2O) were supplied by Wuxi City Yasheng Chemical Co., Ltd. (Wuxi, China). The reagents used in the experiment were all analytically pure, without further purification.

2.2. Methods

2.2.1. Synthesis of Spherical ATP/SiO_2 Gels

The ATP/SiO_2 solution was prepared according to the following steps. Alkaline silica sols, H_2O, and ATP were directly separated in a pot with a mass ratio of 1:1:0.5. Acetic acid was used as a catalyst to adjust the pH value of the solution to 4.5~5.5. Then, the mixture was stirred for approximately 30 min at room temperature. Subsequently, the ATP/SiO_2 solution was obtained by ultrasonic treatment for 15 min. Furthermore, the spherical ATP/SiO_2 gels were prepared by the dripping method. The ATP/SiO_2 solution was dripped into dimethicone oil at 80 °C to form spherical ATP/SiO_2 gels. Finally, the spherical ATP/SiO_2 gels were washed five times with deionized water at 60 °C to remove impurities and then aged in water for 12 h.

2.2.2. Preparation of Spherical ATP/SiO_2 Aerogels Using Different Treatment Techniques

The hydrophobic spherical ATP/SiO_2 gels were obtained by the surface modification process. Trimethoxymethylsilane (MTMS) was employed as a hydrophobic modification agent. The spherical ATP/SiO_2 gels were firstly placed in a mixed MTMS/ethanol solution with a volume ratio of 1:4 for 12 h and then soaked in ethanol solution for 24 h to wash off the excess MTMS and reaction byproducts. Subsequently, the primed samples were dried to hydrophobic spherical ATP/SiO_2 aerogels using the SCD method in an autoclave for 4 h (50 °C, 10 MPa, HELIX 1.1 system, Applied Separations, Inc., Allentown, PA, USA), denoted as HAS-S. The samples without hydrophobic modification were denoted as AS-S.

In addition, the spherical ATP/SiO_2 gels were initially placed in a mixed MTMS/tert-butanol solution with a volume ratio of 1:4 for 12 h and then washed with tert-butanol solution for 24 h to exchange the excess MTMS and reaction byproducts. Finally, the hydrophobic spherical ATP/SiO_2 aerogels were prepared using the FD method (frozen at −15 °C in a fridge for 5 h and then dried at −80 °C for 12 h), denoted as HAS-F.

Moreover, the previous hydrophobic spherical ATP/SiO_2 gels were adequately washed with N-hexane for 24 h to displace the water and byproducts from the samples. The volume of N-hexane was 5 times larger than the volume of the gel. After aging and solvent exchange, the hydrophobic spherical ATP/SiO_2 aerogels were obtained by APD in an air oven (50 °C for 6 h, and 80 °C for 4 h), denoted as HAS-A.

To study the variation in the physicochemical properties of the spherical aerogels, the HAS-A was heat treated at different temperatures (200 °C, 400 °C, 600 °C, and 800 °C) for 3 h in a muffle furnace with a heating rate of 5 °C/min. The 600 °C heat-treated HAS-A is denoted as AS-A.

2.3. *Characterization*

The apparent densities (ρ) of the samples were calculated by $\rho = m/v$, where m and v are the apparent mass and volume. Scanning electron microscope (SEM) images of the samples were carried out using an Ultra-55 (Zeiss, Oberkochen, Germany) at an operating voltage of 15.0 kV. Transmission electron microscope (TEM) images were recorded on a JEOL JEM 2100F microscope at 200 kV. X-ray photoelectron spectroscopy (XPS) was carried out using an Axis Ultra DLD equipped with Al Ka (1486.6 EV). Fourier transform infrared spectroscopy (FT-IR) measurements were recorded on an FT-IR spectrometer (Spectrum 100, Perkin Elmer, Waltham, MA, USA) for KBr pellets. The contact angles were carried out using a JC2000D1 by grinding the aerogels into powder and then pressing them into pieces. The thermogravimetric (TG) analysis was conducted on a Netzsch STA449F5 thermal analyzer (Selb, Germany) in a temperature range of 30 °C to 800 °C with a heating rant of 10 °C/min under flowing air. The BET specific surface area (S_{BET}), pore volume, and pore distribution were measured using a V-sorb 2800P after pretreatment for 6 h at 120 °C. The zeta potential was analyzed and measured using a Malvern Zetasizer Nano ZS90 (Malvern, UK).

2.4. *Adsorption Experiment*

The AS-A was used as an adsorbent for studying the adsorption performance. The MB solution was used to simulate the wastewater, and all adsorption tests were carried out at room temperature. A total of 50 mg AS-A and 100 mL MB solution were mixed in a centrifuge tube and shaken for 100 min (200 rpm/min) at room temperature. After a sufficient adsorption process, the solution was centrifuged for 5 min (5000 r/min), and the concentration of residual dye in the supernatant was analyzed using a UV-Vis spectrometer at 663 nm. The influence of the pH, contact time, and MB concentration were systematically studied to determine the adsorption performance of the AS-A for MB. The removal rate of MB was calculated by a comparison of the initial concentration and the residual concentration after adsorption. The equations for the adsorption efficiency (R) and the adsorption capacity (q_e) of AS-A are shown as follows:

$$R = \frac{C_0 - C_e}{C_0} \times 100\% \quad (1)$$

$$q_e = \frac{(C_0 - C_e)V}{m} \quad (2)$$

where C_0 and C_e represent the initial and residual concentration of MB (mg/L), respectively; V is the volume of the MB solution (L); and m is the mass of AS-A (g).

3. Results and Discussion

3.1. Formation Mechanism of the Spherical ATP/SiO_2 Aerogels

The schematic diagram of the synthesis process for hydrophobic ATP/SiO_2 aerogels is shown in Figure 1a. Because of the incompatibility between water and oily solvent, the hydrophilic ATP/SiO_2 mixed sols can disperse in the oily solvent. When the surface tension of the sol droplet is much greater than the gravity, the isotropic spherical ATP/SiO_2 sol forms in the oil phase [16]. Considering the influence of the temperature on sol-gel kinetics, the oil bath temperature was set at 80 °C for forming the spherical ATP/SiO_2 gel [9]. The final spherical ATP/SiO_2 aerogels were obtained by three drying methods. Figure 1b presents the synthesis mechanism of the spherical hydrophobic ATP/SiO_2 aerogels. Silica sols exist as nanoscale silica particles, which are distributed uniformly in the water. Both silica sol particles and ATP contain many active -OH groups, which can favorably translate into the ATP/SiO_2 gels by a hydrolysis polycondensation reaction under the appropriate conditions of the silica concentration, pH value, and temperature. In addition, the silica sol is bound with ATP via hydrogen bonding, and then the MTMS used as hydrophobic modifier are grafted onto the ATP/SiO_2 gels by the immersion modification method.

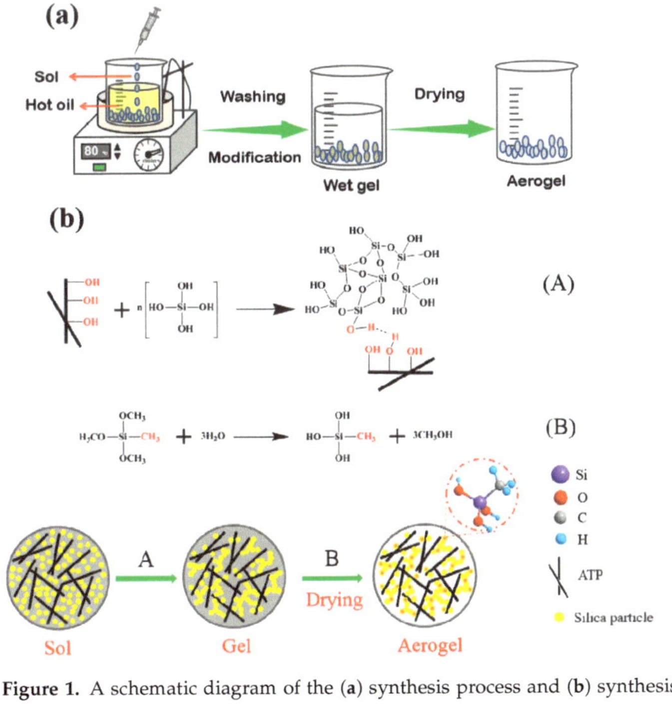

Figure 1. A schematic diagram of the (**a**) synthesis process and (**b**) synthesis mechanism for the spherical hydrophobic ATP/SiO_2 aerogels.

3.2. Effect of Drying Techniques and MTMS Modification on Physical Characteristic and Morphological Properties of Spherical ATP/SiO_2 Aerogels

A comparison of the different drying methods for preparing the spherical ATP/SiO_2 aerogels is shown in Figure 2. All spherical ATP/SiO_2 aerogels exhibited an integrated macroscopic morphology with a diameter of approximately 2.5 ± 0.15 mm (2.65 mm, 2.52 mm, and 2.35 mm for HAS-S, HAS-F, and HAS-A, respectively). By contrast, the shrinkage rate of the HAS-A was larger than the other samples resulting from the drying capillary pressure of the solvent evaporation during the APD process. Table 1 shows the physical parameters of the ATP and spherical ATP/SiO_2 aerogels prepared by different drying methods. The densities of the ATP/SiO_2 aerogel composites (0.46–0.65 g/cm^3) decreased greatly due to the loose three-dimensional skeleton structure instead of the relatively dense structures of the ATP (2.41 g/cm^3). Meanwhile, the densities of the ATP/SiO_2 aerogels modified by MTMS were larger than the samples without hydrophobic modification, which is attributable to the grafting of CH_3-$(SiOH)_3$ groups.

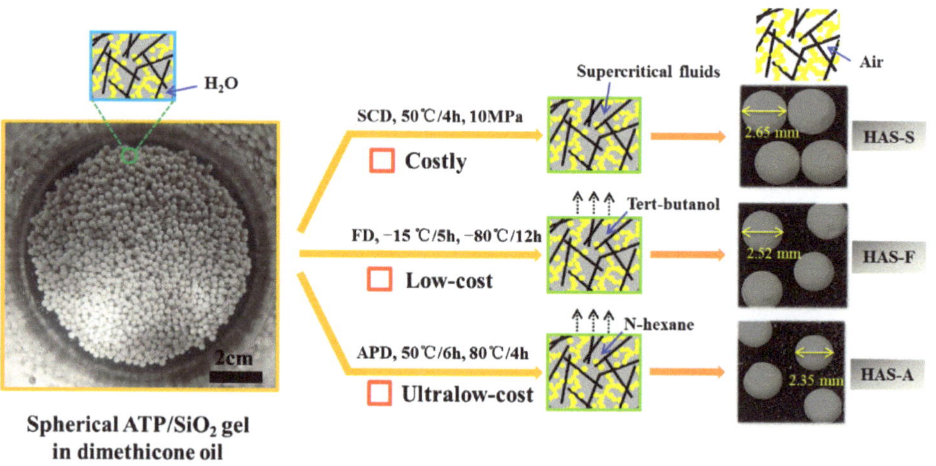

Figure 2. Comparison of the different drying methods for preparing the spherical ATP/SiO_2 aerogels.

Table 1. The physical parameters of the ATP and spherical ATP/SiO_2 aerogels prepared by the different drying methods.

Sample	ρ (g/cm^3)	S_{BET} (m^2/g)	Pore Size (nm)	Pore Volume (cm^3/g)
ATP	2.41	160.5	8.6	0.27
AS-S	0.46	248.7	8.3	0.41
HAS-S	0.53	267.4	7.9	0.37
HAS-F	0.56	241.7	8.1	0.35
HAS-A	0.65	218.5	6.8	0.33

The FT-IR spectra of the as-prepared aerogels are presented in Figure 3a. A broad transmittance appeared at approximately 3200–3700 cm^{-1}, which is attributed to the stretching vibrations of -OH groups [36]. The symmetric stretching and tetrahedra bending vibration of Si-O-Si and Si (or Al)-O appeared at 1035 cm^{-1} and 1100 cm^{-1} [37]. The peaks at approximately 1450 cm^{-1} and 1640 cm^{-1} are assigned to the deformation vibration of the adsorbed water and bound water [32,38]. By contrast, there was a new peak at 780 cm^{-1}, which is normally observed in the hydrophobic group and corresponds to the stretching vibration of -CH_3 in MTMS [39]. Moreover, all of the modified samples showed the symmetric deformation vibration of Si-C bonds and symmetric stretching vibrations of the terminal -CH_3 groups at 2975 cm^{-1} and 1275 cm^{-1},which is proof of the presence of Si-CH_3 [40]. Thus, the FT-IR spectrums have proved that the hydrophobic -CH_3 groups were successfully coupled to the HAS-S, HAS-F, and HAS-A. The XPS measurement (Figure 3b) was subsequently used to quantify the elemental composition of AS-S and HAS-S. The XPS spectra of AS-S and HAS-S show four peaks at 100, 150, 285, and 531 ev, which are assigned to Si 2s, Si 2p, C 1s, and O 1s, respectively [39]. Compared with AS-S, the relative strength of the C 1s peak of the HAS-S increased significantly, while that of the O 1s peak became weak. It can be inferred that with the addition of MTMS, the dehydration condensation reaction between the -OH and CH_3-$(SiOH)_3$ groups resulted in the loss of some O elements and the grafting of -CH_3 groups. The surface wettability of the spherical ATP/SiO_2 aerogels was evaluated by contact angle measurements, which are presented in Figure 3. Due to the large number of –OH, the unmodified AS-S exhibited obvious hydrophilicity with the water contact angle values of 0°. Due to the existence of numerous -CH_3 groups, the spherical ATP/SiO_2 aerogels obtained by MTMS modification possessed excellent hydrophobic performance, corresponding to the values of 142.8°, 136.1°, and 140.7° for HAS-S, HAS-F, and HAS-A, respectively.

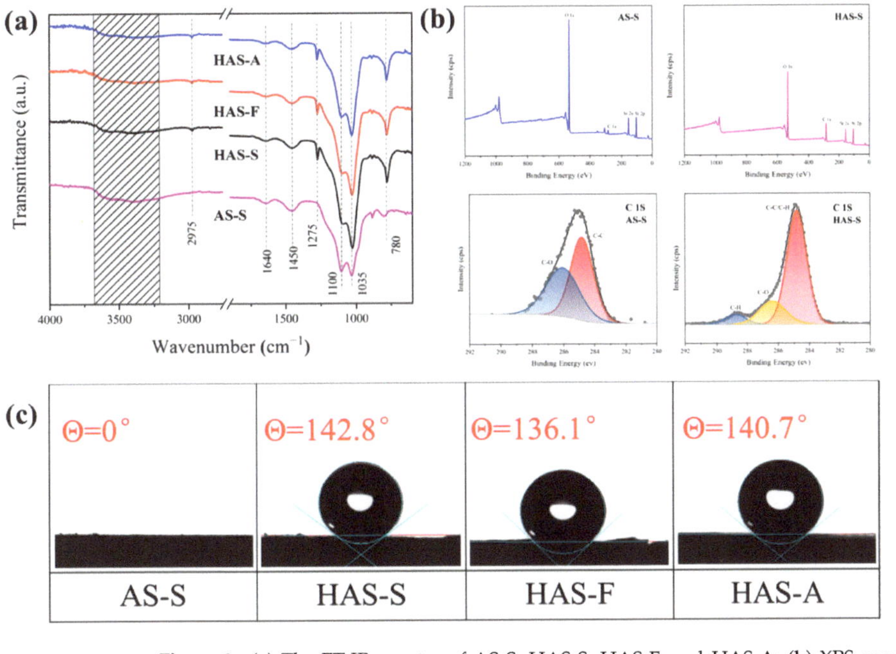

Figure 3. (**a**) The FT-IR spectra of AS-S, HAS-S, HAS-F, and HAS-A; (**b**) XPS spectra and C 1s narrow-scan spectra of AS-S and HAS-S; (**c**) water contact angles of AS-S, HAS-S, HAS-F, and HAS-A.

Figure 4 shows the SEM and TEM images of the prepared spherical ATP/SiO_2 aerogels and pure ATP. As shown in Figure 4e, the pure ATP displays the features of a club-shaped structure with a diameter of approximately 30 nm and partial agglomeration. As shown in Figure 4a–d, all of the ATP/SiO_2 aerogels exhibited inherent three-dimensional porous network structures, which consisted of club-shaped ATP, SiO_2 nanoparticles, and nanopores. As seen from the magnified SEM images, the unordered nano-scaled ATP did not influence the formation of the porous structure of the ATP/SiO_2 aerogels. Additionally, there were some large pores in the HAS-F and densification phenomenon in the HAS-A, which is detrimental to the specific surface areas. By contrast, the AS-S and HAS-S samples exhibited significant homogeneous pore structures. Furthermore, the TEM images (Figure 4f) show that the as-prepared ATP/SiO_2 aerogels using different drying methods displayed random interconnected networks of nanometer-sized SiO_2 aerogel particles and club-shaped ATP, with interconnected amorphous silica particles surrounding the ATP. Meanwhile, the ATP could form hydrogen bonds with silica particles during the reaction, which could strengthen the crosslinked structures of the ATP/SiO_2 aerogels and resist the greater capillary pressure during the drying process. Satisfactorily, compared to other aerogel composites strengthened by coarse fibers, the nanoscale club-shaped ATP was conducive to improving the mechanical performance of the prepared spherical ATP/SiO_2 aerogels without destroying the internal pore structures [19]. The EDS spectrums of AS-S and HAS-S are presented in Figure 5, and the concentration of C, O, and Si elements are presented in Table 2. The apparent concentration of the C element increased from 23.8 wt% (AS-S) to 40.1 wt% (HAS-S). In addition, the homogenous distribution of the C element of HAS-S further proves the successful modification of MTMS.

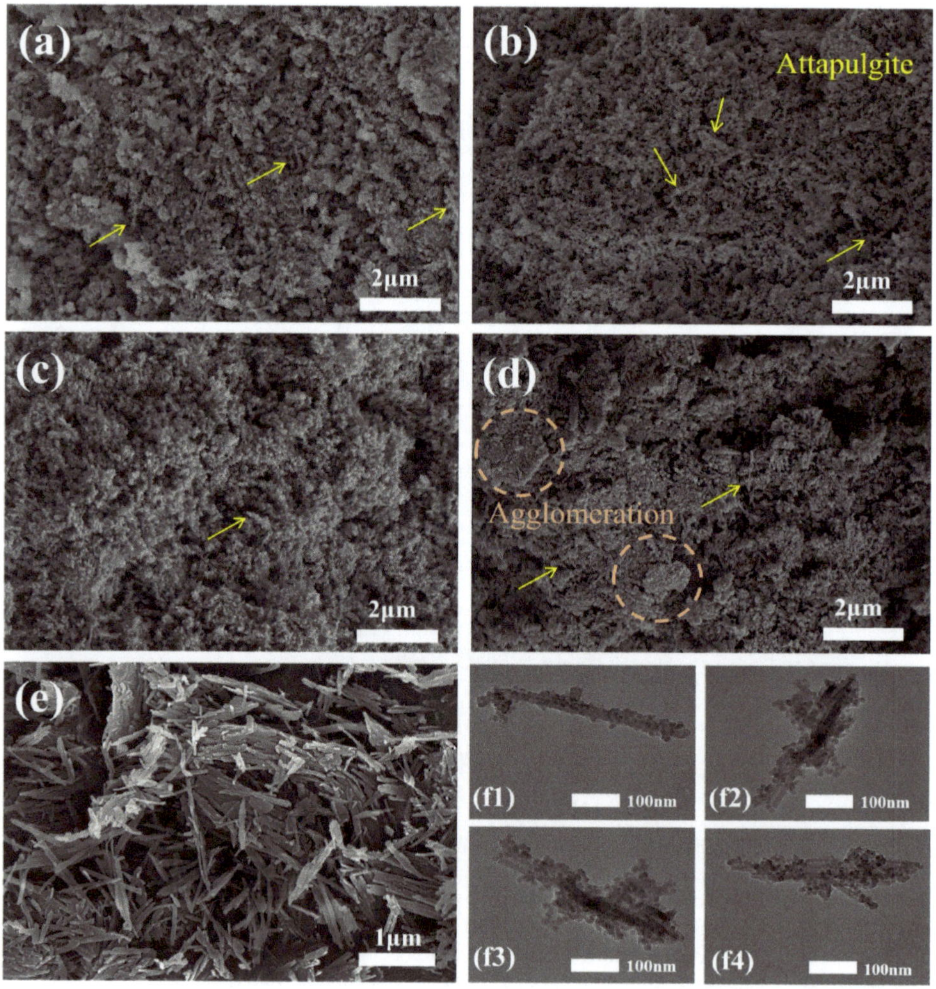

Figure 4. SEM images: (**a**) AS-S; (**b**) HAS-S; (**c**) HAS-F; (**d**) HAS-A; (**e**) ATP. TEM images: (**f1**) AS-S; (**f2**) HAS-S; (**f3**) HAS-F; (**f4**) HAS-A.

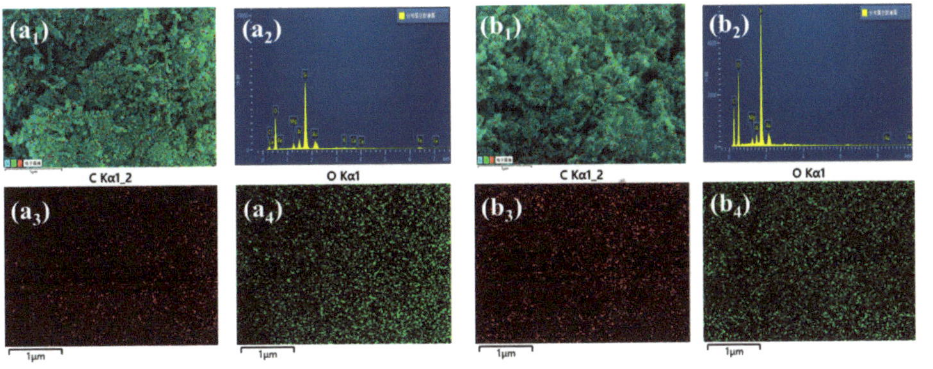

Figure 5. The EDS spectrum of (**a1**–**a4**) AS-S and (**b1**–**b4**) HAS-S.

Table 2. The C, O, Si, and metal element (Mg, Al, K, Ca, Fe) concentration of AS-S and HAS-S.

Sample	C (wt%)	O (wt%)	Si (wt%)	Metal (wt%)
AS-S	23.8	42.9	24.6	8.7
HAS-S	40.1	40.8	17.2	2.3

The N_2 adsorption–desorption isotherms and BJH pore size distribution of the spherical ATP/SiO_2 aerogels are described in Figure 6. All isotherms are Type IV based on the IUPAC classification, reflecting the characteristic of mesoporous materials. The adsorption–desorption curves form closed hysteresis loop, which is generally caused by the capillary condensation in the mesopores. The rapid adsorption process in the low-pressure region (0–0.1) is caused by the micropores inside the aerogel matrix. It is shown in Table 1 that the specific surface area of the as-prepared ATP/SiO_2 aerogels (with the content of ATP over 60 wt%) was much higher than that of pure ATP, resulting from the improvement in the network skeleton structures of the ATP/SiO_2 aerogel composites. Additionally, the specific surface areas of the aerogels obtained by different treatment techniques showed a decreasing trend ($S_{SCD} > S_{FD} > S_{APD}$), corresponding to the values of 248.7–267.4 m^2/g, 241.7 m^2/g, and 218.5 m^2/g, respectively. The HAS-A showed the minimum specific surface area because of the particle agglomeration and closed pores appearing during the APD process (shown in Figure 4d). The HAS-S had higher specific surface areas than AS-S, which can be explained by the fact that the grafting of MTMS improves the hydrophobic property of ATP/SiO_2 aerogel without damaging the nano-porous microstructure. Comparing AS-S to HAS-S, the average pore size and pore volume decreased from 8.3 nm and 0.41 cm^3/g to 7.9 nm and 0.37 cm^3/g, respectively. This is because some organic groups enter the larger pores of spherical ATP/SiO_2 aerogels with the grafting of the CH_3-$(SiOH)_3$ group, which is favorable to increasing the porosity of HAS-S [41]. Furthermore, owing to the high capillary pressure of the solvent evaporation and shrinkage of the pore structure in the APD process, the pore volume of the HAS-A decreased significantly in comparison with the other hydrophobic samples. Compared with HAS-S, HAS-F exhibited lower specific surface area and higher pore volume due to the destruction of the original pore structure during the crystallization and sublimation of the solvent. As revealed in Figure 6b, all spherical ATP/SiO_2 aerogels showed a broad pore size distribution in the range of 2–50 nm. After the hydrophobic modification, the peak of the micropores slightly shifted to higher values (Figure 6c), which can be explained by the fact that the pores in the aerogel are filled with the hydrolysis products of MTMS [41]. This indicates that the hydrophobic modification was beneficial to the microporous structure of the aerogels. Therefore, all spherical ATP/SiO_2 aerogels had a good specific surface area and porous network structure, revealing the feasibility of the multipath preparation via the SCD, FD, and APD techniques.

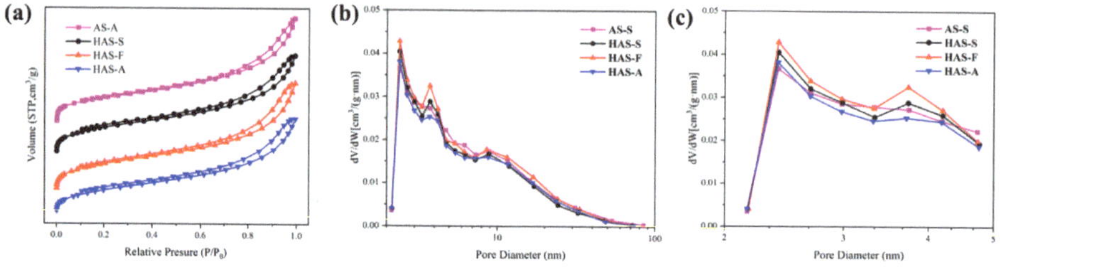

Figure 6. (**a**) The N_2 adsorption–desorption curve and (**b**,**c**) BJH pore size distribution curve of AS-S, HAS-S, HAS-F, and HAS-A.

The TG curves of the prepared spherical ATP/SiO_2 aerogels are displayed in Figure 7, under flowing air. The weight loss rates of the AS-S, HAS-S, HAS-F, and HAS-A were 14.62%, 15.72%, 12.85%, and 13.67%, respectively. The weight loss stage was mainly divided into three parts. The first stage, consisting of the volatilization of the residual solvent and the adsorbed water, appeared at 50–200 °C. In this stage, the weight loss rate of the hydrophobic samples was lower than AS-S (4.51%), which reflects the hydrophobicity of the MTMS modification. Considering the residual solvent, water molecules, and CO_2 during the SCD process, the weight loss rates of AS-S (4.51%) and HAS-S (3.87%) were significantly higher than those of HAS-F (2.59%) and HAS-A (2.17%). In the second stage between 200 °C and 500 °C, the weight loss was assigned to the elimination of the crystal water in the ATP and the condensation of Si-OH. The weight loss in the third stage (over 500 °C) was mainly attributed to the oxidative decomposition of Si-CH_3 and the removal of structural water in the ATP [27]. It is noted that a characteristic temperature point appeared in the hydrophobic sample at approximately 630 °C, which implies the complete decomposition of the Si-CH_3 group [42].

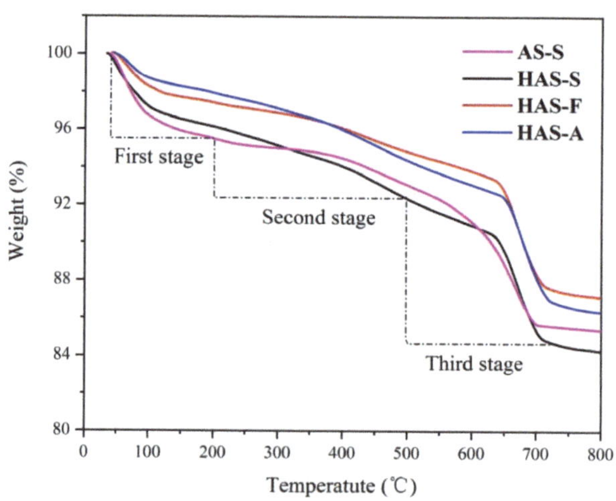

Figure 7. TG curves of AS-S, HAS-S, HAS-F, and HAS-A.

3.3. *Effect of Heat Treatment on Physical Characteristic and Morphological Properties of HAS-A*

The influence of the heat treatment temperatures on the morphology, microstructure, and pore structure is shown in Figures 8 and 9. Heat treatment is a simple and efficient way to modify the crosslinking structure and chemistry of aerogels [43]. The 600 °C heat-treated HAS-A is denoted as AS-A. As the calcination temperature increased, the appearance of the color of the HAS-A gradually changed from off-white to brick red (Figure 8a), which is mainly caused by the component of ATP. It is found from Figure 8b–e that the HAS-A heat treated at 400 °C showed a highly homogeneous porous structure. Due to the pyrolysis of -CH_3 and condensation among -OH, there was a large amount of silica particles aggregated inside the aerogel, and the uniform pore structure was damaged after calcination at 600 °C. A denser silica structure of the as-prepared HAS-A heat treated at 800 °C was clearly observed. All the curves in Figure 9a are type IV isotherms, suggesting that mesopores are still present in the aerogel despite the high temperature heat treatment. Consequentially, the pore size distributions of the heat-treated samples are shown in Figure 9b,c. From 200 °C to 400 °C, the peak of the micropores increased and then decreased after 600 °C. This is because at 200–400 °C, the excess organic matter inside the pores of the HAS-A decomposed, resulting in extra pores. At 600–800 °C, the aerogel particles inside the HAS-A will be aggregated and block the pore channels [44]. The pore structure of the HAS-A at different calcination temperatures is shown in Table 3. The specific surface area increased dramatically from 231.5 m^2/g to 337.7 m^2/g, with the treatment temperature increasing from 200 °C to 400 °C. Simultaneously, it was observed that the total pore volume changed from 0.40 cm^3/g to 0.43 cm^3/g, while the average pore size changed from 11.3 nm to 10.6 nm. However, owing to the oxidation of -CH_3 groups and condensation between -OH, a further increase in the calcination temperature was not favorable for the specific surface area and total pore volume.

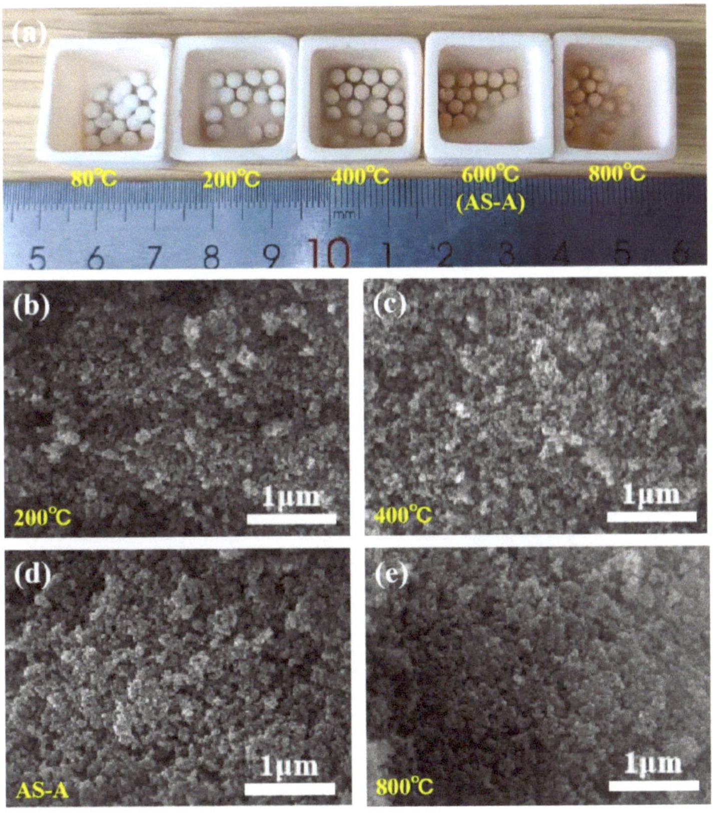

Figure 8. (**a**) Photographs and (**b**–**e**) SEM images of the HAS-A with different heat treatment temperatures.

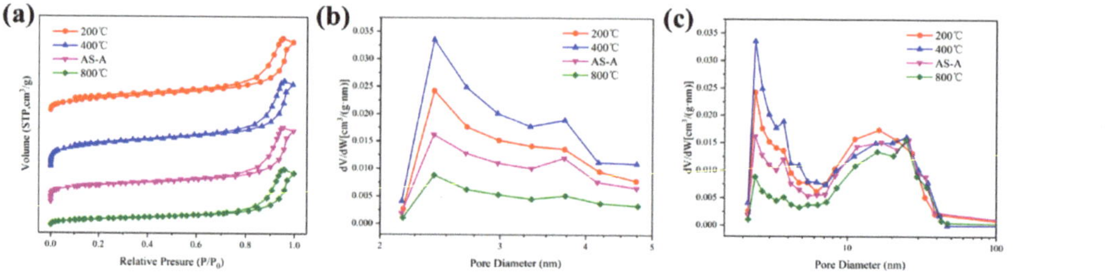

Figure 9. (**a**) N_2 adsorption–desorption and (**b**,**c**) BJH pore size distribution curve of the HAS-A at different calcination temperatures.

Table 3. Pore structure of the HAS-A after calcination.

Sample	S_{BET} (m^2/g)	Pore Size (nm)	Pore Volume (cm^3/g)
200 °C	231.5	11.3	0.40
400 °C	337.7	10.6	0.43
AS-A	246.6	12.4	0.40
800 °C	107.1	13.9	0.33

Figure 10a shows the XRD pattern of the heat-treated HAS-A. The characteristic peaks at 2θ = 27.6° and 30.9° were assigned to the (240) and (400) crystal planes of the ATP [27]. After a 600 °C calcination, the characteristic peak at 30.9° decreased significantly, indicating that the crystal structure of the ATP had been disrupted. After a 800 °C calcination, the characteristic peak at 30.9° gradually disappeared, indicating the complete destruction of the ATP crystal structure. A broad peak at 2θ = 21.5° was clearly identified in all samples, indicating that the silica aerogels preserve their original amorphous structure despite high temperature calcination [44]. The FT-IR spectra of the HAS-A heat treated at different temperatures are depicted in Figure 10b. The -CH_3 groups are observed at the absorption peaks of 780 cm^{-1}, 1275 cm^{-1}, and 2970 cm^{-1}. Those bands of the $-CH_3$ groups disappeared after a 600 °C calcination, indicating the pyrolysis of -CH_3 [42]. Moreover, Figure 10c shows the water contact angles of the HAS-A treated at different temperatures. The hydrophobicity of the HAS-A remained stable at 500 °C (200 °C, 400 °C, and 500 °C, corresponding to the values of 140.4°, 136.7°, and 131.2°, respectively), while the HAS-A converted to hydrophilic AS-A completely at 600 °C, with a water contact angle value of 0°, which is in accordance with the representation results in the TG (Figure 7) and FT-IR (Figure 10b). The transition from hydrophobic to hydrophilic in the spherical ATP/SiO_2 aerogels is conducive to the adsorption of soluble dyes in aqueous solutions. Therefore, the AS-A were used as adsorbents for studying the adsorption performance.

3.4. Adsorption Studies

To observe a visible change in the MB [7], Figure 11a exhibits the absorbance curves of MB at different concentrations. The adsorption capacity of the AS-A at different pH values, times, and MB concentrations is discussed as follows.

3.4.1. Effect of the Solution pH on MB Removal by AS-A

In this experiment, the pH was adapted from 2 to 11, and 50 mg of AS-A was used to adsorb MB at 50 mg/L for 100 min. The result in Figure 11b shows that the removal efficiency improved with the increase in the value of pH. The MB preferred to adsorb on AS-A under alkaline conditions, and the adsorption efficiency could reach 93.2% at a pH value of nine. Under acidic conditions, the adsorption capacity of AS-A is limited. This is due to the presence of excess hydrogen ions under acidic conditions, which leads to the protonation of functional groups on the AS-A surface during the adsorption process. Under alkaline conditions, AS-A has more adsorption sites on its surface, and the adsorption efficiency of MB will increase. Consequently, the pH value for the subsequent experiments was set to nine.

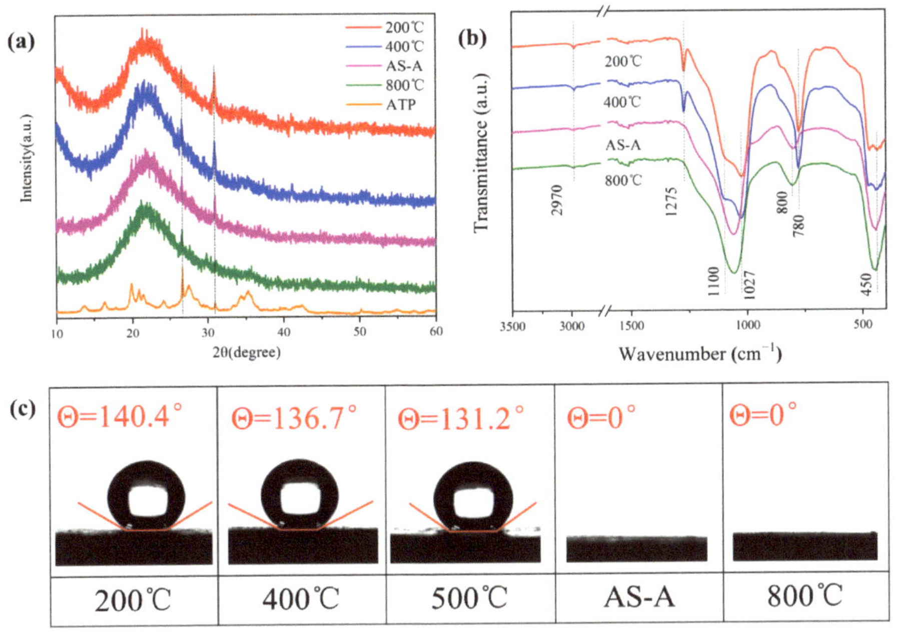

Figure 10. (**a**) The XRD patterns, (**b**) FT-IR spectra, and (**c**) water contact angles of the HAS-A with different heat treatment temperatures.

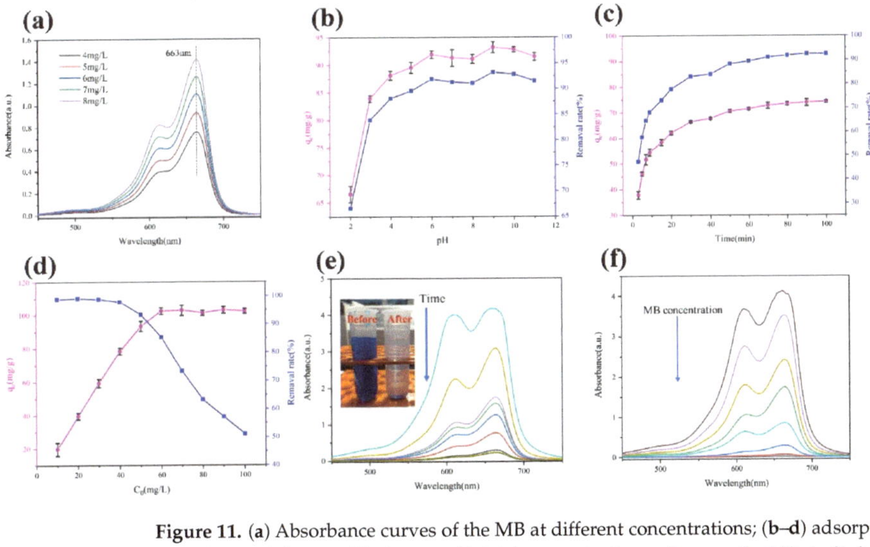

Figure 11. (**a**) Absorbance curves of the MB at different concentrations; (**b**–**d**) adsorption efficiency of the AS-A at different pH, time, and initial concentration values on the MB; (**e**,**f**) absorption spectra with the time and initial concentration of the MB.

3.4.2. Effect of Adsorption Time on MB Removal

In this part, the adsorption quantities from 0 to 100 min were studied to explore the adsorption equilibrium time. The pH value and MB concentration were set to 9 and 40 mg/L, respectively. As shown in Figure 11c, the adsorption capacity increased rapidly to 69.2% within 10 min. Figure 11e presents the spectra of the absorbance of MB with time. After 90 min, the adsorption equilibrium was achieved. The corresponding adsorption efficiency and capacity of the MB were 92.45% and 73.96 mg/g, respectively. After another 10 min, there was no noticeable change in the adsorption rate. Therefore, in subsequent experiments, the adsorption time was set to 100 min.

3.4.3. Effect of Initial Concentration on MB Removal

The effect of the initial MB concentration (10–100 mg/L) on the adsorption efficiency is shown in Figure 11d. Figure 11f shows the variation in the absorption spectra for different initial concentrations of MB. The adsorption capacity increased and then gradually stabilized with the increasing MB concentration. When the MB concentration reached 60 mg/L, the adsorption capacity reached 102.50 mg/g. As the concentration of MB was further increased, there were no significant changes in the adsorption capacity. This indicates that 50 mg of AS-A could not provide enough adsorption sites to remove the excess MB from the solution.

3.4.4. Adsorption Kinetics

The adsorption kinetics is closely related to the contact time, which can reflect the adsorption mechanism of the absorbent. Kinetic models were used to evaluate the experimental data:

Pseudo-first-order equation:

$$\ln(q_e - q_t) = \ln q_e - k_1 t$$

Pseudo-second-order equation:

$$\frac{t}{q_t} = \frac{1}{k_2 q_e^2} + \frac{1}{q_e} t$$

where k_1 and k_2 (g/(mg·min)) represent the pseudo-first-order and pseudo-second-order rate constants, respectively; q_e (mg/g) is the equilibrium adsorption quantity; q_t (mg/g) is the quantity adsorbed at different times ("t").

The model fitting results and the calculated kinetic parameters are shown in Figure 12 and Table 4. Compared with the pseudo-first-order, the experimental data confirm that the adsorption process was consistent with the pseudo-second-order model with high R^2 values of 0.99955. This suggests that the adsorption efficiency is highly dependent on the quantity of unoccupied active sites [45]. The q_e values of the pseudo-second-order model (76.90 mg/g) were found to be essentially the same as the experimental results (73.96 mg/g), indicating that the pseudo-second-order kinetic model can be employed for the adsorption of MB on AS-A.

Table 4. Kinetic parameters of MB adsorption on AS-A.

Pseudo-First-Order Model			**Pseudo-Second-Order Mode**		
q_e **(mg/g)**	k_1	$\mathbf{R^2}$	q_e **(mg/g)**	k_2	$\mathbf{R^2}$
39.98	0.08003	0.94153	76.90	0.00321	0.99955

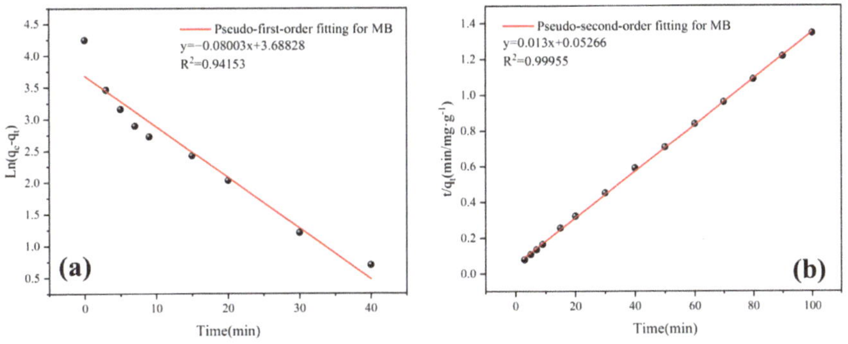

Figure 12. (**a**) Pseudo-first-order equation; (**b**) pseudo-second-order equation.

3.4.5. Adsorption Isotherms

The equilibrium adsorption isotherms play an important role in the interaction between the adsorbent and the adsorbate, which can explore the maximum adsorption capacity of the adsorbent. The Langmuir equation and the Freundlich equation are shown as follows:

Langmuir equation:

$$\frac{C_e}{q_e} = \frac{1}{q_{max}} C_e + \frac{1}{k_L q_{max}}$$

Freundlich equation:

$$\ln q_e = \ln k_F + \frac{1}{n} \ln C_e \quad (3)$$

where q_e (mg/g) is the adsorption amount at adsorption equilibrium; q_{max} (mg/g) stands for the maximum adsorption capacity; C_e (mg/L) represents the equilibrium concentration; k_L and k_F are the constants of the Langmuir isotherm equation and Freundlich isotherm equation; and $1/n$ is the adsorption intensity parameter. The fitting results of the Langmuir and Freundlich equations are shown in Figure 13 and Table 5.

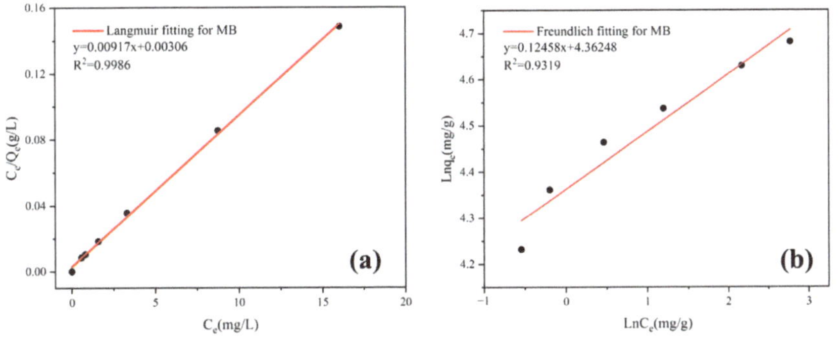

Figure 13. (**a**) Langmuir isotherm model; (**b**) Freundlich isotherm model.

Table 5. Constants for the adsorption isotherms of MB on AS-A.

Langmuir Equation			**Freundlich Equation**		
q_{max} (mg/g)	k_L	R^2	k_F	$1/n$	R^2
109.05	2.997	0.9986	78.45	0.12458	0.9319

The Langmuir model fits the experimental data best, with R^2 values of 0.9986, which proves that the sorption process of MB on AS-A belongs to monolayer adsorption [45]. The maximum adsorption capacity derived from the Langmuir adsorption model was 109.05 mg/g, which closely corresponds to the experimental data (102.50 mg/g). The adsorption performance of the AS-A was compared with the other reported adsorbents (Table 6). The AS-A obtained by heat treatment at 600 °C had excellent adsorption performance on MB in comparison with the other adsorbents. Although it has no advantage in terms of specific surface area, the combination of attapulgite and silica aerogel greatly improved the adsorption of the composite. This reveals that AS-A is a potential adsorbent for dye wastewater.

Table 6. Comparison of MB adsorption capacities with different adsorbents.

Adsorbent	Treatment Method	S_{BET} (m^2/g)	q_{max} (mg/g)	Reference
SiO_2(AG)	Heat treated at 500 °C for 3 h	902.0	9.53	Yi et al. 2019 [10]
HSA	Heat treated at 600 °C for 3 h	888.7	51.16	Wei et al. 2018 [11]
SAs	Mg^{2+} soaked	468.2	40.40	Yang et al. 2020 [12]
Attapulgite	Heat treated at 700 °C for 4 h	-	78.11	Chen et al. 2011 [27]
Ag/SiO_2	Ag decorated	208.0	55.00	Hu et al. 2019 [46]
MPCMs	Heat treated at 700 °C for 30 min	480.3	56.44	Zhang et al. 2021 [47]
ZIF-8	Fe decorated	329.9	9.09	Mahmoodi et al. 2021 [48]
$CaCO_3$@STA/PAM/TOCN	-	39.2	101.01	Li et al. 2022 [49]
CCGA	-	225.2	96.10	Shimizu et al. 2022 [50]
CNCS/SiO_2 aerogel	-	440.7	190.85	Ruan et al. 2022 [51]
AS-A	Heat treated at 600 °C for 3 h	246.6	102.50	This work

3.4.6. Recycling Studies

The regenerative adsorption of the adsorbent is a vital factor in measuring its suitability in practical applications. In the recycling test, AS-A was heat treated at 600 °C for 3 h in a muffle furnace to eliminate the adsorbed dye. As shown in Figure 14a, the adsorption capacity of AS-A remained at 82.32% after five adsorption cycles, indicating that the adsorption of MB on AS-A was stable for multiple cycles. After the regeneration, the disappearance of MB aromatic group at 1603 cm^{-1} illustrates that MB can be completely removed by calcination at 600 °C (Figure 14b) [52].

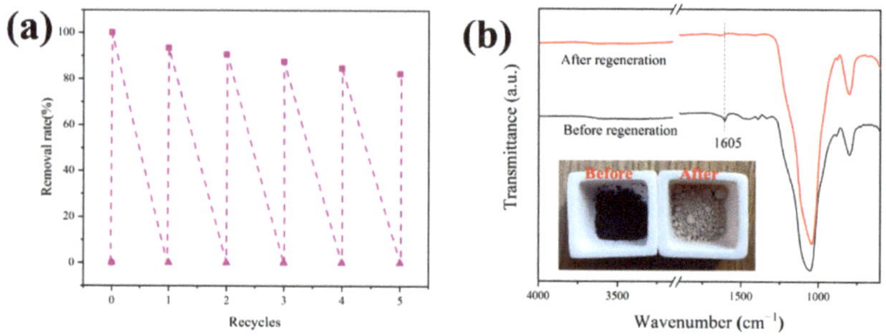

Figure 14. (**a**) Removal rate of AS-A during five cycles (C_0 = 50 mg/L, V = 100 mL, m = 50 mg, pH = 9, and t = 100 min); (**b**) FT-IR spectra of AS-A before and after regeneration.

3.4.7. Adsorption Mechanism

The zeta potential of the AS-A was negative in the pH range of 5–9 (Figure 15a), indicating that the surface of the AS-A exhibited negative charges. In addition, the amount of negative charge on the AS-A surface tended to increase with the increase in the pH. As shown in Figure 15b, MB is a typical cationic dye with a positive charge [53]. The negatively charged AS-A had a significant adsorption effect on the positively charged MB, resulting in electrostatic attraction between MB and AS-A. Therefore, the adsorption mechanism of AS-A on MB is mainly an electrostatic effect. The surface negative charge of AS-A has great attraction to MB with positive charge, which leads to an excellent adsorption performance on MB.

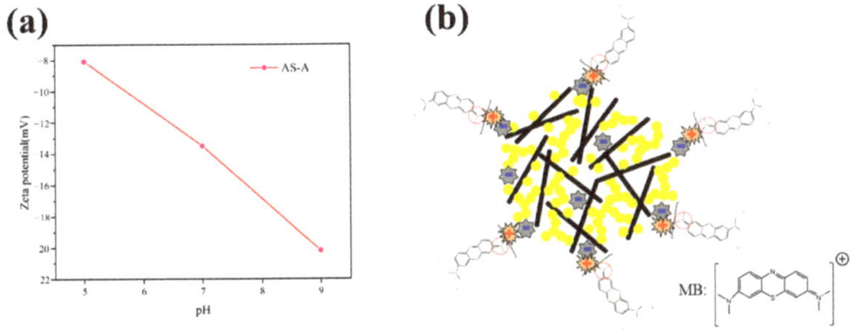

Figure 15. (**a**) Zeta potential of AS-A; (**b**) mechanisms of MB adsorption on AS-A.

4. Conclusions

In this work, the spherical ATP/SiO_2 aerogels were successfully synthesized by three different drying techniques (SCD, FD, and APD). The spherical ATP/SiO_2 aerogels dried with three techniques possessed a complete macromorphology and homogeneous porous network structures. The control of the size of the obtained spherical particles will be included in the next step of the study. The club-shaped ATP was well dispersed in the aerogel matrix without agglomeration, and the silica particles were closely attached to the surface of the ATP. The average pore diameter and specific surface area of the spherical ATP/SiO_2 aerogels prepared by three drying techniques were in the range of 6.8–8.6 nm and 218.5–267.4 m^2/g, respectively. The BET specific surface area of the HAS-A after the 400 °C heat treatment sharply increased to 333.7 m^2/g. Compared to the other adsorbents, AS-A showed a higher adsorption capacity of 102.50 mg/g for MB. After five cycles of regeneration, the adsorption efficiency of AS-A could still maintain 82.32%, exhibiting a good reusability. Therefore, these porous spherical ATP/SiO_2 aerogels with multipath preparation and excellent adsorption performance are expected to have applications in dye wastewater treatment.

Author Contributions: Z.Z., Conceptualization, Data curation, Investigation, Methodology, and Writing—original draft; S.W., Conceptualization and Methodology; Q.Y., Investigation and Writing—review and editing; J.G., Software and Supervision; Y.Z., Supervision, Writing—review and editing, and Funding acquisition; S.C., Supervision and Resources; X.S., Supervision and Formal analysis. All authors have read and agreed to the published version of the manuscript.

Funding: This work was supported by the Key Research and Development Project of Jiangsu Province (BE2021134 and BE2019734), National Natural Science Foundation of China (52102361), and Natural Science Foundation of Jiangsu Province (BK20200827 and BK20200711).

Institutional Review Board Statement: Not applicable.

Informed Consent Statement: Not applicable.

Data Availability Statement: The data that support the findings of this study are available from the corresponding author upon reasonable request.

Acknowledgments: The authors gratefully acknowledge support from the Key Research and Development Project of Jiangsu Province (BE2021134 and BE2019734), National Natural Science Foundation of China (52102361), Natural Science Foundation of Jiangsu Province (BK20200827 and BK20200711), Key Laboratory of Advanced Functional Composites Technology (6142906210508), and the Priority Academic Program Development of Jiangsu Higher Education Institutions (PAPD), China. Any opinions, findings, conclusions, or recommendations expressed in this paper are those of the authors and do not necessarily reflect the views of these programs.

Conflicts of Interest: The authors declare no conflict of interest.

References

1. Shindhal, T.; Rakholiya, P.; Varjani, S.; Pandey, A.; Ngo, H.H.; Guo, W.S.; Ng, H.Y.; Taherzadeh, M.J. A critical review on advances in the practices and perspectives for the treatment of dye industry wastewater. *Bioengineered* **2021**, *12*, 70–87. [CrossRef]
2. Moradihamedani, P. Recent advances in dye removal from wastewater by membrane technology: A review. *Polym. Bull.* **2022**, *79*, 2603–2631. [CrossRef]
3. Yaseen, D.A.; Scholz, M. Textile dye wastewater characteristics and constituents of synthetic effluents: A critical review. *Int. J. Environ. Sci. Technol.* **2019**, *16*, 1193–1226. [CrossRef]
4. Rafaqat, S.; Ali, N.; Torres, C.; Rittmann, B. Recent progress in treatment of dyes wastewater using microbial-electro-Fenton technology. *RSC Adv.* **2022**, *12*, 17104–17137. [CrossRef]
5. Li, W.; Mu, B.N.; Yang, Y.Q. Feasibility of industrial-scale treatment of dye wastewater via bio-adsorption technology. *Bioresour. Technol.* **2019**, *277*, 157–170. [CrossRef]
6. Chang, Z.T.; Chen, Y.J.; Tang, S.X.; Yang, J.Y.; Chen, Y.; Chen, S.S.; Li, P.W.; Yang, Z.M. Construction of chitosan/polyacrylate/graphene oxide composite physical hydrogel by semi-dissolution/acidification/sol-gel transition method and its simultaneous cationic and anionic dye adsorption properties. *Carbohydr. Polym.* **2020**, *229*, 115431. [CrossRef] [PubMed]
7. Krishna Kumar, A.S.; Warchol, J.; Matusik, J.; Tseng, W.-L.; Rajesh, N.; Bajda, T. Heavy metal and organic dye removal via a hybrid porous hexagonal boron nitride-based magnetic aerogel. *NPJ Clean Water* **2022**, *5*, 24. [CrossRef]
8. Yao, C.; Dong, X.; Gao, G.; Sha, F.; Xu, D. Microstructure and Adsorption Properties of MTMS/TEOS Co-precursor Silica Aerogels Dried at Ambient Pressure. *J. Non-Cryst. Solids* **2021**, *562*, 120778. [CrossRef]
9. Jiang, X.; Kong, Y.; Zhao, Z.Y.; Shen, X.D. Spherical amine grafted silica aerogels for CO_2 capture. *RSC Adv.* **2020**, *10*, 25911–25917. [CrossRef] [PubMed]
10. Yi, Z.G.; Tang, Q.; Jiang, T.; Cheng, Y. Adsorption performance of hydrophobic/hydrophilic silica aerogel for low concentration organic pollutant in aqueous solution. *Nanotechnol. Rev.* **2019**, *8*, 266–274. [CrossRef]
11. Wei, W.; Hu, H.H.; Ji, X.L.; Yan, Z.X.; Sun, W.; Xie, J.M. Selective adsorption of organic dyes by porous hydrophilic silica aerogels from aqueous system. *Water Sci. Technol.* **2018**, *78*, 402–414. [CrossRef] [PubMed]
12. Yang, X.H.; Wu, Z.X.; Chen, H.F.; Du, Q.X.; Yu, L.; Zhang, R.Y.; Zhou, Y. A Facile Preparation of Ambient Pressure-Dried Hydrophilic Silica Aerogels and Their Application in Aqueous Dye Removal. *Front. Mater.* **2020**, *7*, 152. [CrossRef]
13. Zu, G.Q.; Shimizu, T.; Kanamori, K.; Zhu, Y.; Maeno, A.; Kaji, H.; Shen, J.; Nakanishi, K. Transparent, Superflexible Doubly Cross-Linked Polyvinylpolymethylsiloxane Aerogel Superinsulators via Ambient Pressure Drying. *ACS Nano* **2018**, *12*, 521–532. [CrossRef] [PubMed]
14. Jadhav, S.B.; Makki, A.; Hajjar, D.; Sarawade, P.B. Synthesis of light weight recron fiber-reinforced sodium silicate based silica aerogel blankets at an ambient pressure for thermal protection. *J. Porous Mat.* **2022**, *29*, 957–969. [CrossRef]
15. Shi, W.; Ching, Y.C.; Chuah, C.H. Preparation of aerogel beads and microspheres based on chitosan and cellulose for drug delivery: A review. *Int. J. Biol. Macromol.* **2021**, *170*, 751–767. [CrossRef]
16. Zong, S.K.; Wei, W.; Jiang, Z.F.; Yan, Z.X.; Zhu, J.J.; Xie, J.M. Characterization and comparison of uniform hydrophilic/hydrophobic transparent silica aerogel beads: Skeleton strength and surface modification. *RSC Adv.* **2015**, *5*, 55579–55587. [CrossRef]
17. Tian, J.Q.; Shafi, S.; Tan, H.J.; Zhao, Y.P. Mechanical and thermal-insulating performance of silica aerogel enhanced jointly with glass fiber and fumed silica by a facile compressing technique. *Chem. Phys. Lett.* **2020**, *739*, 136950. [CrossRef]
18. Slosarczyk, A. Carbon Fiber-Silica Aerogel Composite with Enhanced Structural and Mechanical Properties Based on Water Glass and Ambient Pressure Drying. *Nanomaterials* **2021**, *11*, 258. [CrossRef]
19. Linhares, T.; de Amorim, M.T.P.; Duraes, L. Silica aerogel composites with embedded fibres: A review on their preparation, properties and applications. *J. Mater. Chem. A* **2019**, *7*, 22768–22802. [CrossRef]
20. Zhu, J.D.; Zhao, F.X.; Xiong, R.J.; Peng, T.P.; Ma, Y.Y.; Hu, J.; Xie, L.; Jiang, C.W. Thermal insulation and flame retardancy of attapulgite reinforced gelatin-based composite aerogel with enhanced strength properties. *Compos. Part A Appl. Sci. Manuf.* **2020**, *138*, 106040. [CrossRef]
21. Finlay, K.A.; Gawryla, M.D.; Schiraldi, D.A. Effects of Fiber Reinforcement on Clay Aerogel Composites. *Materials* **2015**, *8*, 5440–5451. [CrossRef]
22. Li, X.L.; Wang, Q.P.; Li, H.L.; Ji, H.M.; Sun, X.H.; He, J. Effect of sepiolite fiber on the structure and properties of the sepiolite/silica aerogel composite. *J. Sol-Gel Sci. Techn.* **2013**, *67*, 646–653. [CrossRef]
23. Cui, M.K.; Mu, P.; Shen, Y.Q.; Zhu, G.R.; Luo, L.; Li, J. Three-dimensional attapulgite with sandwich-like architecture used for multifunctional water remediation. *Sep. Purif. Technol.* **2020**, *235*, 116210. [CrossRef]
24. Mahadik, D.B.; Rao, A.V.; Kumar, R.; Ingale, S.V.; Wagh, P.B.; Gupta, S.C. Reduction of processing time by mechanical shaking of the ambient pressure dried TEOS based silica aerogel granules. *J. Porous Mater.* **2012**, *19*, 87–94. [CrossRef]
25. Xu, L.; Liu, Y.N.; Wang, J.G.; Tang, Y.; Zhang, Z. Selective adsorption of Pb^{2+} and Cu^{2+} on amino-modified attapulgite: Kinetic, thermal dynamic and DFT studies. *J. Hazard. Mater.* **2021**, *404*, 124140. [CrossRef] [PubMed]

26. Zhu, L.X.; Liu, P.; Wang, A.Q. High Clay-Content Attapulgite/Poly(acrylic acid) Nanocomposite Hydrogel via Surface-Initiated Redox Radical Polymerization with Modified Attapulgite Nanorods as Initiator and Cross-Linker. *Ind. Eng. Chem. Res.* **2014**, *53*, 2067–2071. [CrossRef]
27. Chen, H.; Zhao, J.; Zhong, A.G.; Jin, Y.X. Removal capacity and adsorption mechanism of heat-treated palygorskite clay for methylene blue. *Chem. Eng. J.* **2011**, *174*, 143–150. [CrossRef]
28. Zhang, Y.L.; Dang, J.Y.; Su, X.S.; Jin, H.C. Removal of total petroleum hydrocarbon from water using attapulgite fiber silica aerogel composite. *Desalin. Water Treat.* **2016**, *57*, 17463–17472. [CrossRef]
29. Shaari, N.; Kamarudin, S.K. Current status, opportunities, and challenges in fuel cell catalytic application of aerogels. *Int. J. Energy Res.* **2019**, *43*, 2447–2467. [CrossRef]
30. Novak, Z.; Knez, Z. Diffusion of methanol-liquid CO_2 and methanol-supercritical CO_2 in silica aerogels. *J. Non-Cryst. Solids* **1997**, *221*, 163–169. [CrossRef]
31. Pan, Y.L.; Cheng, X.D.; Zhou, T.; Gong, L.L.; Zhang, H.P. Spray freeze-dried monolithic silica aerogel based on water-glass with thermal superinsulating properties. *Mater. Lett.* **2018**, *229*, 265–268. [CrossRef]
32. Li, J.; Lei, Y.; Xu, D.D.; Liu, F.H.; Li, J.W.; Sun, A.H.; Guo, J.J.; Xu, G.J. Improved mechanical and thermal insulation properties of monolithic attapulgite nanofiber/silica aerogel composites dried at ambient pressure. *J. Sol-Gel Sci. Techn.* **2017**, *82*, 702–711. [CrossRef]
33. Ganbavle, V.V.; Kalekar, A.S.; Harale, N.S.; Patil, S.S.; Dhere, S.L. Rapid synthesis of ambient pressure dried tetraethoxysilane based silica aerogels. *J. Sol-Gel Sci. Techn.* **2021**, *97*, 5–10. [CrossRef]
34. Cok, S.S.; Koc, F.; Gizli, N. Lightweight and highly hydrophobic silica aerogels dried in ambient pressure for an efficient oil/organic solvent adsorption. *J. Hazard. Mater.* **2021**, *408*, 124858.
35. Cok, S.S.; Gizli, N. Hydrophobic silica aerogels synthesized in ambient conditions by preserving the pore structure via two-step silylation. *Ceram. Int.* **2020**, *46*, 27789–27799. [CrossRef]
36. Liu, Y.M.; Yan, W.Q.; Zhong, Y.; Wu, Z.W.; Shang, S.S.; Wu, X.D.; Shen, X.D.; Cui, S. Synthesis and characterization of amino-grafted attapulgite/graphene oxide nanocomposites and their adsorption for Pb(II) removal. *J. Nanopart. Res.* **2022**, *24*, 28. [CrossRef]
37. Cai, Y.F.; Xue, J.Y.; Polya, D.A. A Fourier transform infrared spectroscopic study of Mg-rich, Mg-poor and acid leached palygorskites. *Spectrochim. Acta A.* **2007**, *66*, 282–288. [CrossRef]
38. Zhang, J.; Chen, A.S.; Wang, L.H.; Li, X.A.; Huang, W. Striving Toward Visible Light Photocatalytic Water Splitting Based on Natural Silicate Clay Mineral: The Interface Modification of Attapulgite at the Atomic-Molecular Level. *ACS Sustain. Chem. Eng.* **2016**, *4*, 4601–4607. [CrossRef]
39. Shang, S.S.; Ye, X.; Jiang, X.; You, Q.; Zhong, Y.; Wu, X.D.; Cui, S. Preparation and characterization of cellulose/attapulgite composite aerogels with high strength and hydrophobicity. *J. Non-Cryst. Solids* **2021**, *569*, 120922. [CrossRef]
40. Yu, H.J.; Liang, X.F.; Wang, J.X.; Wang, M.M.; Yang, S.Y. Preparation and characterization of hydrophobic silica aerogel sphere products by co-precursor method. *Solid State Sci.* **2015**, *48*, 155–162. [CrossRef]
41. Zhao, Y.F.; Zhong, K.; Liu, W.; Cui, S.; Zhong, Y.; Jiang, S.J. Preparation and oil adsorption properties of hydrophobic microcrystalline cellulose aerogel. *Cellulose* **2020**, *27*, 7663–7675. [CrossRef]
42. Zhou, T.; Cheng, X.; Pan, Y.; Li, C.; Gong, L.; Zhang, H. Mechanical performance and thermal stability of glass fiber reinforced silica aerogel composites based on co-precursor method by freeze drying. *Appl. Surf. Sci.* **2018**, *437*, 321–328. [CrossRef]
43. Sarawade, P.B.; Shao, G.N.; Quang, D.V.; Kim, H.T. Effect of various structure directing agents on the physicochemical properties of the silica aerogels prepared at an ambient pressure. *Appl. Surf. Sci.* **2013**, *287*, 84–90. [CrossRef]
44. He, S.; Huang, Y.; Chen, G.; Feng, M.; Dai, H.; Yuan, B.; Chen, X. Effect of heat treatment on hydrophobic silica aerogel. *J. Hazard. Mater.* **2019**, *362*, 294–302. [CrossRef] [PubMed]
45. Ye, X.; Shang, S.; Zhao, Y.; Cui, S.; Zhong, Y.; Huang, L. Ultra-efficient adsorption of copper ions in chitosan–montmorillonite composite aerogel at wastewater treatment. *Cellulose* **2021**, *28*, 7201–7212. [CrossRef]
46. Hu, M.Q.; Yan, X.L.; Hu, X.Y.; Feng, R.; Zhou, M. Synthesis of silver decorated silica nanoparticles with rough surfaces as adsorbent and catalyst for methylene blue removal. *J. Sol-Gel Sci. Techn.* **2019**, *89*, 754–763. [CrossRef]
47. Zhang, H.; Chen, L.; Li, L.F.; Yang, Y.Z.; Liu, X.G. Magnetic porous carbon microspheres synthesized by simultaneous activation and magnetization for removing methylene blue. *J. Porous Mat.* **2017**, *24*, 341–353. [CrossRef]
48. Mahmoodi, M.; Javanbakht, V. Fabrication of Zn-based magnetic zeolitic imidazolate framework bionanocomposite using basil seed mucilage for removal of azo cationic and anionic dyes from aqueous solution. *J. Biol. Macromol.* **2021**, *167*, 1076–1090. [CrossRef]
49. Li, P.H.; Yang, C.; Xu, X.W.; Miao, C.; He, T.J.; Jiang, B.; Wu, W.J. Preparation of Bio-Based Aerogel and Its Adsorption Properties for Organic Dyes. *Gels* **2022**, *8*, 755. [CrossRef] [PubMed]
50. Shimizu, T.; De Silva, K.K.H.; Hara, M.; Yoshimura, M. Facile synthesis of carbon nanotubes and cellulose nanofiber incorporated graphene aerogels for selective organic dye adsorption. *Appl. Surf. Sci.* **2022**, *600*, 154098. [CrossRef]
51. Ruan, C.; Ma, Y.; Shi, G.; He, C.; Du, C.; Jin, X.; Liu, X.; He, S.; Huang, Y. Self-assembly cellulose nanocrystals/SiO_2 composite aerogel under freeze-drying: Adsorption towards dye contaminant. *Appl. Surf. Sci.* **2022**, *592*, 153280. [CrossRef]

52. Canete, S.J.; Zhang, Z.; Kong, L.; Schlegel, V.L.; Plantz, B.A.; Dowben, P.A.; Lai, R.Y. Application of synchrotron FTIR microspectroscopy for determination of spatial distribution of methylene blue conjugated onto a SAM via "click" chemistry. *Chem. Commun.* **2011**, *47*, 11918–11920. [CrossRef] [PubMed]
53. Han, H.K.; Wei, W.; Jiang, Z.F.; Lu, J.W.; Zhu, J.J.; Xie, J.M. Removal of cationic dyes from aqueous solution by adsorption onto hydrophobic/hydrophilic silica aerogel. *Colloid. Surf. A* **2016**, *509*, 539–549. [CrossRef]

materials

MDPI

Article

Porous Structural Properties of K or Na-Co Hexacyanoferrates as Efficient Materials for CO_2 Capture

Paloma M. Frías-Ureña [1], Maximiliano Bárcena-Soto [1], Eulogio Orozco-Guareño [1], Alberto Gutiérrez-Becerra [2], Josué D. Mota-Morales [3], Karina Chavez [1], Víctor Soto [1,4], José A. Rivera-Mayorga [1], José I. Escalante-Vazquez [1] and Sergio Gómez-Salazar [5,*]

1 Departamento de Química, Universidad de Guadalajara (CUCEI), Boulevard Marcelino García Barragán #1421, Esquina Calzada Olímpica, Guadalajara 44430, Mexico
2 Departamento de Ciencias Básicas y Aplicadas, Universidad de Guadalajara (CUTonala), Avenue Nuevo Periférico 555, Tonalá 45425, Mexico
3 Centro de Física Aplicada y Tecnología Avanzada, Universidad Nacional Autónoma de Mexico, Querétaro 76230, Mexico
4 Graduate Program in Materials Science, Departamento de Ingeniería de Proyectos, Universidad de Guadalajara (CUCEI), Boulevard Marcelino García Barragán #1421, Esquina Calzada Olímpica, Guadalajara 44430, Mexico
5 Departamento de Ingeniería Química, Universidad de Guadalajara (CUCEI), Boulevard Marcelino García Barragán #1421, Esquina Calzada Olímpica, Guadalajara 44430, Mexico
* Correspondence: sergio.gomez@cucei.udg.mx

Abstract: The stoichiometry of the components of hexacyanoferrate materials affecting their final porosity properties and applications in CO_2 capture is an issue that is rarely studied. In this work, the effect that stoichiometry of all element components and oxidation states of transition metals has on the structures of mesoporous K or Na-cobalt hexacyanoferrates (CoHCFs) and CO_2 removal is reported. A series of CoHCFs model systems are synthesized using the co-precipitation method with varying amounts of Co ions. CoHCFs are characterized by N_2 adsorption, TGA, FTIR-ATR, XRD, and XPS. N_2 adsorption results reveal a more developed external surface area (72.69–172.18 m^2/g) generated in samples containing mixtures of $K^+/Fe^{2+}/Fe^{3+}$ ions (system III) compared to samples with Na^+/Fe^{2+} ions (systems I, II). TGA results show that the porous structure of CoHCFs is affected by Fe and Co ions oxidation states, the number of water molecules, and alkali ions. The formation of two crystalline cells (FCC and triclinic) is confirmed by XRD results. Fe and Co oxidation states are authenticated by XPS and allow for the confirmation of charges involved in the stabilization of CoCHFs. CO_2 removal capacities (3.04 mmol/g) are comparable with other materials reported. CO_2 adsorption kinetics is fast (3–6 s), making CoHCFs attractive for continuous operations. Q_{st} (24.3 kJ/mol) reveals a physical adsorption process. Regeneration effectiveness for adsorption/desorption cycles indicates ~1.6% loss and selectivity (~47) for gas mixtures (CO_2:N_2 = 15:85). The results of this study demonstrate that the CoHCFs have practical implications in the potential use of CO_2 capture and flue gas separations.

Keywords: hexacyanoferrates; porosity; adsorption; CO_2 capture; isosteric heat of adsorption

Citation: Frías-Ureña, P.M.; Bárcena-Soto, M.; Orozco-Guareño, E.; Gutiérrez-Becerra, A.; Mota-Morales, J.D.; Chavez, K.; Soto, V.; Rivera-Mayorga, J.A.; Escalante-Vazquez, J.I.; Gómez-Salazar, S. Porous Structural Properties of K or Na-Co Hexacyanoferrates as Efficient Materials for CO_2 Capture. *Materials* **2023**, *16*, 608. https://doi.org/10.3390/ma16020608

Academic Editor: Weiqing Zhang

Received: 12 November 2022
Revised: 23 December 2022
Accepted: 5 January 2023
Published: 8 January 2023

1. Introduction

The need to lessen climate change produced by greenhouse gases, such as CO_2, is a reality nowadays. To this end, a great variety of methodologies have been applied in the removal of CO_2. Amongst these, the use of porous adsorbents displaying fast removal velocities, improved uptake capacities, increased affinities for CO_2 in the presence of other gases, and stable regeneration properties have shown promise as an alternative to more conventional methods in the solution to this problem and are critical when selecting the adsorbent [1]. Particularly, the use of Metal-Organic Frameworks (MOFs) possessing highly developed porosities and outstanding textural properties, e.g., high specific surface areas, optimum pore diameters to ease CO_2 diffusion, and tunable chemical and structural properties have appeared as an encouraging alternative [2,3]. For instance, very recently, MOF-74(Ni)-24-140 was reported to present an adsorption capacity of CO_2 of 8.29/6.61 mmol/g at 273/298 K under the normal pressure of 1.0 bar [4], and this exceeded those of materials such as the amine-functionalized carbon nanotubes TEPA IG-MWCNT of 3.09 mmol g^{-1} [5] and MOF-808 functionalized with ethyleneamines (~2.5 mmol g^{-1}) [6]. Before this, a type of MOFs, metal hexacyanoferrate (also known as Prussian Blue Analogs (PBAs)), attracted increasing interest in CO_2 removal applications due to their intricate micro/meso porous structures, highly developed textural properties and the relatively mild and easy to implement synthesis routes [7]. PBAs have been used in CO_2 capture or H_2 storage applications due to their highly porous properties [8,9]. A recent study demonstrated that hexacyanoferrates, e.g., $K_{2x/3}Cu^{II}[Fe^{II}{}_xFe^{III}{}_{1-x}(CN)_6]_{2/3}$, containing alkali ions such as K^+ comply with these requirements [10]. These materials have demonstrated that they possess outstanding CO_2 capture capacities (~3.0 mmol/g CO_2), fast adsorption kinetics, high selectivity indices for gas mixtures such as CO_2/N_2, and highly stable regeneration capacities [1].

PBAs are coordination compounds with a general formula $M^+{}_pM'_q[M''(CN)_6]_r{\cdot}xH_2O$ where M′ and M″ are commonly transition metals that present different oxidation states. Synthesis conditions and the presence of alkali M^+ ions have an influence on the stoichiometric coefficients p, q, r, and x and the structural and porous properties. These coordination compounds possess a crystalline structure [11]. PBAs have novel applications due to their electronic, magnetic, optical, electrochromic, electrocatalytic, ion sensing, and ion storage properties [12]. In this sense, cobalt hexacyanoferrates (CoHCFs) have been extensively studied due to their physicochemical, electrochromism properties, photoinduced reversible magnetization, and reversible thermochromism [13], properties which derive from the bivalent and trivalent oxidation states of iron and cobalt of the compound [14], so they have been used to manufacture modified graphite electrodes [15]. Furthermore, these materials have also been used in technologies applied in the decontamination of radioactive cesium [16]. CoHCFs have a crystalline structure containing ferricyanide vacancies occupied by water molecules [17,18]. Water molecules form hydrogen bridge networks at the interstitial places where alkali ions are also inserted. The stoichiometry of the compound can vary depending on the number of alkali ions present, and network electroneutrality is guaranteed via the adjustment of $[Fe^{III}(CN)_6]^{3-}$ gaps in the structure [19]. When water molecules are removed from the structure, the material becomes porous with the capability of capturing gases such as CO_2 [10]. There are two types of positions where water molecules can be located within the crystalline structure of the CoHCFs, namely, water coordinated to Co ions at vacant sites of the $[Fe^{III}(CN)_6]^{3-}$, and water contained inside every eighth of the crystal unit cell known as zeolitic water [20]. Consequently, different types of pores are present in the CoHCFs analyzed, where the formation of micropores has been attributed to the removal of zeolitic and coordination water, whereas the formation of mesopores has been attributed to the agglomeration of particles [20,21]. The gaps can be interconnected, forming micro tunnels [22], thus improving the textural properties and increasing their porosity.

The effect of porous and textural properties of PBAs on the extent of CO_2 capture has been studied. There are few studies about the effect of incorporated alkali ions on the porous structure of PBAs. For instance, Li et al. [23] recently reported that K^+ ions exposed after water molecules removal from pores/crystals of PBAs produced adsorption sites owing a high affinity for CO_2 molecules with the consequent modification of the porous structure of the material. On the other hand, Karadas et al. [20] reported that available free pore volumes, V_{free} calculated from Monte Carlo simulations on PBAs samples, enabled the adsorption of significant amounts of CO_2 on the PBAs structures. In addition, they also reported that the adsorption of CO_2 on large pores resulting in random gaps would be more favorable, especially at low and moderated pressures. In this regard, there is controversy in the reports that indicate, on the one hand, that adsorption of CO_2 or H_2 is preferential on large pores of PBAs [24], whereas on the other hand, based upon infrared spectroscopic studies on $M_3[Co(CN)_6]_2$, Natesakhawat et al. [25] reported that the adsorption of CO_2 on these materials was unaffected by the differences between large and small pores. In a recently published review, Lopez et al. [26] reported that the formation of open cavities on the 3D frameworks (such as those present in PBAs) supports ion or gas molecule diffusion. These structural properties, combined with high porosity, clear the way for their use in gas capture and storage applications. Regardless, the above statements indicate the importance of studies dealing with the porous structural properties of PBAs.

The aim of this work is to report on the effect that synthesis stoichiometry has on the structural and porous properties of K or Na-CoHCFs and the efficient CO_2 capture. The novelty of this work includes reporting on the effect that stoichiometric amounts, the types of alkali K and Na ions, and the oxidation state of transition metals have on the final structural and porous properties of CoHCFs and the CO_2 adsorption capacities. A battery of instrumental techniques is applied for the characterization of CoHCFs, including N_2 adsorption, TGA, FTIR-ATR, XRD, and XPS.

2. Experimental

2.1. Synthesis of K or Na-Cobalt Hexacyanoferrates (CoHCFs)

The synthesis of CoHCFs was performed using the drop method and a volumetric ratio of $K_{3,4}/Na_4Fe(CN)_6$:$Co(NO_3)_2$ of 6:1. In a typical synthesis, 300 mL of a solution 0.02 M of $[Fe(CN)_6\ ]^{n-}$ was placed in an Erlenmeyer flask under stirring. To this, 50 mL of a solution of $Co(NO_3)_2$ at varying concentrations of 0.035–0.175 M was added dropwise at time intervals of 2–3 s between each drop addition at 25 °C. After the addition of all the nitrate solution, the reacting mixture was agitated for another 5 min and then was left to rest for 24 h to allow the system to stabilize. All CoHCFs solutions were stored in amber bottles to avoid the reduction of the transition metals by sunlight. To obtain CoHCFs powders, solutions were centrifuged for 20 min, the supernatant removed, and the resulting precipitate washed with doubly distilled water. The precipitate was then dried at 60 °C for 48 h. The powder obtained was ground using an Agatha mortar. Stoichiometric coefficients were determined from ICP-MS after acid digestion. The atomic contents were normalized against Fe. Alkali metals were obtained by charge balances. The H_2O content was determined from TGA. Sample identifications are shown in Table 1.

Table 1. Textural properties of CoHCFs samples obtained from N_2 adsorption measurements and stoichiometric coefficients for each sample.

Sample ID	Reactant Molar Ratio	S_{BET} ($m^2\ g^{-1}$)	D_p (Å)	V_t ($cm^3\ g^{-1}$)	S_{ext} ($m^2\ g^{-1}$)	S_{mic} ($m^2\ g^{-1}$)	V_{mic} ($cm^3\ g^{-1}$)	V_{meso} ($cm^3\ g^{-1}$)	Isotherm	Hysteresis	Formula
	$Na_4Fe(CN)_6$: $Co(NO_3)_2$					**System I**					
M1	1.71:2.5	311.91	64.25	0.444	78.80	233.11	0.1205	0.3235	IVa	H1	$Na_{0.61}Co_{1.45}[Fe(CN)_6]*8.6H_2O$
M2	1.71:2	110.83	155.23	0.417	73.06	37.77	0.0200	0.3970	IVa	H1	$Na_{0.97}Co_{1.29}[Fe(CN)_6]*4.9H_2O$
M3	1.71:1.5	122.43	73.48	0.224	105.88	16.54	0.0084	0.2156	IVa	H1	$Na_{1.83}Co_{0.92}[Fe(CN)_6]*4.2H_2O$
M4	1.71:1	109.53	59.12	0.162	97.99	11.54	0.0054	0.1566	IVa	H1	$Na_{2.01}Co_{0.88}[Fe(CN)_6]*4.4H_2O$
M5	1.71:0.5	52.78	46.34	0.061	49.15	3.63	0.0014	0.0596	IVa	H2b	$Na_{2.14}Co_{0.89}[Fe(CN)_6]*5.3H_2O$
	$K_4Fe(CN)_6$: $Co(NO_3)_2$					**System II**					
M6	1.71:2.5	77.63	109.09	0.216	68.13	9.50	0.0044	0.2116	IVa	H1	$K_{1.52}Co_{1.07}[Fe(CN)_6]*1.6H_2O$
M7	1.71:2	81.28	111.04	0.228	68.89	11.38	0.0053	0.2227	IVa	H1	$K_{1.88}Co_{0.90}[Fe(CN)_6]*1.1H_2O$
M8	1.71:1.5	101.41	31.67	0.085	91.65	9.74	0.0042	0.0808	IVa	H1	$K_{2.13}Co_{0.81}[Fe(CN)_6]*2.1H_2O$
M9	1.71:1	5.72	41.40	0.006	5.72	—	0.000074	0.0059	II	H3	$K_{2.4}Co_{0.71}[Fe(CN)_6]*1.9H_2O$
M10	1.71:0.5	0.45	106.69	0.001	0.4720	—	0.000012	0.0010	II	H3	$K_{2.27}Co_{0.79}[Fe(CN)_6]*2.2H_2O$
	$K_3Fe(CN)_6$: $Co(NO_3)_2$					**System III**					
M11	1.71:2.5	479.21	55.56	0.505	170.91	308.30	0.1603	0.3447	IVa	H1	$K_{0.12}Co_{1.52}[Fe(CN)_6]*6.6H_2O$
M12	1.71:2	425.83	54.69	0.452	172.18	253.65	0.1320	0.3200	IVa	H1	$K_{1.03}Co_{1.18}[Fe(CN)_6]*5.7H_2O$
M13	1.71:1.5	362.87	57.74	0.399	159.68	203.18	0.1059	0.2931	IVa	H1	$K_{1.49}Co_{1.04}[Fe(CN)_6]*5H_2O$
M14	1.71:1	283.70	60.94	0.374	146.18	137.51	0.0720	0.3020	IVa	H1	$K_{1.91}Co_{0.89}[Fe(CN)_6]*4.4H_2O$
M15	1.71:0.5	117.99	79.73	0.215	72.69	45.29	0.0238	0.1912	IVa	H1	$K_{2.21}Co_{0.78}[Fe(CN)_6]*3.9H_2O$

2.2. K or Na-CoHCFs Characterization

2.2.1. Chemical Analyses

The contents of iron, cobalt, sodium, and potassium of the synthesized CoHCFs were obtained by ICP-MS (Agilent Technologies, 7500a model) of the digested samples. Digestion was conducted with boiling sulfuric acid. The contents of C and N were obtained by combustion at 1100 °C using a Leco TruSpec Micro CHNS elemental analyzer (St. Joseph, MI, USA).

2.2.2. N_2 Adsorption Measurements

Nitrogen adsorption measurements were conducted using an ASAP 2020 Micromeritics Sorptometer (Norcross, GA, USA) at 77 K on CoHCFs to study their textural properties and the effect of varying amounts of water on the specific surface areas and pore sizes. About 0.3 g of the sample was conditioned under a heat flow at 100 °C for 6 h under vacuum (10 µm Hg) to remove all impurities (e.g., CO_2 and H_2O) previously adsorbed onto the surface of samples that otherwise cause an error in the surface area measurement. The Brunauer-Emmet-Teller (BET) specific surface area was calculated by the BET equation using adsorption data within the domain of the BET equation in the relative pressure range of $0.05 < P/P_o < 0.26$; the total pore volume (V_t) was determined by nitrogen adsorption at $P/P_o = 0.995$; the pore size (D_p), and pore size distribution was obtained from the desorption branch of the isotherm measurements by the BJH method. The t-plot method was used to estimate the micropore volume (V_{mic}), external surface area (S_{ext}), and micropore area (S_{mic}) using the Harkins and Jura equation in the thickness range of 0.35–0.50 nm. Mesopore volume (V_{meso}) was calculated as the difference between the total volume of pores (V_t) and micropore volume (V_{mic}).

2.2.3. Thermogravimetric Analysis (TGA)

Thermogravimetric measurements (TGA) were conducted on CoHCF to study the %wt of water loss as a function of temperature. TGA analyses were performed using a Discovery model analyzer from TA Instruments (Amarillo, TX, USA). The analyses were performed using dried samples. About 10–15 mg of sample was placed in a platinum pan; a temperature scanning in the range of 25–600 °C was used, but a shorter temperature range (25–500°C) was shown in the thermograms to indicate more clearly the weight loss due to water molecules; the heating rate was 10 °C min^{-1} under a nitrogen atmosphere at 60 mL min^{-1}.

2.2.4. FTIR-ATR

Fourier Transformed Infra-Red Attenuated Total Reflection spectra (FTIR-ATR) of the CoHCFs were determined with a Bruker Alpha FTIR-ATR system (Bruker Optics, Billerica, MA, USA) utilizing a 300 Golden Gate diamond ATR Model. Samples were scanned over the range of 4000–450 cm^{-1} at a resolution of 2 cm^{-1}.

2.2.5. X-ray Powder Diffraction Data Collection and Analysis

A Rigaku diffractometer model Ultima IV was used with CuKα radiation at $\lambda = 1.54051$, a 2θ range of 4°–90°, a step size of 0.026, a time step of 30 s, an intensity of 30 mA, and a power of 40 kV. The Rietveld method implemented in the program MAUD (Materials Analysis Using Diffraction) [27] was used for structure analysis and refinement of sample M6, whereas samples M1 and M11 were refined peak by peak.

2.2.6. XPS Analysis

X-ray Photoelectron Spectroscopy (XPS) measurements were conducted on selected samples. XPS data were collected with an XR50 M monochromatic Al Kα (hν = 1486.7 eV) X-ray source, and a Phoibos 150 spectrometer with the one-dimensional detector 1DDLD provided by SPECS (Berlin, Germany) operated at a potential of 10–14 kV. The base pressure in the analyzer was less than 5.1×10^{-9} Torr. XPS spectra were taken using an aluminum anode X-ray source, operating at 250 kW (12.5 kV and 20 mA). High-resolution scans were acquired over the range of 50–70 eV (Co and Fe 2p) with the pass energy adjusted to 15 eV and step 0.1 eV wide. After baseline subtraction, the spectra were deconvoluted using peak fitting analysis with the AAnalyzer software, and peak positions were found. Peak assignments were determined from literature reports.

2.2.7. Surface Morphology of CoCHFs

The surface morphology of CoHCFs was studied by Scanning Electron Microscopy (SEM) using a Tescan Mira3 LMU high-resolution microscope operated at 15 kV with the detector operating at 13.3–13.4 mm distance and Transmission Electron Microscopy (TEM) using a JEOL JEM 1010 100 kV microscope.

2.2.8. CO_2 Adsorption on CoHCFs

CO_2 adsorption and desorption isotherms on CoHCFs were obtained using an ASAP 2020 Micromeritics Sorptometer (Norcross, GA, USA) equipped with an ISO Controller Micromeritics unit for temperature control with the stability of ±0.1 °C. The analysis gas was supplied by INFRA (México City, México) with 99.995% purity. Isotherms were obtained in the temperature range of 273–353 K in the relative pressure interval of 0.00–1.00 P/Po (0–100 kPa). Prior to the measurements, about 80–100 mg of previously dried samples (60 °C, 72 h) were degassed at 150 °C under vacuum for 6 h. At this temperature, only the loss of water is observed, and there is no decomposition of the CN group of the CoHCFs [28]. An equilibration time of 5 s was used since higher times produced excessive acquisition data of the isotherms. The rate of adsorption (ROA) of CO_2 on CoHCFs was measured at three temperatures (273, 298, and 313 K) by recording pressure change per dose vs. time (or measuring the time it took a dose of gas to adsorb) with the integrated ROA function in the ASAP 2020 Micromeritics Sorptometer.

2.2.9. Isosteric Heats of Adsorption, Q_{st}

The isosteric heats of adsorption were calculated with the incorporated function in the MicroActive software of the ASAP 2020. Q_{st} at each adsorption point was assessed using the Clausius-Clapeyron equation [29] in the form:

$$Q_{st} = -R\left(\frac{\partial \ln P}{\partial\left(\frac{1}{T}\right)}\right)_{\theta} \tag{1}$$

where R is universal gas constant, T is the analysis temperature (K), P is the partial pressure of CO_2, and θ is the adsorption quantity or extent to which the sample surface is covered with adsorbate.

2.2.10. Stability and Regenerability of CoHCFs

The stability and regenerability of CoHCFs after CO_2 adsorption/desorption cycles were investigated. Five runs were performed. In the first cycle, samples were degassed for 6 h at 150 °C under vacuum; for the rest of the cycles, samples were conditioned under vacuum at the same temperature but for 2.5 h to avoid the collapse of the crystalline structure of samples. All cycles were run at 298 K.

3. Results and Discussion

3.1. N_2 Adsorption and Textural Properties of CoCHFs

All samples present type IVa isotherms and hysteresis loop type H1 according to the IUPAC classification [30], Figure 1a–c, characteristic of mesoporous materials, except samples M9 and M10, which are type II with H3 type hysteresis loop. In the case of type H1 hysteresis loop, this is characteristic of solids that present narrow ranges of invariable mesopores; the H3 type hysteresis loop is characteristic of non-rigid aggregates of plate-like particles and pore networks comprised of macropores [30]. On the other hand, type II isotherms present characteristics of physisorption of most gases on nonporous adsorbents. In this case, unlimited combined monolayer and multilayer adsorption takes place, thus conferring the shape of this isotherm [31]. Adsorption/desorption isotherms reveal that samples containing K^+ ions and a mixture of Fe^{2+}/Fe^{3+} ions (see Sections 3.3 and 3.5) of system III exhibit a porous structure with a much more developed external surface area, S_{ext}, (72.69–172.18 m^2 g^{-1}) and a total pore volume V_t, (0.215–0.505 cm^3 g^{-1}) more uniform as compared to samples containing Na^+ and Fe^{2+} ions (systems I and II) (S_{ext} of system I is 49.15–105.88 m^2 g^{-1}, S_{ext} of system II is 0.4720–91.65 m^2 g^{-1}, V_t of system I ~ 0.061–0.444 cm^3 g^{-1} and V_t of system II is 0.001–0.228 cm^3 g^{-1}). In the case of the samples of system II, textural parameters are too small compared to the samples of the other two systems. For instance, samples M9 and M10 reveal an S_{BET} of 5.72 and 0.45 m^2 g^{-1}, respectively. These values indicate that these samples are nonporous materials, as can be observed from the corresponding pore size distributions (Figure 1e).

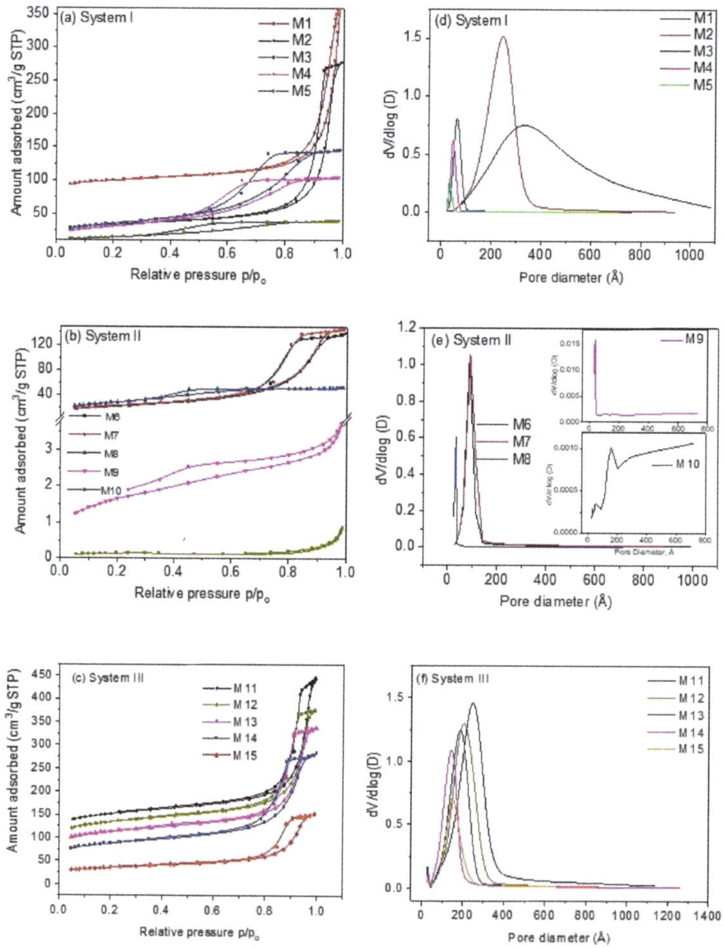

Figure 1. Nitrogen adsorption and desorption isotherms of CoHCFs of (**a**) system I of samples M1–M5; (**b**) system II of samples M6–M10; (**c**) system III of samples M11–M15. Pore size distributions of CoHCFs of (**d**) system I of samples M1–M5; (**e**) system II of samples M6-M-10; (**f**) system III of samples M11–M15.

The specific surface areas decrease as the $Na_4Fe(CN)_6/Co(NO_3)_2$ and $K_3Fe(CN)_6/Co(NO_3)_2$ molar ratios decrease for samples of systems I and III but not for samples of system II, see Table 1. The presence of Na ions, the +2-oxidation state of Fe in system I, and the presence of K^+ ions and both Fe^{2+}/Fe^{3+} oxidation states in system III produce CoHCFs with stoichiometries in which the number of alkali ions and water molecules depend on the reactant initial molar ratios. This behavior can be explained by considering the final number of Co ions in the CoHCFs. In these systems, the number of Co ions in the final stoichiometry is higher (or similar) than the number of Fe ions; thus, it is likely to find Co^{2+} ions not coordinated to C≡N groups, promoting the formation of cubic cells with high specific surface area. This generates a higher number of pores [32], favoring and contributing to the specific surface areas of samples, as is more noticeable in the case of samples M11-M15. Ferro/ferri-cyanide vacancies are filled by coordination water, which, when removed by the heat treatment (100 °C) during the conditioning process, leaves empty spaces generating internal pores inside the material that are joined, forming interconnected tunnels conforming the porous structure of the CoHCFs, as reported by Ojwang et al. [10]. On the other hand, samples M1–M5 depict a range of surface areas S_{BET} of 52.7–311.9 $m^2\ g^{-1}$. This behavior is not observed in system II. It is worth mentioning that even though the only difference between systems I and II is the alkali ion, in the latter, the final stoichiometry shows a lower number of Co ions, promoting a decrease in the specific surface area, which is confirmed by the decrease in the water molecules compared to the other systems. In the case of samples M9 and M10, their specific surface areas are too low (5.72 and 0.45 $m^2\ g^{-1}$, respectively) to fall within the limit of detection of the instrument. These results confirm the impact that stoichiometry has on the final porous properties of the CoHCFs. In the case of the samples of systems I and III, an FCC crystal is obtained (see Section 3.4), and these samples are the ones that present the highest amounts of N_2 adsorbed, whereas the samples of system II present a triclinic cell and the lowest amount of nitrogen adsorbed. These differences can be due to the fact that the formation of porosity in the FCC system is facilitated by the geometry of this system, while in the case of the triclinic system, this geometry prevents the formation of the interconnected pores. Furthermore, when Co concentration is varied during synthesis, the crystalline structure remains, but textural properties, such as porosity, change. The crystalline arrangement obtained in samples of each system is independent of the Co concentration used, given that all samples were synthesized under the same conditions of pressure and temperature. In the case of pore volume, V_t, data from Table 1 reveal that the values for samples within system I are higher in samples synthesized with higher $Co(NO_3)_2$ concentration and present the trend M1 > M2 ... > M5; this trend is also observed in the case of samples of system III, with M11 > M12 .... > M15. However, as in the case of S_{BET}, samples of system II do not show this trend. On the other hand, in the case of D_P, the trend observed is variable for the three systems. In the case of pore size distributions, Figure 1d–f, the distributions observed are very narrow (except for samples M1 and M10) and present a monomodal shape in the mesopore region of 2.5–10.0 nm attributed to the hysteresis loop present and suggesting the formation of a structure with more uniform pores. The uniformity of pore diameters is confirmed by the steep slope of the desorption part of the isotherm. The results obtained from N_2 adsorption measurements provide evidence of the mesoporosity obtained during the synthesis procedure through the formation of interconnected tunnels and cavities formed in the crystalline structure of CoHCFs. The N_2 adsorption measurements present large pore sizes of about 30 to 300 Å. These can be the entrances of interconnected channels through which CO_2 molecules are transported to the smaller pores. These large pores are presumably formed by agglomerated particles and not by part of the crystal structures, which have two intrinsic types of cavities, ones ~5 Å wide formed by the framework where alkali ions and H_2O molecules may reside, and the other ones ~10 Å wide formed by absent $[Fe(CN)_6]^{n-}$ groups, according to Svensson et al. [8]. CO_2 is adsorbed into these cavities. These results suggest that the oxidation states of Co and Fe, the nature of alkali ions, and the water content modify the way particles of CoHCF agglomerate. Consequently, when some of these variables are modified, the adsorption properties of each family change. From the formulas shown in Table 1, it can be observed that the amount of alkali ions increases as the concentration of Co decreases, and this is due to the need of the crystals to satisfy the charge balances.

3.2. TGA Analysis

TGA measurements were performed to determine the amount of water contained in CoHCFs and to see any changes in the chemical structure of the samples when subjected to thermal decomposition. In all cases, a temperature range of 25 to 500 °C was used to ensure that all the water molecules were removed. The thermograms are shown in Figure 2. The weight loss at low temperatures, around 25–150 °C, corresponds to the physically adsorbed or zeolitic water [28]. At ~150–250 °C, the weight loss corresponds to the lattice or coordinated water [28]. The decomposition of the samples with the liberation of C≡N units is carried out at 250–500 °C [33]. The weight loss of water for each sample was obtained from the thermograms using the derivatives of the data (Figures S1–S3). The data were used to determine the corresponding number of water molecules in the formulas of CoHCFs, as shown in Table 1. Although the molar ratio of the reactants varied in the same way in the three systems (Table 1), the obtained stoichiometries do not show the same trend. This behavior is common in CoHCFs, where even the way of mixing the reactants (e.g., dropwise or immediate mixing) has a remarkable effect on the final stoichiometry [14]. Thus, we expect to obtain CoHCFs with diverse stoichiometries due to the reactant excesses used in this work. Nevertheless, it is possible to observe a general behavior between the systems. For instance, the number of water molecules strongly depends on the number of alkali ions in the structure. This behavior is more prominent in systems I and III, where the number of water molecules decreases as the number of alkali ions increases. This behavior suggests that the introduction of more alkali ions into the CoHCFs framework reduces the amount of coordinating water in the unit cell [34]. This behavior is also followed in system II in which the number of alkali ions in the CoHCFs is greater than in the other systems, and it is expected that the number of water molecules is also less than in the other systems, as is shown in Table 1. The number of water molecules in system I and especially in system III may be the reason for the significative textural properties of these samples. On the other hand, the poor textural properties of system II can be explained by the small number of water molecules in the CoHCFs framework.

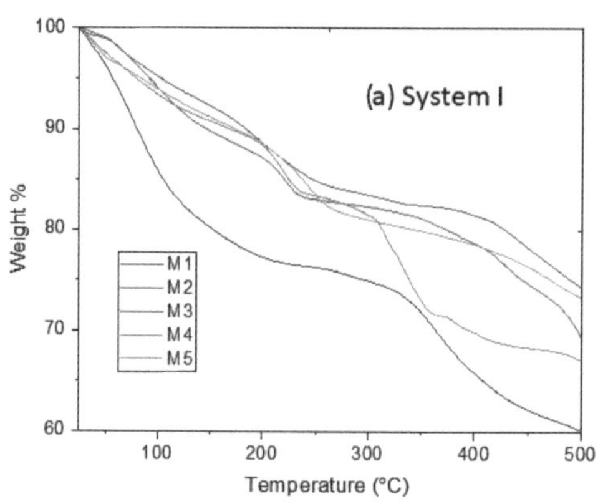

Figure 2. *Cont.*

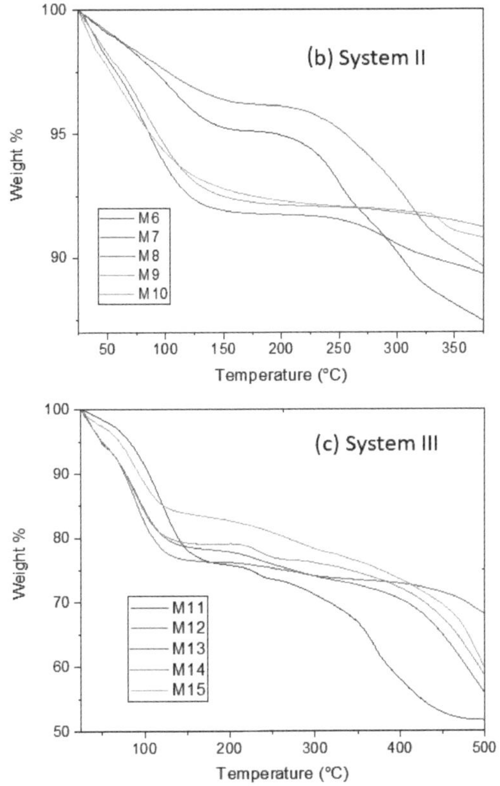

Figure 2. TGA results on CoHCFs for (**a**) system I; (**b**) system II; (**c**) system III.

3.3. FTIR-ATR Study

FTIR is commonly used to characterize CoHCFs [13,14,35–38]. Figure 3a–c shows the FTIR spectra for all the samples. The position of the band due to the CN group in the range of 2200 to 2000 cm^{-1} (shown in spectra of three systems) is useful to qualitatively determine the oxidation states of the metallic ions conforming to the CoHCFs, according to literature reports [13,14,35–38]. On the other hand, bands in the range of 3650 to 3100 cm^{-1} and 1600 to 1630 cm^{-1} indicate the presence of water molecules [35–38]. As expected, the spectra of Figure 3a,b are similar because systems I and II samples were prepared with $[Fe(CN)_6]^{4-}$ ions. In both cases, the presence of a single band at ca. 2077 cm^{-1} indicates that the structure of the CoHCFs is based on the Fe^{2+}–CN–Co^{2+} chain, regardless of the concentration ratio $[X_4[Fe(CN)_6]]/[Co(NO_3)_2]$ (where X= Na or K) used in the synthesis of CoHCFs. This is confirmed by the presence of bands at ca. 590 and 448 cm^{-1} (samples M1–M5 and M6–M10). The presence of Co^{3+} in systems I and II samples could not be detected by FTIR, whereas the presence of this species was detected by XPS in the samples of the three systems (Section 3.5). The main differences between the spectra of these samples are the intensities of the bands due to the presence of water molecules. It has been proposed [35,36] that bands in the range of 3650–3100 cm^{-1} are assigned to the presence of adsorbed water molecules, while bands in the range of 1600–1630 cm^{-1} indicate the presence of water molecules that are trapped within the crystal structure (either coordinated or zeolitic). According to the intensity of these bands, it is evident that CoHCFs prepared with $Na_4[Fe(CN)_6]$ (system I) contain more water molecules than the ones prepared with $K_4[Fe(CN)_6]$ (system II) for all the concentration ratios used, which confirms the results obtained from TGA.

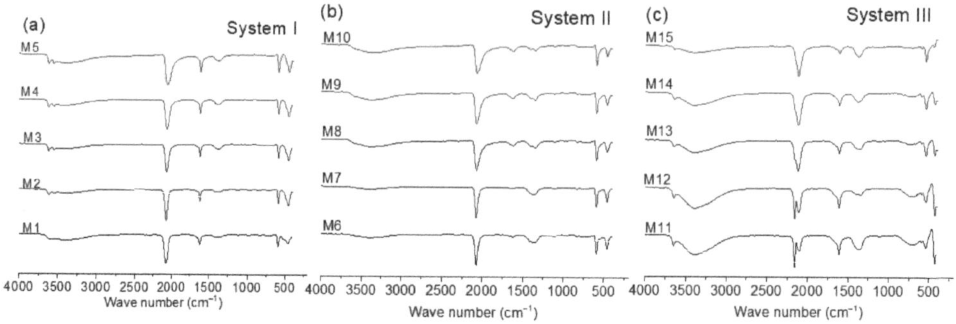

Figure 3. FTIR spectra for CoHCFs prepared by mixing $Co(NO_3)_2$ with (**a**) $Na_4[Fe(CN)_6]$, system I; (**b**) $K_4[Fe(CN)_6]$, system II; (**c**) $K_3[Fe(CN)_6]$, system III.

Unlike the previous cases, Figure 3c (spectra of samples of system III) presents an extraordinary difference in the bands in the range of 2200 to 2000 cm^{-1} as a function of the concentration ratios $[K_3[Fe(CN)_6]/[Co(NO_3)_2]$. At the highest $Co(NO_3)_2$ concentration (sample M11), the presence of two bands at ca. 2160 and 2110 cm^{-1} is evident, which indicates that the CoHCFs synthesized at these conditions are formed by the mixture of two chains Fe^{3+}–CN–Co^{2+} and Fe^{2+}–CN–Co^{3+}, respectively. The presence of Fe^{2+}/Fe^{3+} and Co^{2+}/Co^{3+} ions is also confirmed by the XPS results. The appearance of the Fe^{2+}–CN–Co^{3+} chain is due to the spontaneous electron transport between Fe^{3+} and Co^{2+} ions according to the reaction Fe^{3+}–CN–Co^{2+} $\rightarrow$ Fe^{2+}–CN–Co^{3+} [13,14]. The mixture of chains is also observed by the bands at 544 and 425 cm^{-1} and a small band at 590 cm^{-1} [39]. For CoHCFs prepared at lower $Co(NO_3)_2$ concentrations, the relative intensity of the band assigned to the Fe^{3+}–CN–Co^{2+} chain decreases, while the intensity of the band assigned to the Fe^{2+}–CN–Co^{3+} chain increases. This behavior suggests that when fewer Co ions are available, the electron transport inside the structure of the CoHCFs increases. This can be explained by considering that both chains, Fe^{3+}–CN–Co^{2+} and Fe^{2+}–CN–Co^{3+}, can be expressed as $Co^{2+}{}_3[Fe^{3+}(CN)_6]_2$ and $Co^{3+}{}_4[Fe^{2+}(CN)_6]_3$, respectively. In the former, 33% of the Co^{2+} ions are not coordinated with the C≡N group, while in the latter, this percentage decreases to 25%. Consequently, when the concentration of Co^{2+} ions decreases, the formation of the chain Fe^{2+}–CN–Co^{3+} is favored [13,14,37,38]. This can also be confirmed by the relative intensity of the bands due to the presence of water molecules. To maintain electroneutrality, the Co ions not coordinated to the C≡N group are coordinated with water molecules. Then, for the Fe^{3+}–CN–Co^{2+} chain, where there are more Co^{2+} ions available, more water molecules will be in the CoHCFs structure [13,14,37,38]. When the concentration of the Fe^{2+}–CN–Co^{3+} chain increases, there is a decrease in the water molecules needed to stabilize the crystal. Thus, a decrease in the relative intensity of the bands due to the presence of water is observed as the band also decreases due to the chain Fe^{3+}–CN–Co^{2+}.

3.4. XRD Analysis

The crystalline structure of the as-prepared CoHCFs was determined by XRD (Figure 4). Crystallographic data and structural parameters obtained after refinement are presented in Table 2 for samples M1, M6, and M11 and Table S1. For the analysis of crystalline structure, similar results are obtained for samples within each system; because of this, the results of diffractograms corresponding to the first sample of each system are presented for representativeness (i.e., samples M1, M6, and M11, Figure 4a). It can be observed that either the Fe oxidation state or the type of alkali ion (Na^+ or K^+) influences the formation of the different types of structures. Sample M1 has Na^+ and Fe^{2+} ions; sample M6 has K^+ and Fe^{2+} ions; and sample M11 has K^+ and a mixture of Fe^{2+} and Fe^{3+} ions. The selected samples of systems I and III form FCC cells with a cell parameter a = 10.220(9) Å for sample M1, a = 10.267(1)Å for sample M11 (Table 2), whereas samples from system II form triclinic cells (Table 2) with a = 10.056(2) Å, b = 10.078(2) Å, c = 10.070(3) Å, α = 89.96(9)°, β = 88.47(2)°, γ = 90.39(9)° for sample M6. Some impurities were detected in sample M11, whose origin could not be assigned; however, the identified peaks corresponded to the FCC phase. Samples M1 and **M11** depict a spatial group *Fm-3m*, whereas sample M6 presents a *P1* spatial group and is consistent with reports from others [40,41]. Crystallite sizes obtained by the Scherrer equation are as follows: M1 = 193.88 Å, M6 = 295.72 Å, and M11 = 307.18 Å. These differences may be attributed to the intrinsic nucleation and growth rates in each CoHCF. Further experimentation should be carried out to prove this surmise. To better appreciate the relationship of metal concentrations during the synthesis procedure on the final crystalline phase of samples, two samples are selected for the analysis, namely M1 and M11. If the 200-lattice plane is magnified from sample M11 (Figure 4b), the diffraction peak is shifted to higher angle values, i.e., from 2θ = 17.26 to 2θ = 17.39 compared to sample M1 because more labile K^+ ions replace less labile Na^+ ions caused by the difference in their hydrated radii. These results are consistent with reports from the literature [42].

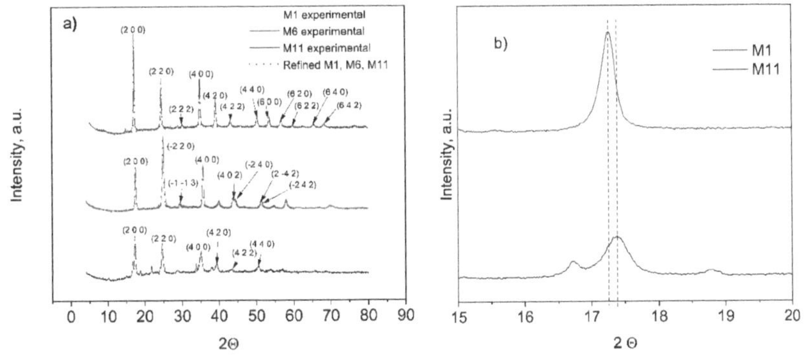

Figure 4. (**a**) Experimental (dots) and refined (continuous lines) powder X-ray diffraction data results for CoHCFs M1, M6, and M11; (**b**) partial magnification of the 200-lattice plane of M1 and M11 samples.

Table 2. Crystallographic data and structure refinement parameters were obtained after the refinement of powder X-ray diffraction measurements on CoCHFs M1, M6, and M11.

Parameter	M1	M6	M11
Space group	Fm-3m	P1	Fm-3m
Crystal system	FCC	Triclinic	FCC
a (Å)	10.220(9)	10.056(2)	10.267(1)
b (Å)	10.220(9)	10.078(2)	10.267(1)
c (Å)	10.220(9)	10.070(3)	10.267(1)
α (deg)	90	89.96(9)	90
β (deg)	90	88.47(2)	90
γ (deg)	90	90.39(9)	90
Crystal size (Å)	193.88	295.72	307.18
Atoms per unit cell	$Na_8Co_4Fe_4C_{24}N_{24}$	$K_8Co_4Fe_4C_{24}N_{24}$	$K_4Co_4Fe_4C_{24}N_{24}$

3.5. XPS Study

XPS was used to authenticate the presence of transition metals to confirm their oxidation state of the CoHCFs. Figure 5a shows the wide scan spectra for selected samples (M1, M6, and M11). The signals correspond to C 1s, O 1s, N 1s, Na 1s, K 2p, Fe 2p, and Co 2p; Figure 5b–h shows high-resolution core level spectra corresponding to these signals; Table S2 summarizes the results of deconvoluted peak assignments. The C 1s spectra of samples M1 and M6 are fitted with five peaks for sample M1 and four peaks for samples M6 and M11 (Figure 5b), respectively. The two peaks at 283.50 and 283.55 eV in samples M1 and M6 are assigned to the C-Fe^{2+} bond by Vannenberg [43] of the CoHCFs, while sample M11 shows two peaks located at 283.55 and 284.03 eV attributed to C-Fe^{2+} [43] and C-Fe^{3+} [43] bonds by Vannenberg [42]. This result indicates electron activity by charge transfer within sample M11. Peaks at 283.48, 284.53, 285.4, and 287.42 eV in sample M1 are assigned to the CO-Co [44] bond, C (elemental) [45], C from the environment [43] and to $Co(CN)_x$, respectively. Sample M6 shows peaks at 284.47 and 285.60 eV corresponding to C (elemental) [45] and C from the environment [43], respectively. These last two assignments correspond to secondary species not removed and formed during the synthesis process and/or due to contact of the sample with the surroundings. Figure 5c shows the O 1s core level spectra. In sample M1, signals are fitted with two peaks, sample M6 needs four peaks, and sample M11 requires three peaks. The three samples present peaks at 535.4 eV [46], 532.8 eV [47], and 532.4 eV [48], all assigned to water molecules of samples M1, M6, and M11, respectively. In addition, sample M1 presents another peak at 531.4 eV corresponding to $Co(NO_3)_2$ [49]. Sample M6 shows peaks at 530 eV assigned to CoO [48] and at 531.3 eV corresponding to $Co(NO_3)_x$ [50]. Sample M11 presents two peaks, one at 530.91 eV assigned to Fe_xO_y [49,51] and the other one at 531.88 eV attributed to $Co(NO_3)_x$ [52]. The core level spectra of N 1s are shown in Figure 5d. The signal for sample M1 requires three peaks, while samples M6 and M11 need four peaks each for fitting. Sample M1 depicts a peak at 396.53 eV assigned to the N-Co^{2+} bond [49], another peak at 396.95 eV corresponding to the N-Co^{3+} bond [43], and a third peak at 400.88 eV assigned to $Co(NO_3)_x$ [49]. In the case of sample M6, peaks are present at 397.30 eV of the N-Co^{2+} bond [43], at 397.76 eV assigned to the N-Co^{3+} bond [43] and small peaks at 401.79 and 406.7 eV corresponding to traces of $Co(NO_3)_x$ [49] and KNO_3 [53], respectively. Fittings of sample M11 present peaks at 397.21 eV assigned to the N-Co^{2+} bond [43], another peak at 397.7 eV corresponding to the N-Co^{3+} bond [43], and a third peak at 396.55 eV assigned to a secondary compound of the type Co_xN_y [54]. Figure 5e shows the Na 1s core level spectrum of sample M1. One peak at 1071.34 eV is required to fit the data and is attributed to Na of the compound $Na_nCo^{2+,\,3+}[Fe(CN)_6]$ [43]. On the other hand, Figure 5f shows the K 2p core level spectra of samples M6 and M11, respectively. Both samples show two peaks at 292.41 and 292.62 eV, both assigned to K in the compound $K_nCo^{2+,3+}[Fe(CN)_6]$ [43]. Some satellites are present in these samples. Figure 5g shows the Fe 2p spectra resolved in individual peaks. Sample M1 requires one peak at 707.34 eV corresponding to the C-Fe^{2+} bond [55]; the rest of the peaks correspond to Auger-Co and satellites; sample M6 shows two peaks at 709.23 and 709.49 eV, both assigned to the C-Fe^{2+} bond [55] and another peak at 708.23 eV assigned to Fe° [56] overlapped with a satellite; the rest of the peaks correspond to Auger-Co and satellites. Sample M11 depicts three main peaks, one located at 712.72 eV corresponding to the C-Fe^{2+} bond [55], the second one at 715.27 eV assigned to the C-Fe^{3+} bond [55], and the third one at 711.7 eV assigned to Fe_xO_y [57]. The XPS results for Fe 2p indicate that when CoHCFs are synthesized with Fe^{2+} ion in ferrocyanide, iron maintains a stable oxidation state. Conversely, when the Fe^{3+} ion in ferricyanide is used during the synthesis, iron presents electron activity by changing its oxidation state. Figure 5h shows the Co 2p core level spectra resolved in individual peaks. Sample M1 depicts two important peaks, the first at 780.68 eV corresponding to the N-Co^{2+} bond [49,58] and the second at 783.59 eV assigned to the N-Co^{3+} bond [13]. Sample M6 presents two relevant peaks, the first at 781.90 eV corresponding to the N-Co^{2+} bond [49,58] and the second at 783.58 eV assigned to the N-Co^{3+} bond [13]. Sample M11 shows two peaks at 785.54 and 786.36 eV assigned to N-Co^{2+} [49,58] and N-Co^{3+} [13], respectively. These results indicate that Co also presents spontaneous electron transfer reflected in changes to its oxidation state. The XPS results indicated that Fe^{2+}, Co^{2+}, and Co^{3+} were detected in samples M1 and M6 only, while in the case of sample M11, both Fe and Co presented the two oxidation states. The XPS technique was used to obtain the formulas for samples with the highest concentration of Co by separating the oxidation states of Fe and Co and the results obtained were:

$M1 = Na_{0.61}Co^{II}_{0.95}Co^{III}_{0.5}\left[Fe^{II}(CN)_6\right] * 8.6H_2O$, $M6 = K_{1.52}Co^{II}_{0.71}Co^{III}_{0.36}\left[Fe^{II}(CN)_6\right] * 1.6H_2O$, $M11 = K_{0.12}Co^{II}_{1}Co^{III}_{0.52}\left[Fe^{II}_{0.67}Fe^{III}_{0.33}(CN)_6\right] * 6.6H_2O$. In conclusion, even with the large number of assignations ascribed to the different bonds within the CoHCFs, those with correspondence to two or more spectra of different elements are studied and provide information confirming the different oxidation states Fe and Co acquired during the synthesis procedure as well as the different bonds involved in the structure of the CoHCFs that ultimately will have an impact on the CO_2 capture.

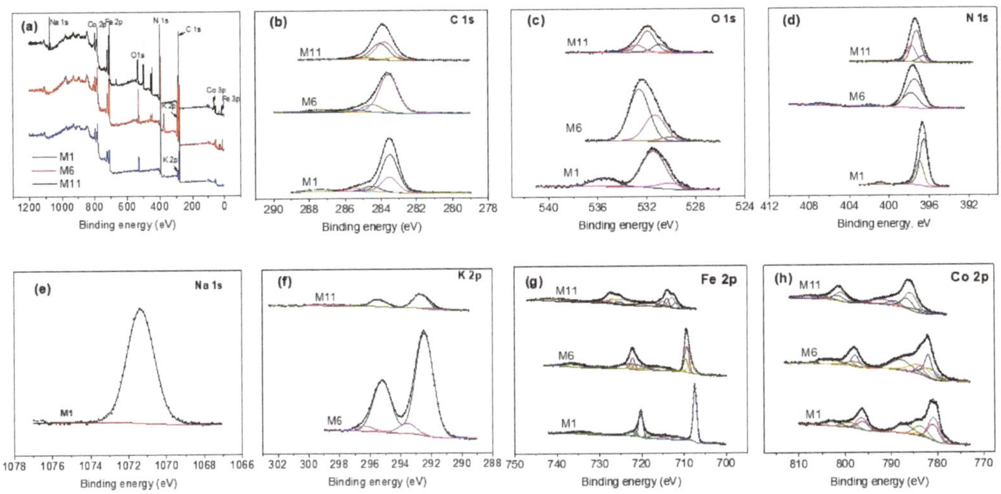

Figure 5. XPS core-level spectra on CoHCFs M1, M6, M11 (**a**) survey spectra; (**b**) C 1s; (**c**) O 1s; (**d**) N 1s; (**e**) Na 1s; (**f**) K 2p; (**g**) Fe 2p; (**h**) Co 2p.

3.6. Surface Morphology of CoCHFs

The surface morphologies of the CoHCFs materials were examined by TEM (Figure 6) and by SEM (Figure 7) for samples M4 and M11, respectively. These samples were selected randomly for this study. It is demonstrated that the particles of CoHCF materials are typical nanocubes with particle sizes in the range of 80–130 nm (Figure 6a–d). For sample M11, aggregation of nanocubes is clearly observed (Figure 7a,b), suggesting that high molar ratios of $K_3Fe(CN)_6$:$Co(NO_3)_2$ lead to the aggregation of the precipitated CoHCF. However, from the magnification of SEM images, it can be seen that CoHCFs have rough surfaces.

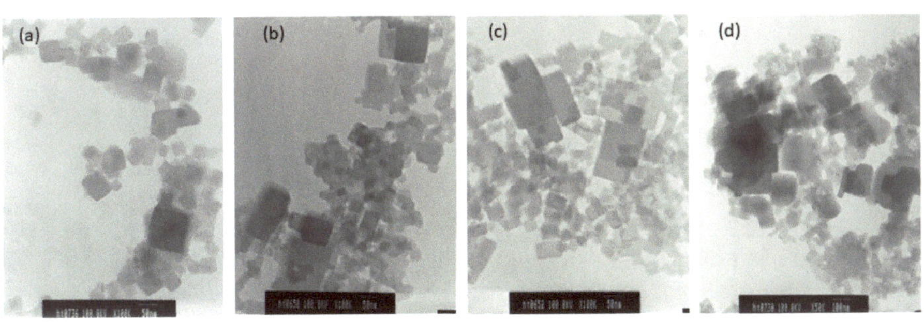

Figure 6. TEM micrographs of CoHCFs of sample M4 keeping fixed the molar ratio of $Na_4Fe(CN)_6$: $Co(NO_3)_2$ at 1.71:1, but using different amounts of reactants (**a**) 10 % wt, (**b**) 15 %wt, (**c**) 20 %wt, (**d**) 30 %wt.

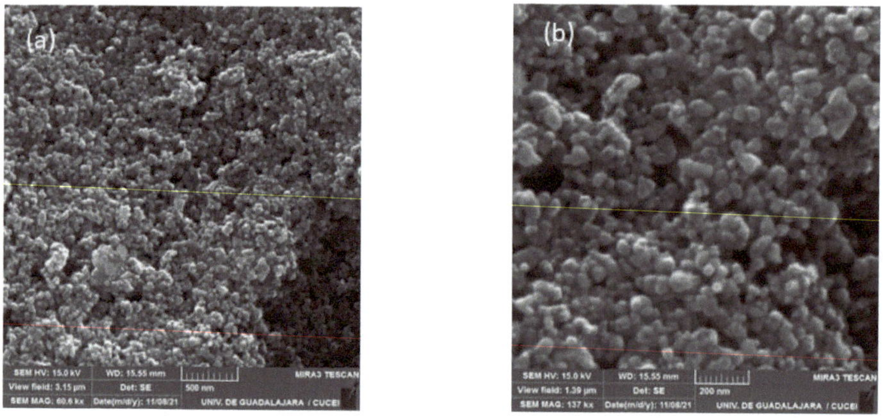

Figure 7. SEM micrographs of sample M11 at different magnifications (**a**) 500 nm and (**b**) 200 nm.

3.7. CO_2 Adsorption on CoHCFs

3.7.1. Equilibrium Adsorption Isotherms

The results of CO_2 adsorption equilibrium measurements at 273 K and 760 mmHg on three representative samples of CoHCFs (M1, M6, and M11) are depicted in Figure 8a. As it is expected from the results of N_2 adsorption and XPS measurements, sample M11 shows the highest CO_2 adsorption capacity (3.04 mmol g^{-1}), followed by sample M1 (2.45 mmol g^{-1}) and sample M6 (0.31 mmol g^{-1}). Consequently, sample M11 is used to further perform all CO_2 characterization experiments so that the diffusion processes of CO_2 can be facilitated and the adsorption sites can be easily accessed by CO_2 molecules. The high adsorption capacity presented by sample M11 is comparable with capacities reported for CoHCFs [59]. The CO_2 adsorption and desorption isotherms at temperatures of 273–353 K on sample M11 are depicted in Figure 8b. The isotherms correspond to Type I b according to the IUPAC classification [30] and exhibit strong desorption hysteresis. These hysteresis results can be attributed to either of the following two reasons. On the one hand, Zhao et al. [60] report that the desorption hysteresis can be due to the diffusion of the gas through the intra-crystalline structure caused by the entanglement of CO_2 molecules inside the vacancies of the CoHCFs upon desorption; further admission of CO_2 to the material continues taking place at the onset of the desorption, and this is reflected by an increase in gas removal. On the other hand, Riascos-Rodriguez et al. [61] report that this type of hysteresis can be attributed to the expansion of the material structure, which appears when the filling of pores by the gas takes place when the material is in its relaxed state.

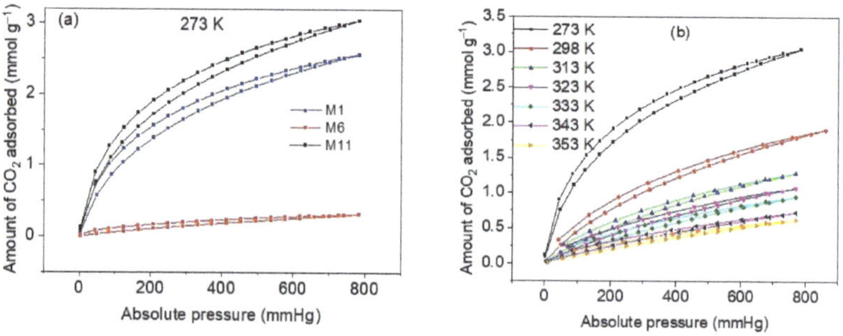

Figure 8. (**a**) CO_2 adsorption capacities on samples M1, M6 and M11 of CoHCFs at 273 K; (**b**) CO_2 adsorption isotherms at 273–353 K, on sample M11 of CoHCFs.

3.7.2. Kinetic Measurements

To be a good adsorbent, the material must possess fast adsorption kinetics as one of the most important properties, according to Serna-Guerrero et al. [62]. This is relevant, especially in dynamic operations (e.g., in fixed bed adsorption separations) where the adsorbent ought to stand up high solute flows as related to its rate of adsorption [62]. As a result, a pivotal step in CO_2 adsorption studies should include kinetic measurements. Figure 9a depicts the kinetic profiles at temperatures 273, 298, and 313 K and low absolute pressures. These results indicate the strong dependence of the adsorption rate on the temperature. For increased temperatures, CO_2 uptakes are faster due to boosted kinetics, but the adsorption capacities diminish due to thermodynamic constraints [63]. This is consistent with the exothermal condition of the adsorption process. Conversely, the equilibrium is achieved at shorter times and higher temperatures, as can be observed by the sharper curves, and this is in line with rapidly increased diffusion and mass transfer phenomena. In general, very fast kinetics can be observed to occur within the first 10 s of CO_2 capture at 273 K, whereas between ~60 and ~80% of CO_2 is adsorbed within the first 6 s in the case of 298 K and is much faster in the case of 313 K, to within 3 s of contact time.

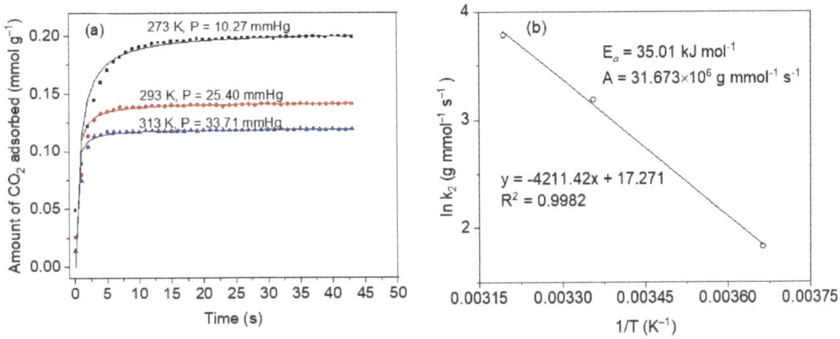

Figure 9. (**a**) CO_2 adsorption kinetics at 273, 298 and 313 K, on sample M11 of CoHCFs; symbols represent experimental data and lines are for eye guidance; pressures up to 33.71 mmHg (4.5 kPa); (**b**) Arrhenius plot obtained from pseudo-second-order model constants.

Using the Arrhenius equation [62], the activation energy (E_a) of CO_2 adsorption on CoHCFs is determined employing the mass transfer coefficients determined from the pseudo-second-order model at the three different temperatures studied. The Arrhenius equation is given by:

$$k_f = Aexp\left(-\frac{E_a}{RT}\right) \tag{2}$$

where: k_f (g mmol^{-1} s^{-1}) is the mass-transfer constant, A (g mmol^{-1} s^{-1}) is the Arrhenius factor, T is the adsorption temperature (K), R (8.314 J mol^{-1} K^{-1}) is the universal gas constant, E_a (J mol^{-1}) is the activation energy of adsorption. Parameters k_f and E_a are determined from the slope and intercept of a plot of ln k_f vs. 1/T, respectively.

An activation energy E_a = 35.01 kJ mol^{-1} and an Arrhenius factor A = 31.673 × 10^6 g mmol^{-1} s^{-1}, Figure 9b. A positive value of E_a indicates that the adsorption rate increases as the temperature increases. This value is within the range of a physical CO_2 adsorption process on CoHCFs (<40 kJ mol^{-1}) [64].

3.7.3. Effect of Adsorption Temperature on CO_2 Uptake

One of the operating conditions that can significantly affect the magnitude of CO_2 uptake by CoHCFs is the temperature. The results of the comparative study of CO_2 removal by CoHCFs (sample M11) are depicted in Figure 10a. An adsorption capacity of 2.72 mmol g^{-1} is observed at 273 K, which is approximately 4.4 times higher than the one corresponding to 353 K (0.62 mmol g^{-1}). The CO_2 adsorption capacities decrease from 2.72 to 0.62 mmol g^{-1} as the temperature increases. This can be explained by the fact that the interactions between the cyano groups of CoHCFs and CO_2 molecules are exothermic; the strong observed temperature dependence corresponds to a predominant kinetic diffusion-controlled process rather than a thermodynamic one. The increase in temperature prevents CO_2 molecules from being transferred from the bulk to the surface of CoHCFs dictated by the kinetic hindrance. This, in turn, causes a diminution of the number of available adsorption sites at elevated temperatures.

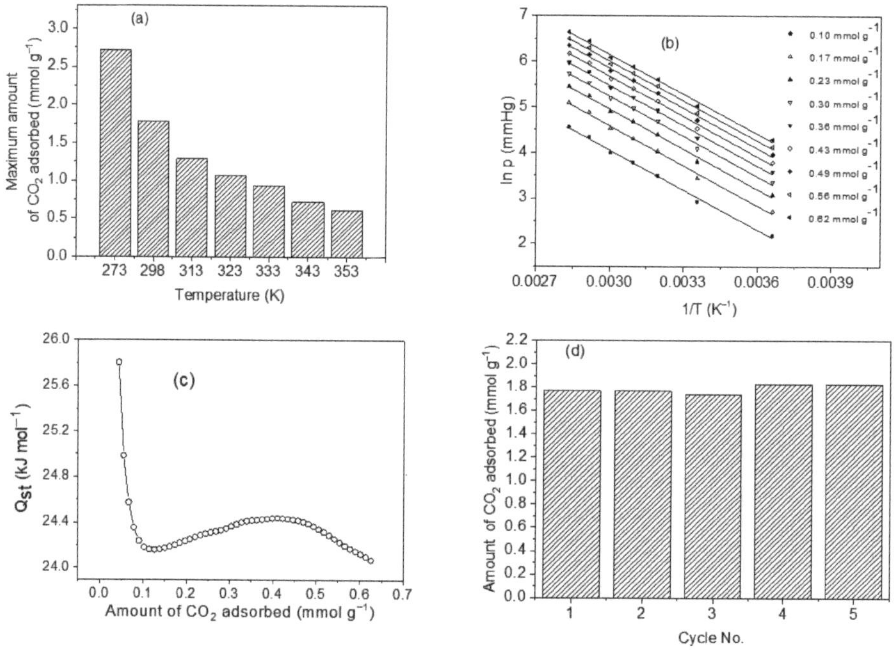

Figure 10. (**a**) CO_2 adsorption capacity of sample M11 of CoHCFs at a temperature range of 273–353 K; (**b**) isosteres of CO_2 adsorption as a function of temperature and amount adsorbed on sample M11 of system III of CoHCF; (**c**) change of isosteric heat of adsorption with the amount adsorbed for sample M11 of system III of CoHCFs using CO_2 adsorption from 0 to 101 kPa; (**d**) recovery of sample M11 of CoHCFs after five cycles of CO_2 adsorption/desorption.

3.7.4. Isosteric Heats of Adsorption (Q_{st}) of CO_2 on CoHCFs

The heat released upon an adsorption process at certain coverage is defined as the isosteric heat of adsorption, Q_{st} [10,60]. This is a thermodynamic property indicating the strength of bonding between the molecules of a gas and the surface of a solid [10]. It can be used to reflect the affinities of guest gas molecules to the adsorbent. In general, a high value of Q_{st} is advantageous at low pressures, but a very large value of Q_{st} produces an elevated energy demand for regeneration purposes of the solid adsorbent [65]; a value of Q_{st} neither too large nor too small is suitable for CO_2 removal at low pressures. Plots of ln P_{CO2} vs. 1/T shown in Figure 10b at constant coverage (called isosteres) in the range of 0.10–0.62 mmol g^{-1} and in the temperature range of 273–353 K were used to assess Q_{st} from the slopes. The isosteres present negative slopes being concordant with an exothermic CO_2 adsorption process. It has been reported [66] that by monitoring the magnitude and changes of Q_{st} throughout the CO_2 uptake process, it is possible to obtain information regarding the interactions at the molecular level involving CO_2 molecules and the adsorbent and to elucidate the energetic diversity of the adsorbent surface [66]. Figure 10c shows the results of the calculated values of Q_{st} as a function of the amount adsorbed. We can observe a range of Q_{st} of 24.07–25.80 kJ mol^{-1} and an average value of 24.34 kJ mol^{-1}, which falls within the range of physical adsorption [67]. The trend observed in Q_{st} with the amount adsorbed is that it first decreases to a minimum, then slightly increases, and finally decreases again. This is attributed to a high degree of surface heterogeneity and changes in the sorbate/sorbent and sorbate/sorbate interplays [68]. The main adsorption sites involved in the CO_2 uptake include K and Na ions since these have basic characteristics to bind acidic CO_2 molecules. In contrast, Co and Fe atoms are classified as Lewis acids that do not bind CO_2 molecules easily due to the acidic nature of this gas. It was reported very recently [24,59] that the K^+ ions exposed after the removal of water molecules from the crystallites of MOFs and successfully implanted in the MOFs produce adsorption sites with a strong affinity towards CO_2 molecules through a symbiotic effect, and the K^+ ions act effectively as gas confinement sites improving the CO_2-framework affinity, thus enhancing the values of Q_{st}. Q_{st} values show a decreasing trend as CO_2 surface covering increases indicating that the CoHCFs are distinguished by a high degree of heterogeneity and showing differences in the adsorbate–adsorbent, and adsorbate–adsorbate interplays. Generally, different classes of adsorptive molecular forces (e.g., Van der Waals, hydrophobicity, or chemical bonds) correlate Q_{st} to the interplays between CO_2 molecules and adsorbent surface [69]. The trends of Q_{st} observed in Figure 10c indicate that at the initial adsorption phases, a substantial amount of unoccupied adsorption sites is exposed on the surface of CoHCFs, and more molecules of CO_2 can contact the surface of the adsorbent and the strength of adsorptive forces between CO_2 molecules and the adsorbent is large, thus resulting in a low activation energy and high Q_{st}. As the CO_2 uptake process continues, an increased amount of occupied adsorption sites is present, and the adsorption of more CO_2 molecules is burdensome [70]. In addition to this, adsorbate–adsorbate interactions start to take place, and attraction–repulsion London dispersion forces start acting between already adsorbed CO_2 molecules [68]. Consequently, when the removal of CO_2 continues, activation energy soars, and Q_{st} declines [70].

3.7.5. Regenerability

The stability of adsorbent, after adsorption/desorption cycles and uptake capacity, is important as part of the properties of CoHCFs if they are to be used in industrial applications for CO_2 capture [10]. In this study, adsorption/desorption cycle tests on sample M11 of CoHCFs were conducted at 298 K, and the results are shown in Figure 10d. The adsorption remains at ca 1.77 mmol g^{-1} during the first two cycles (100 and 99.9% adsorption efficiency). In the third cycle, the adsorption decreases to 1.74 mmol g^{-1} (98.4%), representing only ~1.6% loss. This marginal loss can be attributed to the possible blocking of pores by CO_2 molecules to the adsorption sites and which can be removed along with water molecules as the repeated adsorption/desorption cycles were performed, whereas adsorption capacity increased to just 1.83 mmol g^{-1} in the fourth and fifth cycles, respectively. This slight increase in adsorption capacities in the last two cycles can be attributed to the further removal of water molecules from the CoHCFs lattices that possibly prevailed trapped in the interconnected channels of the materials during the conditioning process. These results are indicative that chemical functionalities and textural properties (i.e., S_{BET} and V_t) remain unchanged and that the CoHCFs are reversible adsorbents for CO_2 capture.

3.7.6. Selectivity

The selectivity of CoHCFs for a mixture of gases was studied in this work. Generally, flue-discharged gases in industrial or power plants contain ~15% CO_2 with mostly balanced N_2. Consequently, measurements were conducted at 298 K, at a pressure range of 1–760 mmHg, and CO_2:N_2 = 15:85 using pure component adsorption isotherms of CO_2 and N_2 to mimic industrial flue gases. The CO_2 and N_2 gases are assumed as ideal gases, and their adsorptive properties are taken individually [71]. Selectivities are calculated using the

Ideal Adsorption Solution Theory (IAST) [72]. This method uses the following Equation to calculate selectivities [73]:

$$S_{CO_2/N_2} = \frac{\frac{q_{CO_2}}{q_{N_2}}}{\frac{p_{CO_2}}{p_{N_2}}} \quad (3)$$

where S_{CO_2/N_2} is the selectivity of CO_2 over N_2, q_{CO_2} and q_{N_2} represent the amounts of CO_2 and N_2 adsorbed at the absolute pressures of p_{CO_2} and p_{N_2}, respectively.

Figure 11a compares the two adsorption isotherms on sample M11 of CoHCF at 298 K. It shows a significant difference in the amount of adsorbed CO_2 compared to that of N_2. At 760 mmHg, the amounts of CO_2 and N_2 adsorbed are 1.79 mmol g^{-1} and 0.10 mmol g^{-1}, respectively. This is a significant difference reflected in the selectivity of CO_2 over N_2. Figure 11b depicts selectivity calculation results and reveals that a maximum selectivity of ~47 is reached at a pressure of 30.7 mmHg, and it starts to slowly decrease after 35 mmHg until a minimum of 16.7 is attained at 760 mmHg. The highly enhanced selectivity of CO_2 over N_2 has been attributed to small pore sizes in CoHCFs and PBAs and powerful dipole–quadrupole interactions acting in the extremely polarizing exposed metal sites cladding the cavities, which are able to distinguish between these two gases. Since CoHCFs are capable of selectively separating gases (CO_2 over N_2) from a mixture of gases, they can be potentially used in flue gas industrial applications.

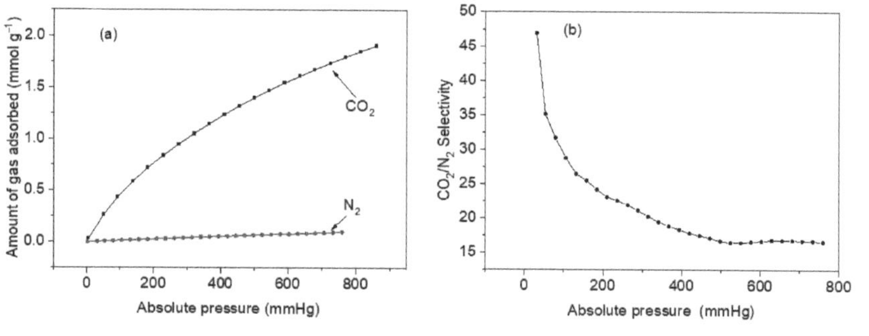

Figure 11. (**a**) CO_2 and N_2 adsorption isotherms at 25 °C and pressure between 1 and 760 mmHg for sample M11 of CoHCFs; (**b**) adsorption selectivity curve of CO_2/N_2 at the adsorption temperature of 25 °C, pressure range of 1–760 mmHg, and CO_2:N_2 = 15:85 for sample M11 of CoHCFs.

4. Conclusions

Three K or Na-CoHCFs systems were synthesized with different stoichiometries by the co-precipitation method at mild reaction conditions to establish which stoichiometry is more efficient in the removal of CO_2. The stoichiometry of samples from system III produced the best material for CO_2 uptake. The three proposed systems were studied through a battery of analytical techniques, including N_2 adsorption, TGA, FTIR, XRD, and XPS. Based on the results of N_2 adsorption measurements, this work revealed the effect that alkali ions such as K^+ or Na^+ have on the final porous structure of the CoHCFs. It was found that samples containing K^+ ions, a mixture of Fe^{2+}/Fe^{3+} ions, and a higher concentration of Co^{2+}/Co^{3+} ions in the CoHCFs structure in system III exhibited a porous structure with a much more developed external specific surface area, S_{ext} (72.69–172.18 m^2 g^{-1}) compared with samples containing Na^+ and Fe^{2+} ions (systems I and II). The specific surface areas decreased as the $Na_4[Fe(CN)_6]/Co(NO_3)_2$ and $K_4[Fe(CN)_6]/Co(NO_3)_2$ molar ratios decreased for samples of systems I and III but not for samples of system II, and it was attributed to the fact that high molar ratios of $Na_4[Fe(CN)_6]/Co(NO_3)_2$ and $K_4[Fe(CN)_6]/Co(NO_3)_2$ favored the formation of CoHCFs cubic cells with high BET surface area. The results from N_2 adsorption measurements provided evidence of the mesoporosity gained during the synthesis, along with the formation of interconnected micro tunnels and cavities formed in the crystalline structure of CoHCFs. TGA study revealed that the oxidation states of Fe, Co, and the types of alkali ions affected the number of water molecules contained in the samples and, consequently, the final porous structure of the material. XRD results allowed us to determine the formation of two different crystalline cells, FCC for systems I and III and triclinic for system II. These differences were attributed to the fact that samples in system I have Na^+ and Fe^{2+} ions, system II samples contain K^+ and Fe^{2+} ions, and system III samples have K^+ and a mixture of Fe^{2+} and Fe^{3+} ions. XPS studies confirmed the different oxidation states of Fe and Co obtained during the synthesis of CoHCFs and allowed us to authenticate the charges involved in the stabilization of some of the CoHCFs structures. The CO_2 adsorption isotherms indicated that the CoHCFs provided removal capacities for CO_2 with a maximum CO_2 removal capacity of 3.04 mmol g^{-1}, comparable with other sorbents presented in the literature. The low calculated isosteric heat of adsorption of ca. 24.3 kJ mol^{-1} provides support that these materials can be used in batch and continuous CO_2 removal operations due to low amounts of heat released that can make the adsorption process more controllable. The values of isosteric heats observed are attributed to the presence of K^+ ions exposed after the removal of water molecules from the crystallites of K-CoHCFs, creating adsorption sites with a strong affinity towards CO_2 molecules, and the K^+ ions act as gas confinement sites improving the CO_2 affinity for the CoHCFs, thus enhancing the values of Q_{st}. The CoHCFs showed significantly fast CO_2 adsorption kinetics (3 to 6 s to reach equilibrium), a property that makes them attractive for continuous adsorption operations. CoHCFs showed excellent efficiency towards adsorption/desorption cycles with a loss of ~1.6% of adsorption capacity. CoHCFs presented excellent selectivity for gas mixtures with selectivity factors of ca. 47 of CO_2 over N_2, thus demonstrating the potential application in flue gas operations. The limitations of this work include the CO_2 removal efficiencies presented by the shapes of the porous structure and particle clustering composing the structures of the CoHCFS, which is, in turn, reflected in diffusional restrictions. As a possible prospect to help overcome these limitations, the synthesis of new CoHCFs is proposed using other synthesis routes, such as inverse microemulsion techniques that would help disperse the CoHCFs nanoparticles that are expected to reduce diffusional resistances during the CO_2 uptake process.

Supplementary Materials: The following are available online at https://www.mdpi.com/article/10.3390/ma16020608/s1, Figure S1. TGA plots (as Weight % loss and first derivative) of samples M1–M5 of CoHCFs, Figure S2. TGA plots (as Weight % loss and first derivative) of samples M6–M10 of CoHCFs, Figure S3. TGA plots (as Weight % loss and first derivative) of samples M11–M15 of CoHCFs, Table S1. XRD Atomic coordinates of sample M6, Table S2. XPS binding energy assignments of samples M1, M6 and M11 of CoHCFs.

Author Contributions: P.M.F.-U.: sample synthesis, carried out the BET, CO_2 capture, experiments, XPS data processing, conceived the original idea and writing. M.B.-S.: supervised the project. E.O.-G.: carried out TGA measurements and data processing. J.D.M.-M.: carried out XRD measurements. K.C.: performed XRD data refinement. V.S.: performed and interpreted XRD data. J.A.R.-M.: performed XPS data processing and assignments. J.I.E.-V.: supervised the project. A.G.-B.: performed FTIR measurements, TGA data interpretation, and writing. S.G.-S.: conceived the original idea, supervised project, and performed writing. All authors have read and agreed to the published version of the manuscript.

Funding: This work was supported by the University of Guadalajara through the program PRO-SNI 2021.

Institutional Review Board Statement: Not applicable.

Informed Consent Statement: Not applicable.

Data Availability Statement: Not applicable.

Acknowledgments: Paloma M. Frias-Ureña is grateful for the scholarship No. 45388 from Mexico's National Council of Science and Technology, CONACyT.

Conflicts of Interest: The authors declare no conflict of interest.

References

1. Hedin, N.; Andersson, L.; Bergström, L.; Yan, J. Adsorbents for the post-combustion capture of CO_2 using rapid temperature swing or vacuum swing adsorption. *Appl. Energy* **2013**, *104*, 418–433. [CrossRef]
2. Llewellyn, P.L.; Bourrelly, S.; Serre, C.; Vimont, A.; Daturi, M.; Hamon, L.; De Weireld, G.; Chang, J.; Hong, D.; Hwang, Y.K.; et al. High Uptakes of CO_2 and CH_4 in Mesoporous Metal—Organic Frameworks MIL-100 and MIL-101. *Langmuir* **2008**, *24*, 7245–7250. [CrossRef] [PubMed]
3. Eddaoudi, M.; Li, H.; Yaghi, O.M. Highly porous and stable metal-organic frameworks: Structure design and sorption properties. *J. Am. Chem. Soc.* **2000**, *122*, 1391–1397. [CrossRef]
4. Lei, L.; Yan, C.; Changwei, C.; Mohammadreza, K.; Zeyu, J.; Chi, H. Taming structure and modulating carbon dioxide (CO_2) adsorption isosteric heat of nickel-based metal organic framework (MOF-74(Ni)) for remarkable CO_2 capture. *J. Colloid Interface Sci.* **2022**, *612*, 132–145. [CrossRef]
5. Liu, Q.; Shi, Y.; Zheng, S.; Ning, L.; Ye, Q.; Tao, M.; He, Y. Amine-functionalized low-cost industrial grade multi-walled carbon nanotubes for the capture of carbon dioxide. *J. Energy Chem.* **2014**, *23*, 111–118. [CrossRef]
6. Jun, H.J.; Yoo, D.K.; Jhung, S.H. Metal-organic framework (MOF-808) functionalized with ethyleneamines: Selective adsorbent to capture CO_2 under low pressure. *J. CO2 Util.* **2022**, *58*, 101932. [CrossRef]
7. Ogilvie, S.H.; Duyker, S.G.; Southon, P.D.; Peterson, V.K.; Kepert, C.J. Identification of bridged CO_2 binding in a Prussian blue analogue using neutron powder diffraction. *Chem. Commun.* **2013**, *49*, 9404–9406. [CrossRef]
8. Svensson, G.; Grins, J.; Eklöf, D.; Eriksson, L.; Wardecki, D.; Thoral, C.; Bodoignet, L. Influence of the presence of different alkali cations and the amount of $Fe(CN)_6$ vacancies on CO_2 adsorption on copper hexacyanoferrates. *Materials* **2019**, *12*, 3371. [CrossRef]
9. Krap, C.P.; Balmaseda, J.; Del Castillo, L.F.; Zamora, B.; Reguera, E. Hydrogen storage in prussian blue analogues: H_2 interaction with the metal found at the cavity surface. *Energy Fuels* **2010**, *24*, 581–589. [CrossRef]
10. Ojwang, D.O.; Grins, J.; Svensson, G. The adsorption kinetics of CO_2 on copper hexacyanoferrate studied by thermogravimetric analysis. *Microporous Mesoporous Mater.* **2018**, *272*, 70–78. [CrossRef]
11. Shriver, D.F.; Shriver, S.A.; Anderson, S.E. Ligand Field Strength of the Nitrogen End of Cyanide and Structures of Cubic Cyanide Polymers. *Inorg. Chem.* **1965**, *4*, 725–730. [CrossRef]
12. Zamponi, S.; Giorgetti, M.; Berrettoni, M.; Kulesza, P.J.; Cox, J.A.; Kijak, A.M. Cobalt hexacyanoferrate in PAMAM-doped silica matrix: 1. Solid state electrochemistry and thermochromism. *Electrochim. Acta* **2005**, *51*, 118–124. [CrossRef]
13. Lezna, R.O.; Romagnoli, R.; De Tacconi, N.R.; Rajeshwar, K. Cobalt hexacyanoferrate: Compound stoichiometry, infrared spectroelectrochemistry, and photoinduced electron transfer. *J. Phys. Chem. B* **2002**, *106*, 3612–3621. [CrossRef]
14. Berrettoni, M.; Giorgetti, M.; Zamponi, S.; Conti, P.; Ranganathan, D.; Zanotto, A.; Saladino, M.L.; Caponetti, E. Synthesis and characterization of nanostructured cobalt hexacyanoferrate. *J. Phys. Chem. C* **2010**, *114*, 6401–6407. [CrossRef]
15. Jassal, V.; Shanker, U.; Shankar, S. Synthesis, Characterization and Applications of Nano-structured Metal Hexacyanoferrates: A Review. *J. Environ. Anal. Chem.* **2015**, *2*, 2. [CrossRef]
16. Kawamoto, T.; Tanaka, H.; Hakuta, Y.; Takahashi, A.; Parajuli, D.; Minami, K.; Yasutaka, T.; Uchida, T. Radioactive cesium decontamination technology for ash. *Synth. Engl. Ed.* **2016**, *9*, 139–154. [CrossRef]
17. Sato, O.; Einaga, Y.; Fujishima, A.; Hashimoto, K. Photoinduced long-range magnetic ordering of a cobalt-iron cyanide. *Inorg. Chem.* **1999**, *38*, 4405–4412. [CrossRef]
18. De Tacconi, N.R.; Rajeshwar, K.; Lezna, R.O. Metal hexacyanoferrates: Electrosynthesis, in situ characterization, and applications. *Chem. Mater.* **2003**, *15*, 3046–3062. [CrossRef]
19. Aguilà, D.; Prado, Y.; Koumousi, E.S.; Mathonière, C.; Clérac, R. Switchable Fe/Co Prussian blue networks and molecular analogues. *Chem. Soc. Rev.* **2016**, *45*, 203–224. [CrossRef]
20. Karadas, F.; El-Faki, H.; Deniz, E.; Yavuz, C.T.; Aparicio, S.; Atilhan, M. CO_2 adsorption studies on Prussian blue analogues. *Microporous Mesoporous Mater.* **2012**, *162*, 91–97. [CrossRef]
21. Roque, J.; Reguera, E.; Balmaseda, J.; Rodríguez-Hernández, J.; Reguera, L.; del Castillo, L.F. Porous hexacyanocobaltates(III): Role of the metal on the framework properties. *Microporous Mesoporous Mater.* **2007**, *103*, 57–71. [CrossRef]
22. Boudjema, L.; Long, J.; Salles, F.; Larionova, J.; Guari, Y.; Trens, P. A Switch in the Hydrophobic/Hydrophilic Gas-Adsorption Character of Prussian Blue Analogues: An Affinity Control for Smart Gas Sorption. *Chem.—A Eur. J.* **2019**, *25*, 479–484. [CrossRef]
23. Li, N.; Chang, Z.; Huang, H.; Feng, R.; He, W.W.; Zhong, M.; Madden, D.G.; Zaworotko, M.J.; Bu, X.H. Specific K^+ Binding Sites as CO_2 Traps in a Porous MOF for Enhanced CO_2 Selective Sorption. *Small* **2019**, *15*, 1900426. [CrossRef] [PubMed]
24. Kaye, S.S.; Long, J.R. Hydrogen storage in the dehydrated prussian blue analogues $M_3[Co(CN)_6]_2$ (M = Mn, Fe, Co, Ni, Cu, Zn). *J. Am. Chem. Soc.* **2005**, *127*, 6506–6507. [CrossRef]
25. Natesakhawat, S.; Gulp, J.T.; Matranga, C.; Bockrath, B. Adsorption properties of hydrogen and carbon dioxide in Prussian blue analogues $M_3[Co(CN)_6]_2$, M = Co, Zn. *J. Phys. Chem. C* **2007**, *111*, 1055–1060. [CrossRef]
26. López, Y.C.; Ortega, G.A.; Reguera, E. Microporous prussian blue analogs and their application for environmental remediation: A deeper look from the structure-property-functionality perspective. *Microporous Mesoporous Mater.* **2022**, *333*, 111755. [CrossRef]
27. Lutterotti, L.; Bortolotti, M.; Ischia, G.; Lonardelli, I.; Wenk, H.-R. Rietveld texture analysis from diffraction images. *Z. Krist. Suppl.* **2007**, *26*, 125–130. [CrossRef]

28. Li, W.J.; Han, C.; Cheng, G.; Chou, S.L.; Liu, H.K.; Dou, S.X. Chemical Properties, Structural Properties, and Energy Storage Applications of Prussian Blue Analogues. *Small* **2019**, *15*, 1900470. [CrossRef]
29. Wark, K. *Generalized Thermodynamic Relationships*, 5th ed.; McGraw-Hill: New York, NY, USA, 1988; ISBN 978-0-07-068286-3.
30. Thommes, M.; Kaneko, K.; Neimark, A.V.; Olivier, J.P.; Rodriguez-Reinoso, F.; Rouquerol, J.; Sing, K.S.W. Physisorption of gases, with special reference to the evaluation of surface area and pore size distribution (IUPAC Technical Report). *Pure Appl. Chem.* **2015**, *87*, 1051–1069. [CrossRef]
31. Thommes, M.; Cychosz, K.A. Physical Adsorption Characterization of Nanoporous Materials: Progress and Challenges. *Adsorption* **2014**, *20*, 233–250. [CrossRef]
32. Soek, R.N.; Schmidt, A.; Winnischofer, H.; Vidotti, M. Anisotropic behavior of layer-by-layer films using highly disordered copper hexacyanoferrate(II) nanoparticles. *Appl. Surf. Sci.* **2016**, *378*, 253–258. [CrossRef]
33. Ji, Z.; Han, B.; Liang, H.; Zhou, C.; Gao, Q.; Xia, K.; Wu, J. On the Mechanism of the Improved Operation Voltage of Rhombohedral Nickel Hexacyanoferrate as Cathodes for Sodium-Ion Batteries. *ACS Appl. Mater. Interfaces* **2016**, *8*, 33619–33625. [CrossRef]
34. Li, W.J.; Chou, S.L.; Wang, J.Z.; Kang, Y.M.; Wang, J.L.; Liu, Y.; Gu, Q.F.; Liu, H.K.; Dou, S.X. Facile method to synthesize na-enriched $Na_{1+x}FeFe(CN)_6$ frameworks as cathode with superior electrochemical performance for sodium-ion batteries. *Chem. Mater.* **2015**, *27*, 1997–2003. [CrossRef]
35. Gerber, S.J.; Erasmus, E. Surfactant-stabilized nano-metal hexacyanoferrates with electrocatalytic and heterogeneous catalytic applications. *Transit. Met. Chem.* **2018**, *43*, 409–420. [CrossRef]
36. Gerber, S.J.; Erasmus, E. Electronic effects of metal hexacyanoferrates: An XPS and FTIR study. *Mater. Chem. Phys.* **2018**, *203*, 73–81. [CrossRef]
37. Reguera, L.; Reguera, E.; Balmaseda, J.; Rodriguez-Hernandez, J.; Yee-Madeira, H. Mixed valences system in cobalt iron cyanide. Microporous structure stability. *J. Porous Mater.* **2008**, *15*, 719–729. [CrossRef]
38. Martínez-García, R.; Knobel, M.; Balmaseda, J.; Yee-Madeira, H.; Reguera, E. Mixed valence states in cobalt iron cyanide. *J. Phys. Chem. Solids* **2007**, *68*, 290–298. [CrossRef]
39. Ng, C.W.; Ding, J.; Shi, Y.; Gan, L.M. Structure and magnetic properties of copper(II) hexacyanoferrate(III) compound. *J. Phys. Chem. Solids* **2001**, *62*, 767–775. [CrossRef]
40. Mink, J.; Stirling, A.; Ojwang, D.O.; Svensson, G.; Mihály, J.; Németh, C.; Drees, M.; Hajba, L. Vibrational properties and bonding analysis of copper hexacyanoferrate complexes in solid state. *Appl. Spectrosc. Rev.* **2019**, *54*, 369–424. [CrossRef]
41. Cattermull, J.; Sada, K.; Hurlbutt, K.; Cassidy, S.J.; Pasta, M.; Goodwin, A.L. Uncovering the Interplay of Competing Distortions in the Prussian Blue Analogue $K_2Cu[Fe(CN)_6]$. *Chem. Mater.* **2022**, *34*, 5000–5008. [CrossRef]
42. Qiu, Y.; Lin, Y.; Yang, H.; Wang, L. Ni-doped cobalt hexacyanoferrate microcubes as battery-type electrodes for aqueous electrolyte-based electrochemical supercapacitors. *J. Alloys Compd.* **2019**, *806*, 1315–1322. [CrossRef]
43. Vannerberg, N.G. The ESCA-spectra of sodium and potassium cyanide and of the sodium and potassium salts of the hexacyanometallates of the first transition metal series. *Chem. Scr.* **1976**, *9*, 122–126.
44. Wesner, D.A.; Linden, G.; Bonzel, H.P. Alkali promotion on cobalt: Surface analysis of the effects of potassium on carbon monoxide adsorption and Fischer-Tropsch reaction. *Appl. Surf. Sci.* **1986**, *26*, 335–356. [CrossRef]
45. Belton, D.N.; Schmieg, S.J. Electron spectroscopic identification of carbon species formed during diamond growth. *J. Vac. Sci. Technol. A Vac. Surf. Film.* **1990**, *8*, 2353–2362. [CrossRef]
46. Schulze, P.D.; Shaffer, S.L.; Hance, R.L.; Utley, D.L. Adsorption of water on rhenium studied by XPS. *J. Vac. Sci. Technol. A Vac. Surf. Film.* **1983**, *1*, 97–99. [CrossRef]
47. Russat, J. Characterization of Polyamic Acid/Polyimide Films in the Nanometric Thickness Range from Spin-deposited Poly amic Acid. *Surf. Interface Anal.* **1988**, *11*, 414–420. [CrossRef]
48. Nefedov, V.I.; Gati, D.; Dzhurinskii, B.F.; Sergushin, N.P.; Salyn, Y.V.; Naumkin, A.V.; Kraut-Vass, A.; Gaarenstroom, S.; Powell, C. NIST X-ray Photoelectron Spectroscopy Database, NIST Standard Reference Database 20, v4.1, 2012. *Zhurnal Neorg. Khimii* **1975**, *20*, 2307–2314. [CrossRef]
49. Strydom, C.A.; Strydom, H.J. X-ray photoelectron spectroscopy studies of some cobalt(II) nitrate complexes. *Inorg. Chim. Acta* **1989**, *159*, 191–195. [CrossRef]
50. Bonnelle, J.P.; Grimblot, J.; D'huysser, A. Influence de la polarisation des liaisons sur les spectres esca des oxydes de cobalt. *J. Electron Spectros. Relat. Phenom.* **1975**, *7*, 151–162. [CrossRef]
51. Paparazzo, E. XPS and Auger Spectroscopy on mixtures of the oxides SiO_2, Al_2O_3, Fe_2O_3, and Cr_2O_3. *J. Electron Spectros. Relat. Phenom.* **1987**, *43*, 97–112. [CrossRef]
52. Porta, P.; Dragone, R.; Fierro, G.; Inversi, M.; Lo Jacono, M.; Moretti, G. Preparation and characterisation of cobalt-copper hydroxysalts and their oxide products of decomposition. *J. Chem. Soc. Faraday Trans.* **1992**, *88*, 311–319. [CrossRef]
53. Nefedov, V.I.; Salyn, Y.V.; Leonhardt, G.; Scheibe, R. A comparison of different spectrometers and charge corrections used in X-ray photoelectron spectroscopy. *J. Electron Spectros. Relat. Phenom.* **1977**, *10*, 121–124. [CrossRef]
54. Biwer, B.M.; Bernasek, S.L. Electron spectroscopic study of the iron surface and its interaction with oxygen and nitrogen. *J. Electron Spectros. Relat. Phenom.* **1986**, *40*, 339–351. [CrossRef]
55. Yatsimirskii, K.B.; Nemoshkalenko, V.V.; Nazarenko, Y.P.; Aleshin, V.G.; Zhilinskaya, V.V.; Tomashevsky, N.A. Use of X-ray photoelectron and Mössbauer spectroscopies in the study of iron pentacyanide complexes. *J. Electron Spectros. Relat. Phenom.* **1977**, *10*, 239–245. [CrossRef]

56. Siriwardane, R.V.; Cook, J.M. Interactions of SO_2 with sodium deposited on silica. *J. Colloid Interface Sci.* **1985**, *108*, 414–422. [CrossRef]
57. Kishi, K. Adsorption of ethylenediamine on clean and oxygen covered Fe/Ni(100) surfaces studied by XPS. *J. Electron Spectros. Relat. Phenom.* **1988**, *46*, 237–247. [CrossRef]
58. Tan, B.J.; Klabunde, K.J.; Sherwood, P.M.A. XPS Studies of Solvated Metal Atom Dispersed Catalysts. Evidence for Layered Cobalt—Manganese Particles on Alumina and Silica. *J. Am. Chem. Soc.* **1991**, *113*, 855–861. [CrossRef]
59. Andonova, S.; Akbari, S.S.; Karadas, F.; Spassova, I.; Paneva, D.; Hadjiivanov, K. Structure and properties of KNi-hexacyanoferrate Prussian Blue Analogues for efficient CO_2 capture: Host-guest interaction chemistry and dynamics of CO_2 adsorption. *J. CO2 Util.* **2021**, *50*, 101593. [CrossRef]
60. Zhao, J.; Xie, K.; Singh, R.; Xiao, G.; Gu, Q.; Zhao, Q.; Li, G.; Xiao, P.; Webley, P.A. Li^+/ZSM-25 Zeolite as a CO_2 Capture Adsorbent with High Selectivity and Improved Adsorption Kinetics, Showing CO_2-Induced Framework Expansion. *J. Phys. Chem. C* **2018**, *122*, 18933. [CrossRef]
61. Riascos-Rodríguez, K.; Schroeder, A.J.; Arend, M.R.; Evans, P.G.; Hernández-Maldonado, A.J. Hysteretic adsorption of CO_2 onto a $Cu_2(pzdc)_2(bpy)$ porous coordination polymer and concomitant framework distortion. *Dalt. Trans.* **2014**, *43*, 10877–10884. [CrossRef]
62. Serna-Guerrero, R.; Sayari, A. Modeling adsorption of CO_2 on amine-functionalized mesoporous silica. 2: Kinetics and breakthrough curves. *Chem. Eng. J.* **2010**, *161*, 182–190. [CrossRef]
63. Ammendola, P.; Raganati, F.; Chirone, R.; Miccio, F. Fixed bed adsorption as affected by thermodynamics and kinetics: Yellow tuff for CO_2 capture. *Powder Technol.* **2020**, *373*, 446–458. [CrossRef]
64. Creamer, A.E.; Gao, B.; Zhang, M. Carbon dioxide capture using biochar produced from sugarcane bagasse and hickory wood. *Chem. Eng. J.* **2014**, *249*, 174–179. [CrossRef]
65. Xu, C.; Dinka, E.; Hedin, N. Hydrophobic porous polyketimines for the capture of CO_2. *Chempluschem* **2016**, *81*, 58–63. [CrossRef]
66. Shafeeyan, M.S.; Daud, W.M.A.W.; Shamiri, A.; Aghamohammadi, N. Adsorption equilibrium of carbon dioxide on ammonia-modified activated carbon. *Chem. Eng. Res. Des.* **2015**, *104*, 42–52. [CrossRef]
67. Zhou, X.; Yi, H.; Tang, X.; Deng, H.; Liu, H. Thermodynamics for the adsorption of SO_2, NO and CO_2 from flue gas on activated carbon fiber. *Chem. Eng. J.* **2012**, *200–202*, 399–404. [CrossRef]
68. Singh, V.K.; Anil Kumar, E. Measurement and analysis of adsorption isotherms of CO_2 on activated carbon. *Appl. Therm. Eng.* **2016**, *97*, 77–86. [CrossRef]
69. Heydari-Gorji, A.; Sayari, A. CO_2 capture on polyethylenimine-impregnated hydrophobic mesoporous silica: Experimental and kinetic modeling. *Chem. Eng. J.* **2011**, *173*, 72–79. [CrossRef]
70. Himeno, S.; Tomita, T.; Suzuki, K.; Yoshida, S. Characterization and selectivity for methane and carbon dioxide adsorption on the all-silica DD3R zeolite. *Microporous Mesoporous Mater.* **2007**, *98*, 62–69. [CrossRef]
71. Jang, E.; Choi, S.W.; Lee, K.B. Effect of carbonization temperature on the physical properties and CO_2 adsorption behavior of petroleum coke-derived porous carbon. *Fuel* **2019**, *248*, 85–92. [CrossRef]
72. Myers, A.L.; Prausnitz, J.M. Prausnitz, Thermodynamics of Mixed-Gas Adsorption. *AIChE J.* **1965**, *11*, 121–127. [CrossRef]
73. Manyà, J.J.; González, B.; Azuara, M.; Arner, G. Ultra-microporous adsorbents prepared from vine shoots-derived biochar with high CO_2 uptake and CO_2/N_2 selectivity. *Chem. Eng. J.* **2018**, *345*, 631–639. [CrossRef]

materials

MDPI

Article

Porous Carbon Sponge from White-Rot Fungus *Phanerochaete chrysosporium* for the Removal of Oils and Organic Solvents

Yue Gong [1,†], Lejie Pan [1,†], Huahui Yuan [1], Juncheng Li [1], Xin Li [1], Qian Chen [1], Yue Yuan [1,*], Xian Wu [2] and Sheng-Tao Yang [1,2,*]

1 Key Laboratory of Pollution Control Chemistry and Environmental Functional Materials for Qinghai-Tibet Plateau of the National Ethnic Affairs Commission, School of Chemistry and Environment, Southwest Minzu University, Chengdu 610041, China

2 Key Laboratory of General Chemistry of the National Ethnic Affairs Commission, School of Chemistry and Environment, Southwest Minzu University, Chengdu 610041, China

* Correspondence: smileyuan1987@163.com (Y.Y.); yangst@swun.edu.cn (S.-T.Y.); Tel.: +86-28-8570-9707 (Y.Y. & S.-T.Y.)

† These authors contributed equally to this work.

Abstract: Oil leakage incidentally occurs and leads to environmental disasters. Because of their porous and hydrophobic characteristics, graphene sponges are often studied as an oil adsorbent to repair oil spills at sea. Graphene materials are very expensive, and their biological toxicity has been given serious concerns; however, the easier preparation and eco-friendly, biomass-derived porous carbon materials can be used as an alternative to graphene materials. In this study, we prepared a porous carbon sponge (PCS) for oil and organic solvent removal by carbonizing white-rot fungus *Phanerochaete chrysosporium*, a fast-growing microorganism for the production of lignin-degrading enzymes and the environmental remediation. *P. chrysosporium* fungus balls were converted into black PCS by carbonization at high temperatures, where PCS was light (density of 56 g/L), hydrophobic (contact angle of 115°) and porous. According to the results of BET and XPS analysis, the surface area of PCS was 14.43 m^2/g, and the carbon in PCS is mainly sp^2 carbon. PCS could adsorb pure oils and organic solvents within seconds. The adsorption capacities of PCS were 20.7 g/g for gasoline, 30.1 g/g for peanut oil, 27.7 g/g for toluene, 18.5 g/g for dodecane, 32.5 g/g for chloroform, 27.1 g/g for tetrahydrofuran, 23.7 g/g for acetone and 13.7 g/g for ethanol. According to the reusability study, there was no obvious capacity loss after recycling up to 10 cycles. Our results indicated that white-rot fungi could be adopted as a cheap carbon resource for oil and organic solvent removal.

Keywords: biomass carbon; carbonization; oil removal; fungi; environmental remediation

Citation: Gong, Y.; Pan, L.; Yuan, H.; Li, J.; Li, X.; Chen, Q.; Yuan, Y.; Wu, X.; Yang, S.-T. Porous Carbon Sponge from White-Rot Fungus *Phanerochaete chrysosporium* for the Removal of Oils and Organic Solvents. *Materials* **2023**, *16*, 534. https://doi.org/10.3390/ma16020534

Academic Editor: Elisabetta Loffredo

Received: 22 November 2022
Revised: 26 December 2022
Accepted: 28 December 2022
Published: 5 January 2023

1. Introduction

Oils and organic solvents are important pollutants nowadays. The accidental discharge of oils inevitably would induce ecological disaster [1–3]. Due to the lower density, oils could float on water, thus blocking the air and light, which leads to the deaths of aquatic organisms. In addition, the biotransformation of oils is usually very slow. Great efforts have been dedicated to remediating oil pollution. To circumvent the oil spill events, commonly applied technologies are developed using insulation (booms) and oil gathering (skimmers) [4,5]. Chemical methods usually use dispersants, solidifiers and bio-reducing agents. The in situ burning of the dispersed fuel is also a possibility. Among these approaches, adsorption is a physical method with the advantages of fast remediation, easy operation, the recovery of lost oil, and low environmental impact [6–8].

Porous carbon materials, in particular graphene sponges, attract great interest in oil/water separation due to their huge capacities [9–12]. In 2012, Zhao and coworkers assembled graphene oxide sheets by hydrothermal treatment with thiourea to prepare a graphene sponge. The adsorption capacity of the graphene sponge for diesel oil reached 129 g/g [13]. Bi et al. annealed graphene oxide aerogel to produce a graphene sponge for oil removal, and the capacities were 20–86 g/g [14]. We reported the one-pot hydrothermal preparation of the graphene sponge, and the as-obtained graphene sponge showed high capacities for oils and organic solvents (23–35 g/g) [15]. The adsorption capacity of graphene prepared by the vapor phase reduction method for oil and organic solvent could be improved to 72–224 g/g [16]. The main limit of graphene sponge is the expensive producing cost. Graphene was used to modify the melamine sponge for oil adsorption [17]. Such a cheaper sponge showed a capacity of 111 g/g for heavy oil. Granular activated carbon was prepared to adsorb and degrade the organics in oil recycled from oil sand [18,19]. Cheap alternatives of graphene sponge have highly demanded oil/water separation [6].

Biomass carbon materials could serve as alternatives to graphene sponges in oil removal applications [20,21]. Banana peel and waste paper were adopted to prepare the hybrid aerogel by freezing-cast, freeze-drying and pyrolysis [22]. The hybrid aerogel showed the adsorption capacities of 35–115 g/g for oils and separated various surfactant-stabilized water-in-oil emulsions with high fluxes up to 8550 L/m^2/h. Shaddock peel was converted into carbon aerogel by hydrothermal carbonization, freezing-drying and post-pyrolysis process [23]. The carbon aerogel had huge adsorption capacities of 23–48 g/g for oils and organic solvents with good recyclability. Superhydrophobic *Enteromorpha*-derived carbon aerogels showed high performance in multi-behavioral oil/water separation with a capacity of 140 g/g [24]. Sugarcane residue was a good bioresource for carbon aerogel production [25]. The carbon aerogel had capacities of 17.1–37.7 g/g for organic solvents. Plant fibers could be extracted from the *Platanus orientalis* fruit [26]. After vacuum carbonization, the adsorption capacities of 71.72–172.37 g/g were achieved for oils. Beyond biomass from plants, fungi could be used for the preparation of carbon aerogel, too. The mycelia of fungi allow the formation of a porous structure after carbonization. Yi et al. reported a pilot study on bamboo fungus [27]. The bamboo fungus was freeze-dried, coated with carbon ink and carbonized. The fungus carbon aerogel could adsorb 20–42 g/g of oils. Although fungi are good candidates for this purpose, bamboo fungus as a delicious mushroom is too expensive.

White-rot fungi are microorganisms that produce lignin-degrading enzymes. White-rot fungi grow fast and have been applied in environmental remediation, while the preparation of porous carbon sponge (PCS) by white-rot bacteria and its application in the simultaneous removal of oil and organic solvent are rarely studied [28–31]. In this study, we utilized the fungus balls of the white-rot fungus *Phanerochaete chrysosporium* as the carbon source to prepare PCS for the removal of oils and organic solvents. Fungus balls were homogenized, lyophilized and carbonized to obtain PCS. The as-obtained PCS was characterized by multiple techniques. The adsorption performance of PCS was evaluated in the pure oils and also in the simulated seawater. The regeneration of PCS after adsorbing oils was achieved by evaporation. The implications of the applications of biomass carbon materials in oil removal are discussed.

2. Materials and Methods

2.1. Preparation of PCS

P. chrysosporium strain (MTCC 787) was purchased from Guangdong Microbiology Culture Center and cultivated following the recommended protocols. The recipe of culture medium and cultivation conditions were described in our previous study [28]. After 14 days of cultivation, the fungus balls were collected by filtration to remove the excess culture medium. The fungus mycelia were homogenized and then lyophilized to obtain the white sponge. The white sponge was carbonized under N_2 atmosphere (99.99%, with flowrate of 150 mL/min) in a tubular furnace at 600 °C for 2 h at a heating rate of 5 °C/min. The hydrophilic white sponge was converted into black PCS after the carbonization. PCS was characterized by multiple techniques, including scanning electron microscopy (SEM,

JSM-7500, JEOL, Tokyo, Japan), X-ray photoelectron spectroscopy (XPS, ESCALAB 250XI, Thermo-Fisher, Waltham, USA), infrared spectroscopy (IR, Magna-IR 750, Nicolet, Madison, WI, USA), contact angle measuring device (JC2000D1, Powereach Co., Shanghai, China), X-ray diffraction (XRD, XD-6, Purkinje General Instrument Co., Beijing, China) and Raman spectroscopy (inVia, Renishaw Co., London, UK).

2.2. Removal of Oils and Organic Solvents

To visualize the adsorption performance of PCS, chloroform (2 g) was stained red by Sudanred 5B. The chloroform was placed on a glass plate and added with a piece of PCS. The photographs were taken at time intervals of 16 s. At the end of experiment, the plate was wiped with the PCS to remove the residues completely.

Similarly, the visualization of the removal process was performed in simulated seawater. Gasoline (dyed with Sudan red 5B) was added to simulated seawater. The recipe of simulated sea water was as follows: 25 g of NaCl, 1.14 g of $CaCl_2$, 0.7 g of KCl and 1 L of deionized water. PCS floated on the simulated seawater and was driven by a pipet tip to suck the gasoline. The photographs were taken before and after adding PCS at a time interval of 2 min.

Adsorption capacities of PCS for oils and organic solvents were measured by weighting PCS before and after adsorption. A piece of PCS was weighted before use (m_0). Then, it was sunk into oil or organic solvent (peanut oil, gasoline, toluene, dodecane, chloroform, tetrahydrofuran, acetone, ethanol). After 24 h, the PCS was collected and weighed again (m). The adsorption capacity of PCS was calculated by $(m - m_0)/m_0$ (g/g). All data contained three duplicates and were expressed as the mean with the standard deviation (mean $\pm$ SD).

2.3. Recycling of PCS

The recycling evaluation was performed with dodecane. After full adsorption of dodecane, PCS was dried at 80 °C to evaporate the adsorbed dodecane. The dried PCS was used to adsorb dodecane again for the adsorption capacity measurements following the aforementioned protocol. The adsorption capacities were measured up to 10 cycles.

3. Results and Discussion

3.1. Characterization of PCS

PCS was a black sponge. Under SEM, the pores of PCS could be recognized (Figure 1). The surface of PCS seemed tight and flat. The pores were found at the edge of PCS. The presence of pores was important for the adsorption to accommodate oils and organic solvents. These pores were too big to be detectable in N_2 adsorption/desorption measurements (Figure 2). The contact angle of PCS was 115°, indicating that the hydrophilicity of PCS was moderate. The hydrophilicity was crucial for the adsorption of oils and organic solvents. The N_2 adsorption/desorption isotherm curve followed type IV, suggesting the adsorption hysteresis nature of N_2 on PCS. The surface area of PCS measured by the BET method was 14.43 m^2/g (Figure 2a). The pore width was distributed mostly at 4 nm (Figure 2b) and the pore volume was 0.022384 cm^3/g.

The efficient carbonization was indicated by XPS analysis (Figure 3a). There were 79.77% of C, 17.1% of O and 3.13% of N in atomic ratios. The majority of C atoms were in the form of sp^2 carbon (66.9%) based on the C1s spectrum. The other two components were C-O/sp^3 carbon (30.43%) and C=O/shake-up signal (2.67%). The oxygen-containing groups were reflected in the IR spectrum (Figure 3b), including 3442 cm^{-1} for -OH/-COOH and 1101 cm^{-1} for C-O. The sp^2 carbon was identified at 1637 cm^{-1}. The small peak at 2931 cm^{-1} was attributed to the remnant hydrocarbon -CH_2-/-CH_3. There was no signal at 1720 cm^{-1}, suggesting the lack of C=O groups. The lack of negatively charged carboxyl groups explained the hydrophilicity. The sp^2 carbon was also indicated by the G band (1590 cm^{-1}) of the Raman spectrum (Figure 3c). The sp^3 carbon was indicated by the D band (1340 cm^{-1}) as defects. The sp^2 carbon was in the form of graphene sheets as indicated by the XRD (Figure 3d) band between 13.7–37.8°. The intensity was very low, and the peak was centered at 24.3°, which refers to the 002 crystal face of graphite.

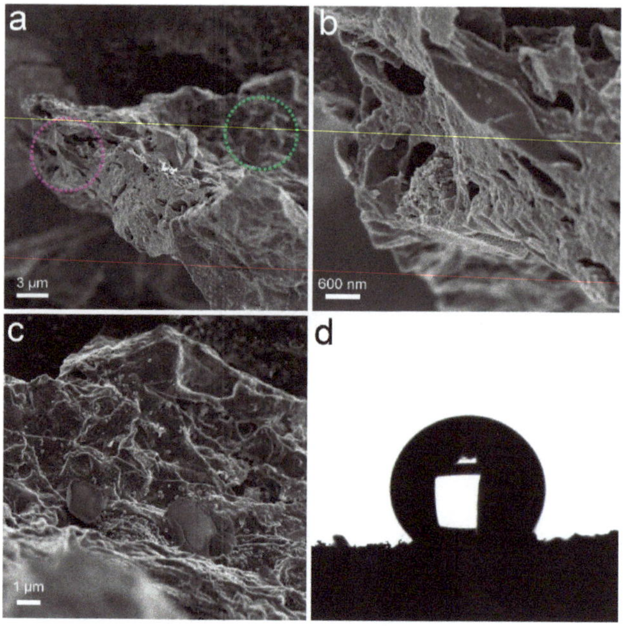

Figure 1. Characterization data of PCS. (**a**–**c**) SEM images; (**b**) Enlargement of the purple cycle; (**c**) Enlargement of the green cycle; (**d**) Contact angle.

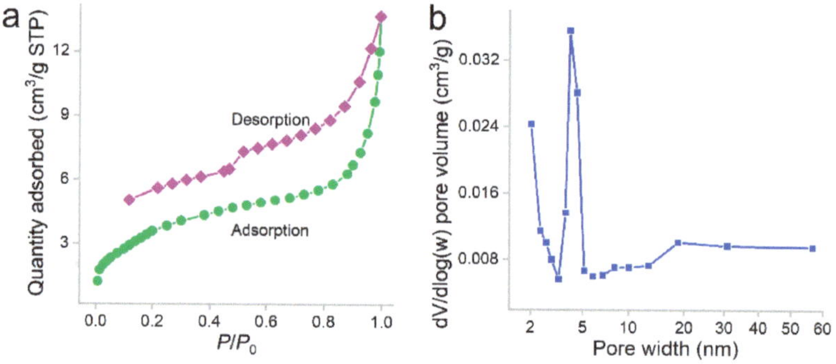

Figure 2. Nitrogen adsorption/desorption isotherms (**a**) and pore size distribution (**b**) of PCS.

3.2. Adsorption of Oils and Organic Solvents by PCS

In our previous studies, we suggested the accommodation of oils and organic solvents in hydrophobic pores was the main reason for oil adsorption by porous carbon materials [16]. PCS had a contact angle of 115° and a porous structure; thus, PCS was supposed to adsorb oils and organic solvents efficiently. To visualize the adsorption performance, the chloroform was stained with Sudan red 5B, and a piece of PCS was added to the chloroform. As shown in Figure 4, the chloroform was quickly adsorbed into PCS. Within 1 min, all chloroform was sucked into PCS. The quick adsorption of pure chloroform by PCS suggested that PCS could be used for oils and organic solvents.

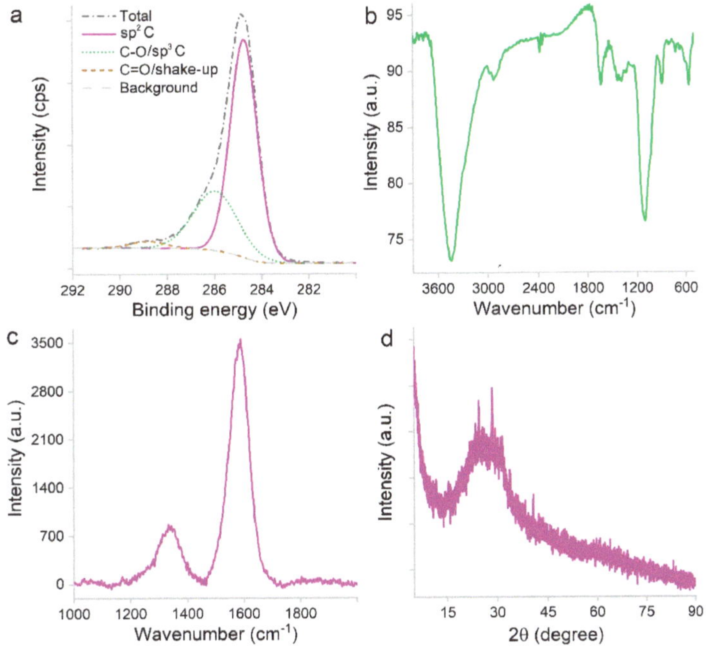

Figure 3. C1s XPS (**a**), IR (**b**), Raman (**c**) and XRD (**d**) spectra of PCS.

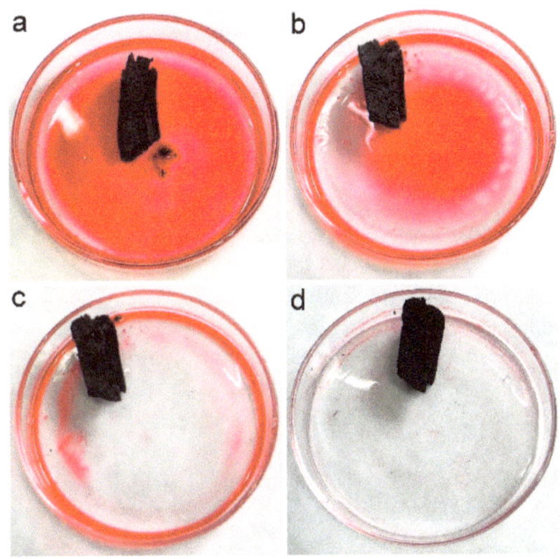

Figure 4. Removal of chloroform (dyed with Sudan red 5B) by PCS. (**a**) chloroform (dyed with Sudan red 5B) before PCS addition (0 s), (**b**) PCS removes chloroform (dyed with Sudan red 5B) for 16 s, (**c**) PCS removes chloroform (dyed with Sudan red 5B) for 32 s, (**d**) PCS removes chloroform (dyed with Sudan red 5B) for 48 s.

A more important situation is that oils are leaked into the water environment. Simulated seawater was dumped with gasoline (dyed with Sudan red 5B) (Figure 5). A piece of PCS was added to the beaker, and it floated on the simulated seawater steadily. The PCS was driven by a pipette tip to achieve complete removal within 2 min. After the adsorption, PCS floated on simulated seawater, which could be easily collected.

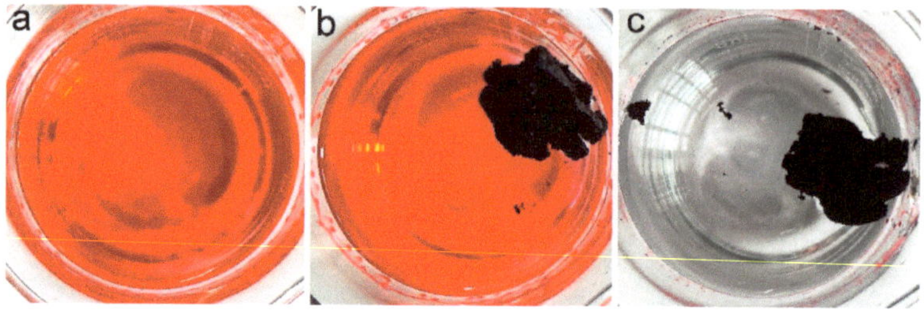

Figure 5. Removal of gasoline (dyed with Sudan red 5B) in simulated seawater by PCS. (**a**) gasoline in simulated seawater (dyed with Sudan red 5B) (0 min), (**b**) PCS starts to remove gasoline (dyed with Sudan red 5B) in simulated seawater (0 min), (**c**) PCS removes gasoline (dyed with Sudan red 5B) in simulated seawater for 2 min.

The adsorption capacities of PCS for several species of oils and organic solvents were measured in batch experiments. PCS was used to adsorb pure liquids, where the capacities were obtained by weighting PCS before and after the adsorption. The capacities of PCS for oils and organic solvents are presented in Figure 6. The adsorption capacity for ethanol was the lowest, namely 13.7 g/g, which should be due to the higher hydrophilicity of ethanol. The highest adsorption capacity was found in chloroform (32.5 g/g). Rest ranged in the range of 18.5–30.0 g/g. Of particular interest was that PCS adsorbed gasoline efficiently (20.7 g/g), suggesting PCS had the potential for the remediation of oil leakage. The performance of PCS was on the same order of magnitude compared to other biomass carbon materials. For example, carbon aerogels from sugarcane residue had adsorption capacities of 17.1–30.7 g/g for organic solvents [25]. Porous tubular carbon fibers from the fruits of *Platanus orientalis* could adsorb 71.72–172.37 times their own weight [24]. Superhydrophobic *Enteromorpha*-derived carbon aerogels were modified by $NH_4H_2PO_4$ and showed adsorption capacities of 62–140 g/g for oil and organic solvents [24]. Hybrid aerogels derived from banana peel and waste paper had high oil sorption capacities of 35–115 times their own weight [22]. Carbon aerogel from shaddock peel had 23–48 times its own weight in selectively adsorption oils and organic solvents [23]. Tubelike aerogels from bamboo fungus showed 20–42 g/g capacities for oils and organic solvents [27]. The adsorption capacities of PCS were competitive to graphene sponges from hydrothermal methods [13–15]. Compared to the literature results (Table 1) [15,16,25,27,32–35], the advantages of white-rot fungi in our study included the following. First, white-rot fungi are common fungus species in nature and grow fast. Second, during the growth of white-rot fungi, they could secrete enzymes and might be applied in pollutant remediation. Third, white-rot fungi have mycelia, and the fabrication of porous structures is easy. Due to the aforementioned merits and the high performance, we concluded that white-rot fungi were good precursors of porous carbon sponge for oil and organic solvent removal.

Table 1. Comparison of performance between PCS and similar sorbents.

Materials	q_m for Oil (g/g)	BET (m^2/g)	Reference
Porous carbon sponge	20.7	14.43	This study
carbon aerogels	17.1–30.7 g/g	342.34 ± 5.59	[25]
Graphene sponge	29.3	—	[15]
Vapor phase reduction graphene sponge	165	—	[16]
Tubelike aerogels	20–42 g/g	129.61 ± 3.96	[27]
Graphene/polyester staple composite	52	—	[32]
Volume-based magnetic foams	95	89.5	[33]
Magnetic biochar	8.77	48.22	[34]
Coconut oil-modified biochar	5.315	443.81	[35]

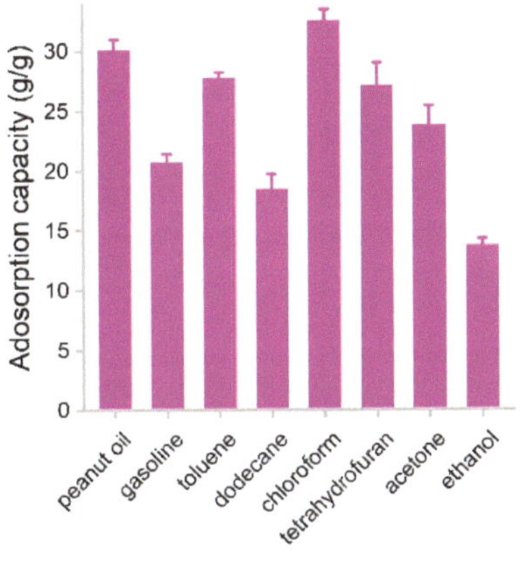

Figure 6. Adsorption capacities of PCS for oils and organic solvents.

3.3. Regeneration of PCS

The recyclability of adsorbent is a crucial parameter that largely determines the operating cost of the adsorption process. The recyclability of hydrophobic adsorbents depends on the maintenance of the porous structure for the accommodation of oils and organic solvents. In our study, PCS showed very good regeneration performance (Figure 7). Using dodecane as the model solvent, the initial capacity of PCS was 19.8 g/g. After adsorption, the dodecane was evaporated at 80 °C. The surface tension of dodecane was low, so during the drying process, the pores could be retained well. The first regeneration resulted in a capacity of 18.8 g/g. The values varied in the range of 17.5–19.2 g/g during the recycling, and the adsorption capacity was almost always maintained above 90%. At the 10th cycle, the capacity of PCS was 19.1 g/g, equaling 96.5% of the initial value. Compared with graphene sponge, the adsorption capacity is down to 77% of the raw material at 10 cycles [15]. The adsorption capacity was reduced to 78.4% of the raw graphene/polyester staple composite at nine cycles [32]. These results suggested that PCS could be well recycled without obvious capacity loss. In future applications, the reuse of PCS in a real environment should be evaluated.

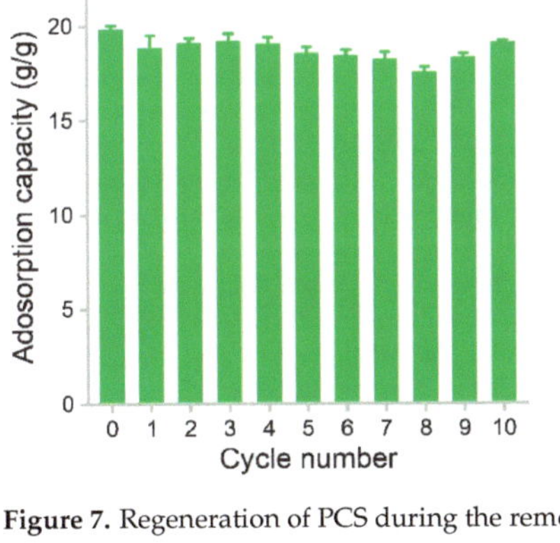

Figure 7. Regeneration of PCS during the removal of dodecane.

4. Conclusions

In summary, a biomass carbon sponge was prepared by the carbonization of fungus balls, which showed high performance in removing oil and organic solvents. The low density, hydrophobic nature and porous structure enabled the large adsorption capacities of PCS for pure oils and organic solvents within seconds. PCS could float on the simulated seawater to suck oils efficiently. After 10 cycles of regeneration, PCS retained the high capacity. Our results collectively indicated that white-rot fungi were good precursors of porous carbon materials. It is hoped that our results will benefit the ongoing exploration of biomass carbon for oil pollution remediation.

Author Contributions: Conceptualization, S.-T.Y.; Methodology, H.Y., J.L., Y.Y. and X.W.; Investigation, Y.G., L.P., X.L. and Q.C.; Writing—original draft, Y.G.; Writing—review & editing, Y.Y. and S.-T.Y.; Supervision, S.-T.Y.; Project administration, Y.Y. and X.W.; Funding acquisition, L.P. and S.-T.Y. All authors have read and agreed to the published version of the manuscript.

Funding: This research was funded by China Natural Science Foundation (No. 21777132) and the Fundamental Research Funds for the Central Universities, Southwest Minzu University (No. ZYN2022002).

Institutional Review Board Statement: Not applicable.

Informed Consent Statement: Not applicable.

Data Availability Statement: Not applicable.

Conflicts of Interest: The authors declare no conflict of interest.

References

1. Wang, Q.; Lü, Y.; Li, Q. A review on submarine oil and gas leakage in near field: Droplets and plume. *Environ. Sci. Pollut. Res.* **2022**, *29*, 8012–8025. [CrossRef] [PubMed]
2. Saadoun, I. Impact of Oil Spills on Marine Life. *Emerg. Pollut. Environ.* **2015**, *4*, 75–103.
3. Barron, M. Ecological impacts of the deepwater Horizon oil spill: Implications for immunotoxicity. *Toxicol. Pathol.* **2012**, *40*, 315–320. [CrossRef] [PubMed]
4. Dhaka, A.; Chattopadhyay, P. A review on physical remediation techniques for treatment of marine oil spills. *J. Environ. Manag.* **2021**, *288*, 112428. [CrossRef]
5. Melo, A.; Hoff, R.; Molognoni, L.; Oliveira, T.; Daguer, H.; Barreto, P. Disasters with oil spills in the oceans: Impacts on food safety and analytical control methods. *Food. Res. Int.* **2022**, *157*, 111366. [CrossRef] [PubMed]
6. Oliveira, L.; Saleem, J.; Bazargan, A.; Duarte, J.; Mckay, G.; Meili, L. Sorption as a rapidly response for oil spill accidents: A material and mechanistic approach. *J. Hazard. Mater.* **2021**, *407*, 124842. [CrossRef]
7. Zhang, T.; Li, Z.; Lü, Y.; Liu, Y.; Yang, D.; Li, Q.; Qiu, F. Recent progress and future prospects of oil-absorbing materials. *Chin. J. Chem. Eng.* **2019**, *27*, 1282–1295. [CrossRef]
8. Yang, W.; Yuen, A.; Li, A.; Lin, B.; Chen, T.; Yang, W.; Lu, H.; Yeoh, G. Recent progress in bio-based aerogel absorbents for oil/water separation. *J. Polym. Res.* **2019**, *26*, 6449–6476. [CrossRef]
9. Zhao, L.; Yang, S.; Yilihamu, A.; Wu, D. Advances in the applications of graphene adsorbents: From water treatment to soil remediation. *Rev. Inorg. Chem.* **2019**, *39*, 47–76. [CrossRef]
10. Bagoole, O.; Rahman, M.; Shah, S.; Hong, H.; Chen, H.; Ghaferi, A.; Younes, H. Functionalized three-dimensional graphene sponges for highly efficient crude and diesel oil adsorption. *Environ. Sci. Pollut. Res.* **2018**, *25*, 23091–23105. [CrossRef]
11. Abbas, Z.; Maimoona, M.; Nada, D. A simple method for developing efficient room temperature reduced graphene oxide-coated polyurethane sponge and cotton for oil-water separation. *Sep. Sci. Technol.* **2022**, *57*, 2596–2605.
12. Yang, S.; Li, J.; Zhen, C.; Li, F.; Sha, S.; Hou, C.; Lu, H.; Wu, J.; Sheng, Z.; Ma, J. Graphene-based melamine sponges with reverse wettability for oil/water separation through absorption and filtration. *J. Environ. Chem. Eng.* **2022**, *10*, 107543. [CrossRef]
13. Zhao, J.; Ren, W.; Cheng, H. Graphene sponge for efficient and repeatable adsorption and desorption of water contaminations. *J. Mater. Chem.* **2012**, *22*, 20197–20202. [CrossRef]
14. Bi, H.; Xie, X.; Yin, K.; Zhou, Y.; Wan, S.; He, L.; Xu, F.; Banhart, F.; Sun, L.; Ruoff, R. Spongy graphene as a highly efficient and recyclable sorbent for oils and organic solvents. *Adv. Funct. Mater.* **2012**, *22*, 4421–4425. [CrossRef]
15. Wu, R.; Yu, B.; Liu, X.; Li, H.; Wang, W.; Chen, L.; Bai, Y.; Ming, Z.; Yang, S. One-pot hydrothermal preparation of graphene sponge for the removal of oils and organic solvents. *Appl. Surf. Sci.* **2016**, *362*, 56–62. [CrossRef]
16. Wu, R.; Yu, B.; Jin, X.; Liu, X.; Bai, Y.; Chen, L.; Ming, Z.; Yang, H.; Yang, S.; Luo, J. Preparation of graphene sponge by vapor phase reduction for oil and organic solvent removal. *Mater. Res. Express.* **2016**, *3*, 105602. [CrossRef]
17. Yang, F.; Hao, L.; Zhu, Y.; Liu, H.; Chen, S.; Shao, Y. Preparation of graphene modified melamine sponge and solar-assisted cleanup of heavy oil spills. *J. Environ. Chem. Eng.* **2022**, *10*, 107779. [CrossRef]
18. Islam, M.S.; Zhang, Y.; McPhedran, K.N.; Liu, Y.; El-Din, M.G. Granular activated carbon for simultaneous adsorption and biodegradation of toxic oil sands process-affected water organic compounds. *J. Environ. Manag.* **2015**, *152*, 49–57. [CrossRef]
19. Islam, M.S.; McPhedran, K.N.; Messele, S.A.; Liu, Y.; El-Din, M.G. Isotherm and kinetic studies on adsorption of oil sands process-affected water organic compounds using granular activated carbon. *Chemosphere* **2018**, *202*, 716–725. [CrossRef]

20. Madhubashani, A.; Giannakoudakis, D.; Amarasinghe, B.; Rajapaksha, A.; Kumara, P.P.; Triantafyllidis, K.; Vithanage, M. Propensity and appraisal of biochar performance in removal of oil spills: A comprehensive review. *Environ. Pollut.* **2021**, *288*, 117676. [CrossRef]
21. Zhang, L.; Liu, Z.; Cui, G.; Chen, L. Biomass-derived materials for electrochemical energy storages. *Prog. Polym. Sci.* **2015**, *43*, 136–164. [CrossRef]
22. Yue, X.; Zhang, T.; Yang, D.; Qiu, F.; Li, Z. Hybrid aerogels derived from banana peel and waste paper for efficient oil absorption and emulsion separation. *J. Clean. Prod.* **2018**, *199*, 411–419. [CrossRef]
23. Liu, H.; Li, A.; Liu, Z.; Tao, Q.; Li, J.; Peng, J.; Liu, Y. Preparation of lightweight and hydrophobic natural biomass-based carbon aerogels for adsorption oils and organic solvents. *J. Porous. Mat.* **2022**, *29*, 1001–1009. [CrossRef]
24. Dan, H.; Ji, K.; Gao, Y.; Yin, W.; Gao, B.; Yue, Q. Fabrication of superhydrophobic Enteromorpha-derived carbon aerogels via $NH_4H_2PO_4$ modification for multi-behavioral oil/water separation. *Sci. Total. Environ.* **2022**, *837*, 155869. [CrossRef]
25. Li, Z.; Lei, S.; Xi, J.; Ye, D.; Hu, W.; Song, L.; Hu, Y.; Cai, W.; Gui, Z. Bio-based multifunctional carbon aerogels from sugarcane residue for organic solvents adsorption and solar-thermal-driven oil removal. *Chem. Eng. J.* **2021**, *426*, 129580. [CrossRef]
26. Yang, K.; Ren, J.; Cui, Y.; Wang, Y.; Shah, T.; Zhang, Q.; Zhang, B. Fabrication of porous tubular carbon fibers from the fruits of *Platanus orientalis* and their high oil adsorption properties. *J. Environ. Chem. Eng.* **2021**, *9*, 105706. [CrossRef]
27. Yi, L.; Xia, Y.; Tan, Z.; Fang, X.; Zhao, L.; Wu, H.; Guo, S. Design of tubelike aerogels with macropores from bamboo fungus for fast oil/water separation. *J. Clean. Prod.* **2020**, *264*, 121558. [CrossRef]
28. Zhang, Q.; Shi, R.; Li, Q.; Maimaiti, T.; Lan, S.; Ouyang, P.; Ouyang, B.; Bai, Y.; Yu, B.; Yang, S. Low toxicity of fluorescent carbon quantum dots to white rot fungus *Phanerochaete chrysosporium*. *J. Environ. Chem. Eng.* **2021**, *9*, 104633. [CrossRef]
29. Zhuo, R.; Fan, F. A comprehensive insight into the application of white rot fungi and their lignocellulolytic enzymes in the removal of organic pollutants. *Sci. Total. Environ.* **2021**, *778*, 146132. [CrossRef]
30. Omoni, V.; Lag-Brotons, A.; Ibeto, C.; Semple, K. Effects of biological pre-treatment of lignocellulosic waste with white-rot fungi on the stimulation of ^{14}C-phenanthrene catabolism in soils. *Int. Biodeter. Biodegr.* **2021**, *165*, 105324. [CrossRef]
31. Huang, D.; Zeng, G.; Feng, C.; Hu, S.; Zhao, M.; Lai, C.; Zhang, Y.; Jiang, X.; Jiang, X.; Liu, H. Mycelial growth and solid-state fermentation of lignocellulosic waste by white-rot fungus *Phanerochaete chrysosporium* under lead stress. *Chemosphere* **2010**, *81*, 1091–1097. [CrossRef] [PubMed]
32. Wu, R.; Yu, B.; Liu, X.; Li, H.; Bai, Y.; Ming, Z.; Chen, L.; Yang, S.; Chang, X. Graphene/polyester staple composite for the removal of oils and organic solvents. *Mater. Res. Express* **2016**, *3*, 065601. [CrossRef]
33. Chen, N.; Pan, Q. Versatile Fabrication of Ultralight Magnetic Foams and Application for Oil-Water Separation. *ACS Nano* **2013**, *7*, 6875–6883. [CrossRef] [PubMed]
34. Sun, X.; Bao, H.; Liu, W.; Luo, C.; Li, Y.; Lu, J. Development of a new hydrophobic magnetic biochar for removing oil spills on the water surface. *Microb. Cell Factories* **2022**, *4*, 1–17. [CrossRef]
35. Gurav, R.; Bhatia, S.; Choi, T.; Choi, Y.; Kim, H.; Song, H.; Park, S.; Lee, H.; Lee, S.; Choi, K.; et al. Adsorptive removal of crude petroleum oil from water using floating pinewood biochar decorated with coconut oil-derived fatty acids. *Sci. Total Environ.* **2021**, *781*, 146636. [CrossRef]

materials

Article

Study of Electric and Magnetic Properties of Iron-Modified MFI Zeolite Prepared by a Mechanochemical Method

Fabian N. Murrieta-Rico [1,*], Joel Antúnez-García [2], Rosario I. Yocupicio-Gaxiola [2], Jonathan Zamora [3], Armando Reyes-Serrato [2], Alexey Pestryakov [4,5,*] and Vitalii Petranovskii [2]

1 Ingeniería Mecatrónica, Universidad Politécnica de Baja California, Mexicali 21376, Mexico
2 Centro de Nanociencias y Nanotecnología, Universidad Nacional Autónoma de México, Ensenada 22800, Mexico
3 Departamento de Ingeniería Metalúrgica, Facultad de Química, Universidad Nacional Autónoma de México, Ciudad de México 04510, Mexico
4 Research School of Chemistry and Applied Biomedical Sciences, Tomsk Polytechnic University, Tomsk 634050, Russia
5 Laboratory of Catalytic and Biomedical Technologies, Sevastopol State University, Sevastopol 299053, Russia
* Correspondence: fnmurrietar@upbc.edu.mx (F.N.M.-R.); pestryakov@tpu.ru (A.P.); Tel.: +52-686-104-2727 (F.N.M.-R.)

Abstract: Zeolites are materials of undeniable importance for science and technology. Since the properties of zeolites can be tuned after the inclusion of additional chemical species into the zeolitic framework, it is necessary to study the nature of zeolites after modification with transition metals to understand the new properties that were obtained, and with this information, novel applications can be proposed. This paper reports a solvent-free approach for the rapid synthesis of zeolites modified with iron and/or iron oxide particles. The samples were characterized, and their electrical and magnetic properties were investigated.

Keywords: zeolite; MFI; iron; mechanochemistry; impedance spectroscopy

check for updates

Citation: Murrieta-Rico, F.N.; Antúnez-García, J.; Yocupicio-Gaxiola, R.I.; Zamora, J.; Reyes-Serrato, A.; Pestryakov, A.; Petranovskii, V. Study of Electric and Magnetic Properties of Iron-Modified MFI Zeolite Prepared by a Mechanochemical Method. *Materials* **2022**, *15*, 7968. https://doi.org/10.3390/ma15227968

Academic Editors: Zhifeng Wang, Weiqing Zhang and Yichao Wang

Received: 23 September 2022
Accepted: 7 November 2022
Published: 11 November 2022

1. Introduction

Zeolites are materials that have plenty of applications in diverse areas of science and technology [1]. Among other things, zeolites are used as adsorbents [2,3], ion exchangers [4], and catalyst supports [5,6]. Moreover, they can serve as sensitive materials for sensors [7]. By composition, zeolites are porous crystalline aluminosilicates that have interconnected channels. To date, there are more than 240 topologically different zeolite structures [8], each of which has a specific distribution of nanometric channels and cavities. The characteristic properties of zeolites make it easy to modify their chemical composition. Ion-exchange cations, located in the voids of their structure to neutralize the excess negative charge of the framework created by Al atoms, are easily exchanged for others. The isomorphic substitution of tetrahedrally coordinated atoms of Al and/or Si in the crystal structure also occurs rather easily [9,10]. Among various zeolite modifiers, iron attracts special attention [11]. Due to the inherent properties of iron and its variable valency, it can play the role of both an ion-exchange cation and/or replace atoms of the framework. In addition, iron easily forms oxide nanoparticles of various compositions on the surface of zeolites [12]. Among all known zeolitic structures, the MFI framework is used as an adsorbent or in the development of catalysts [13], which grants MFI zeolite an important place for research and industrial applications. Accordingly, it is desirable to understand the effects associated with the modification of MFI zeolite with metallic species such as iron.

Fe-containing zeolites are widely used in many catalytic applications, both because of the zeolite properties and due to the redox properties provided by the introduction of Fe compounds. It is noteworthy that after introducing Fe into the zeolite framework, the poor properties of conventional aluminosilicate zeolites, such as lack of active sites, low catalytic activity, poor anti-coking properties, etc., can be effectively improved [11]. The state of Fe in the zeolite, which significantly affects the properties of the obtained materials (stability, acidity, redox properties, etc.) of the Fe-zeolite, significantly depends on the method of introduction. The review [11] briefly outlines the latest developments in the field of synthesis, properties, and applications of Fe-based zeolites. Different species of iron in zeolites are attractive due to the variety of structures and functions, the relative ease of synthesis for a number of methods, economic efficiency due to the wide distribution of Fe in nature and its low cost, environmental friendliness, biocompatibility, and synergistic interaction with many other materials. The magnetic properties of some iron oxide nanoparticles make them much easier to clean and handle. These features make them promising catalysts for various chemical reactions [14].

Iron forms compound in several oxidation states. Some of the iron oxides and hydroxides, especially in the nanometer size range, are of great interest as heterogeneous catalysts, in particular in environmental remediation procedures using an advanced oxidation process that allows the efficient formation of reactive species such as hydroxyl radicals even at room temperature and at atmospheric pressure [13,15].

Catalysts play a crucial role in SCR technology, making it necessary to modify existing catalysts or develop new materials. A wide range of active temperatures from 150 to 650 °C is required, as well as exceptional N_2 selectivity. The selective catalytic reduction of NO_x with NH_3 over zeolite catalysts is recognized as the most practical techniques used in mobile vehicle exhaust to convert NO_x to N_2 [16]. It should be noted that Fe-zeolite is being considered for commercial applications due to its advantages in fast SCR reactions. In addition, Fe-zeolite has been shown to provide higher NO_x conversion under real transient conditions [16].

Heinrich et al. [17] proposed a single-stage method for the preparation of Fe-containing zeolite with maximum ecological purity since the mechanochemical reaction was carried out in the absence of solvents. In contrast to the mentioned work, in the present study, we used not the protonated zeolite H-ZSM5 but its ammonium form NH_4-ZSM5. The aim of our work was to confirm the possibility of the mechanochemical incorporation of iron into the composition of a zeolite sample and to characterize the material obtained by a number of physicochemical methods.

2. Materials and Methods

Zeolite MFI with a nominal SiO_2/Al_2O_3 mole ratio of 50 in ammonium cation form (product CBV 5524G) and $FeCl_3$ were purchased from Zeolyst International (KS, USA) and Sigma-Aldrich (MO, USA), respectively. According to the method of Heinrich et al. [17], the MFI zeolite was intensely ground in a mortar together with $FeCl_3$ in a weight ratio of 2:1. The mixture was then placed in a beaker and stirred in water for 30 min; in this case, the excess of $FeCl_3$ was dissolved. The solid precipitate was filtered off and washed with 500 mL of deionized water. The crystals were dried in an oven at 100 °C for 1 h. The chemical composition of the samples was evaluated by inductively coupled plasma-optical emission spectroscopy (ICP-OES) using a Vista-MPX CCD (CO, USA) simultaneous ICP-OES (Varian) (CO, USA). The samples were characterized by X-ray diffraction (XRD) using Aeris Panalytical (Malvern, UK) equipment with Cu K alpha monochromatic radiation (λ = 0.154056 nm, 40 kV, 15 mA) and UV-Vis spectroscopy using a UV-Vis NIR Cary 5000 (CA, USA) spectrophotometer. Micrographs of the samples were obtained using scanning electron microscopy (SEM) on JEOL JIB-4500 (MA, USA) equipment.

The textural properties of materials were determined by means of the nitrogen adsorption-desorption isotherms at −196 °C with ASAP 2000 equipment from Micromeritics (GA, USA). The samples were previously degassed at 300 °C for 4 h under vacuum before nitrogen adsorption, and two samples were pretreated in N_2 flow at 160 °C and 300 °C. Surface area was measured with the Brunauer–Emmett–Teller (BET) method at relative pressures of $0.05 < P/P0 < 0.30$. Average pore diameter was calculated following the Barrett–Joyner–Halenda method (BJH) using the desorption branch of the N_2 isotherm. The cumulative pore volume (V_P) was obtained according to the amount adsorbed at a relative pressure of P/P0 = 0.99. The micropore surface area (S_{micro}) and the micropore volume (V_{micro}) were obtained with the t-plot method.

The XPS spectra of the samples were recorded using a SPECS® spectrometer (Berlin, Germany) with a PHOIBOS® 150 WAL hemispherical energy analyzer with angular resolution (<0.5°), equipped with an XR 50 X-Ray Al/Mg-X-ray anode and μ-FOCUS 500 X-ray monochromator (Al excitation line) sources.

In order to obtain electrochemical impedance spectroscopy (EIS) data, each sample was ground and compressed into a pellet with a diameter of 1 cm and thickness of 1 mm; the pellet was then placed between two polished copper electrodes, which were connected to E4980A Precision LCR Meter. As a result, for each sample, a data set was obtained corresponding to the magnitude of the total impedance $|Z_T|$ in ohms, phase angle θ in degrees, and frequency of interrogation signal in hertz. Magnetic measurements were performed using a Vibrating Sample Magnetometer (VSM) Squid Magnetometer Quantum Design MPMS® 3 system (LA, USA) with a maximum applied field of 2.0 T at room temperature.

3. Results and Discussion

3.1. X-ray Diffraction

The experimental diffraction patterns obtained for the MFI and MFI-Fe samples are shown in Figure 1. For comparison, an XRD pattern obtained from the IZA database [8] corresponding to the standard MFI zeolite is shown. As can be observed, the experimental diffraction patterns of the original and iron-modified samples reveal the presence of characteristics peaks, corresponding to the reference data: although this is expected for a commercial zeolite (MFI), after modification with iron, this indicates the preservation of the topology of the zeolite structure. However, after the inclusion of iron in the composition of the sample, a slight shift to higher angles at the peak position can be observed. For example, the c lattice parameter undergoes a contraction from 20.15 Å to 20.05 Å with the inclusion of iron. This is consistent with the idea of a slight replacement of Al for Fe in the zeolitic structure, possible as an isomorphous substitution, due to the fact that Fe-O and Si-O bond lengths both are shorter than Al-O [18–20]. Therefore, its inclusion contributes to the contraction of the unit cell. On the other hand, we observe that the original zeolitic structure is retained after the mechanochemical modification with Fe. This fact is important because it is known that for some not-very-stable zeolites, such as LTA, as well as for too long mechanochemical treatment of other, stronger frameworks, the crystal structure of zeolites can be amorphized and destroyed.

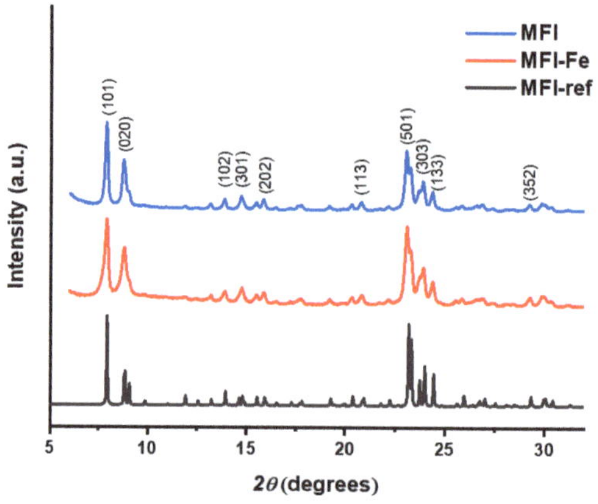

Figure 1. Diffractograms of MFI zeolite, iron-modified zeolite (MFI-Fe), and reference ZSM-5 sample from the IZA database [8] (MFI-ref).

For both samples, the degree of crystallinity or peak-to-noise ratio from XRD is presented in Table 1. In particular, the crystallinity is calculated from XRD by multiplying by 100 the division of area corresponding to crystalline peaks by the area of all peaks. In addition, from the Scherrer Equation (1), the average crystallites size $\overline{D}$ is calculated.

$$D = \frac{K\lambda}{\beta \cos\theta} \tag{1}$$

where D is the crystallite size in nm, $K = 0.9$ is the Scherrer constant, β is the line broadening at half the maximum intensity (FWHM), and $\lambda = 0.15406$ is the wavelength in nm of the X-ray source.

Table 1. Analysis of average crystallite size and crystallinity of MFI and MFI-Fe samples.

Sample	Crystallinity	$\overline{D}$ (nm)
MFI	93.55%	19
MFI-Fe	98.12%	15

3.2. Scanning Electron Microscopy and Energy Dispersive Spectroscopy

The morphology of MFI and MFI-Fe samples is presented in Figure 2. According to SEM micrographs, it can be noted that at a resolution of 5 μm, both samples, zeolite prior (Figure 2a) and after modification with iron chloride (Figure 2c), presents a quite similar grain morphology, but there are zones (see yellow circles in Figure 2a) that present a higher agglomeration on the MFI zeolite. Although in Figure 2c grain agglomeration is observed, agglomerated zones in Figure 2a could be attributed to large grains with small particles attached to them, while in Figure 2c, most of the grains are clearly defined. After a higher magnification, a similar level of homogeneity is observed for both samples (Figure 2b,c). In general, it can be noted from SEM micrographs that the average particle size slightly decreased as a result of mechanochemical modification using $FeCl_3$.

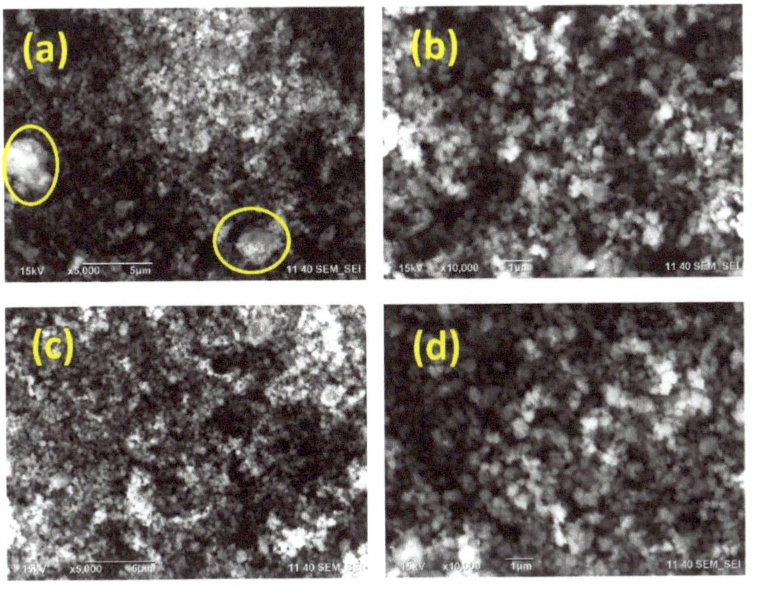

Figure 2. Micrographs of zeolites: MFI with a resolution of 5 μm (**a**) and 1 μm (**b**); and MFI-Fe with a resolution of 5 μm (**c**) and 1 μm (**d**).

The chemical composition of both samples was evaluated from EDS data. The normalized spectra of both samples are shown in Figure 3, which demonstrate the presence of elements typical for zeolites, such as Si, O, and Al. However, there are also substantive differences. Although Na is not present in the original sample, trace amounts of Na appear in the MFI-Fe sample after the mechanochemical process. It is assumed that this is a consequence of the presence of Na in the mortar material, and as a result of the grinding process, sodium could be integrated into the resulting material. Since EDS is a surface method of analysis, the appearance of a larger relative amount of Al on the surface, where characteristic X-rays are generated, could be due to the appearance of surface aluminum in an environment other than tetrahedral, namely, surface Al in an octahedral arrangement due to dealumination at the moment of crystal destruction. Although for MFI-Fe, the Fe peaks should appear at 0.705 and 6.398 eV for the Lα and Kα lines, respectively, its apparent absence could be attributed to low iron content. In general, terms, taking into account that the EDS data are generated by the characteristic X-ray radiation, which is emitted after atoms under study are impacted by electrons at a given penetration depth, from Figure 3, we can conclude that, after mechanochemical modification, the oxygen atoms of the MFI zeolite framework are rearranged and, as a consequence, the Al atoms are more exposed. This is caused by the level of particle agglomeration, as well as a decrease in the average crystallite size. In addition, since the O emission is increased, it could be concluded that oxides are present in greater amounts in the MFI-Fe sample. In other words, after the mechanochemical process, amorphization of a part of Al occurs, which is deposited on the surface. This raises the level of Al and O signals.

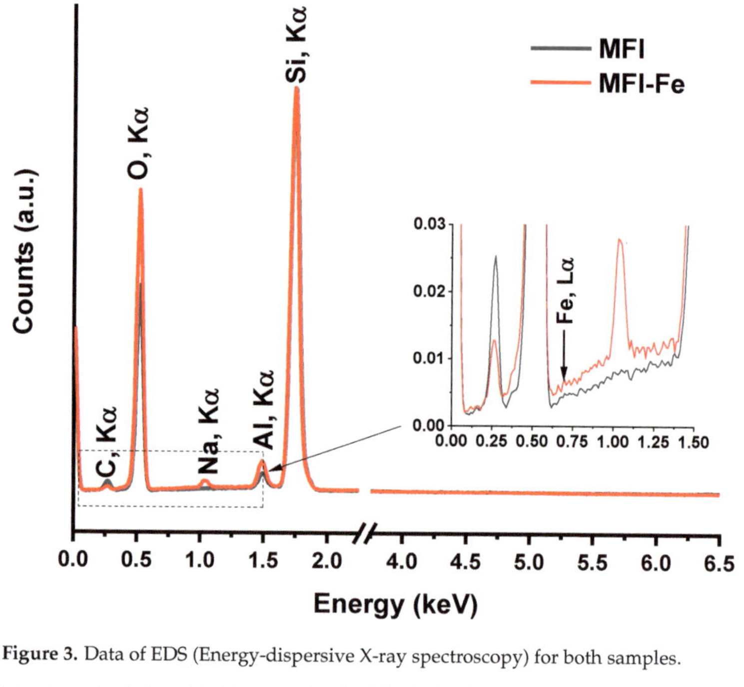

Figure 3. Data of EDS (Energy-dispersive X-ray spectroscopy) for both samples.

3.3. Inductively Coupled Plasma—Optical Emission Spectrometry

The results of the ICP-MS analysis are reported in Table 2. The SiO_2/Al_2O_3 molar ratio was slightly above that specified by the supplier (SiO_2/Al_2O_3). In passing, we note that the commercial sample of MFI contains an impurity of iron.

Table 2. Elemental composition data from ICP-OES for MFI and MFI-Fe zeolites.

Sample	Al_2O_3 % Mole	SiO_2 % Mole	Fe_2O_3 % Mole	NH_4 % Mole	SiO_2/Al_2O_3 Mole Ratio	SiO_2/Fe_2O_3 Mole Ratio	$SiO_2/(Al_2O_3 + Fe_2O_3)$ Mole Ratio
MFI	1.52	95.34	0.09	3.05	62.54	1045.49	59.01
MFI-Fe	1.55	94.67	0.70	3.09	61.25	136.02	42.23

We cannot enter into a discussion of the reasons for such contamination; however, very often, the reagents used for synthesis are contaminated with traces of iron, which can change the composition of the final material. After our mechanochemical treatment by grinding in a mortar with ferric chloride, an increase in the concentration of iron by order of magnitude is observed. Moreover, a slight decrease in the amount of silicon in the material after the iron introduction was evident.

3.4. BET Analysis

After BET analysis, N_2 adsorption isotherms were obtained (Figure 4). As is observed similar behavior is observed for both samples, which indicates that the main textural properties of the material are retained after modification. However, there is a difference in the absorbed quantity, which can be attributed to the difference in surface area as well as pore volume.

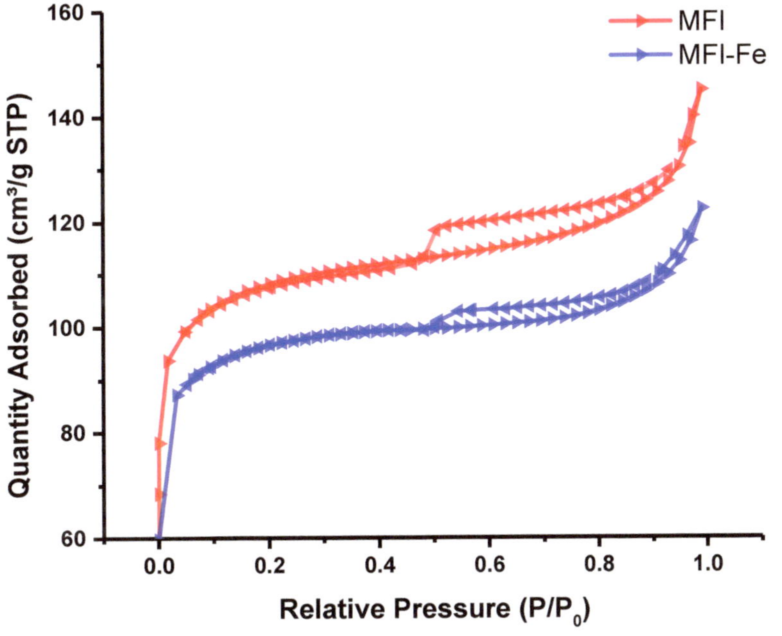

Figure 4. N_2 adsorption-desorption isotherms for MFI and MFI-Fe zeolites.

As can be seen from Table 3, the surface area of the MFI sample is larger than that of MFI-Fe, which indicates that there is a decrement in surface area after modification and, as a result, a lower adsorption capacity is observed. Thus, these measurements confirm the data of XRD analysis that the porous structure of the MPI zeolite is generally preserved during mechanochemical activation, and the observed decrease in the available pore volume is associated with the formation of nanoparticles of iron-containing compounds.

Table 3. Pore textural properties of MFI and MFI-Fe zeolites.

Sample	V_p (cm^3/g)	S_{BET} (m^2/g)
MFI	0.14	322
MFI-Fe	0.11	286

3.5. XPS Analysis

Finally, the XPS spectra for the MFI and MFI-Fe samples were obtained and analyzed. Previously, when analyzing the XRD data, it was noted that the experimental diffraction patterns indicate the preservation of the topology of the zeolite structure. However, a slight shift in the position of the peaks towards larger angles gave grounds to assume that there was a partial incorporation of Fe into the crystalline zeolite structure, which led to the compression of the unit cell. Analysis of the SEM data (Figure 2) showed that both samples have very similar grain morphology. The conducted EDS analysis did not reveal Fe peaks, which should appear at 0.705 and 6.398 eV for the Lα and Kα lines, respectively. Since the elemental composition data obtained by the ICP-OES method showed the presence of iron, the EDS results were explained by the fact that this is a surface analysis technique, in contrast to the ICP-OES method. The N_2 adsorption-desorption isotherms for the MFI and MFI-Fe zeolites confirm the data of X-ray diffraction analysis that the porous structure of the MFI zeolite is generally preserved upon mechanochemical activation. The decrease in the available pore volume observed, in this case, is due to the fact that during mechanochemical treatment, nanoparticles of iron-containing compounds are formed in the voids of zeolite crystals due to ions diffusing into the channels of the zeolite structure.

Figure 5 shows high-resolution XPS spectra taken for elements such as Al, Si, and Fe. These data fully confirm the sum of observations obtained by other methods. The positions of the Si and Al peaks (103.01 and 74.47 eV, respectively) do not change since the structure of the MFI zeolite is preserved. There is no Fe peak in the spectra, which once again confirms the main result of this work. During mechanochemical activation, there is no "spreading" of iron compounds over the surface of zeolite crystals. Iron ions, on the contrary, penetrate into their volume, where some of them are introduced into the crystal structure, as evidenced by XRD data, and some are hydrolyzed with the formation of oxide-hydroxide nanoparticles in the channels, which is confirmed by adsorption data. These iron compounds are responsible for the magnetic properties of the samples.

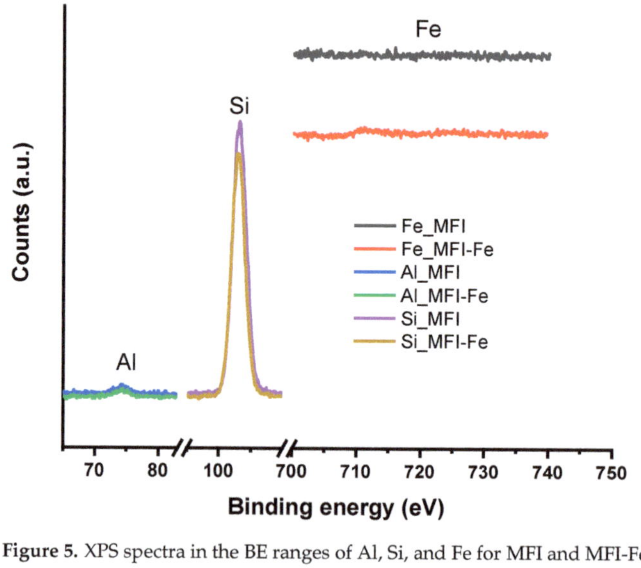

Figure 5. XPS spectra in the BE ranges of Al, Si, and Fe for MFI and MFI-Fe zeolites.

3.6. Band-Gap Analysis

The Band gap of zeolitic nanoparticles was obtained using the UV–Vis spectra of the samples under study in absorbance mode and following the Tauc model:

$$(ahv)^{1/n} = A(hv - E_g) \quad (2)$$

where h is Planck's constant, v is the frequency of the photon, α is the adsorption coefficient, E_g is the band gap, and A is the slope of the Tauc graph in the linear region. The data of linear fitting are presented in Figure 6. The band gap of the MFI zeolite is modified after the inclusion of Fe. In fact, the band gap of the MFI-Fe sample lies between the band gaps of MFI and $FeCl_3$.

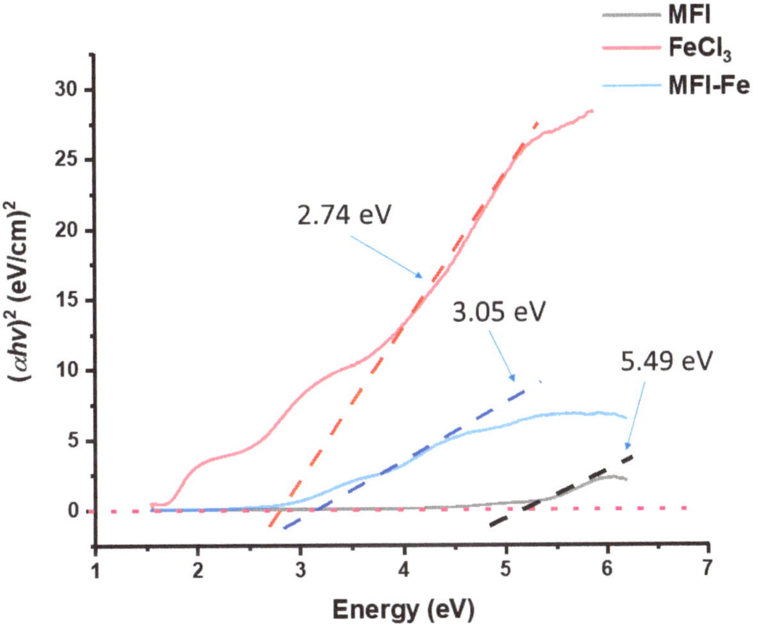

Figure 6. Tauc plot for band-gap characterization of pure MFI zeolite, iron-chloride, and iron-modified MFI zeolite.

3.7. Electrochemical Impedance Spectroscopy

It can be seen from the EIS data that MFI and MFI-Fe have almost the same frequency response (Figure 7a,b); at low frequencies (Figure 7c), the DC conductivity σ_{dc} is higher for the MFI-Fe sample; in contrast, at high-frequency values, the AC conductivity σ_{ac} is greater for the MFI sample. These plots show how the conductivity of the MFI zeolite can be tuned after the addition of Fe, which is observed in the Argand diagram (Figure 7d). According to Jonscher's power law for, the universal behavior of conductivity is given by

$$\sigma_{ac} = \sigma_{dc} + A_j\omega^n, \quad (3)$$

where A_j is the pre-exponential factor, ω is the angular frequency in rad/s, and n is the frequency exponent. Using the data in Figure 7c, the parameters of Equation (3) (σ_{dc}, A_j, n) can be fitted, and the parameters of the elements in Joshcher's power law are found for each sample.

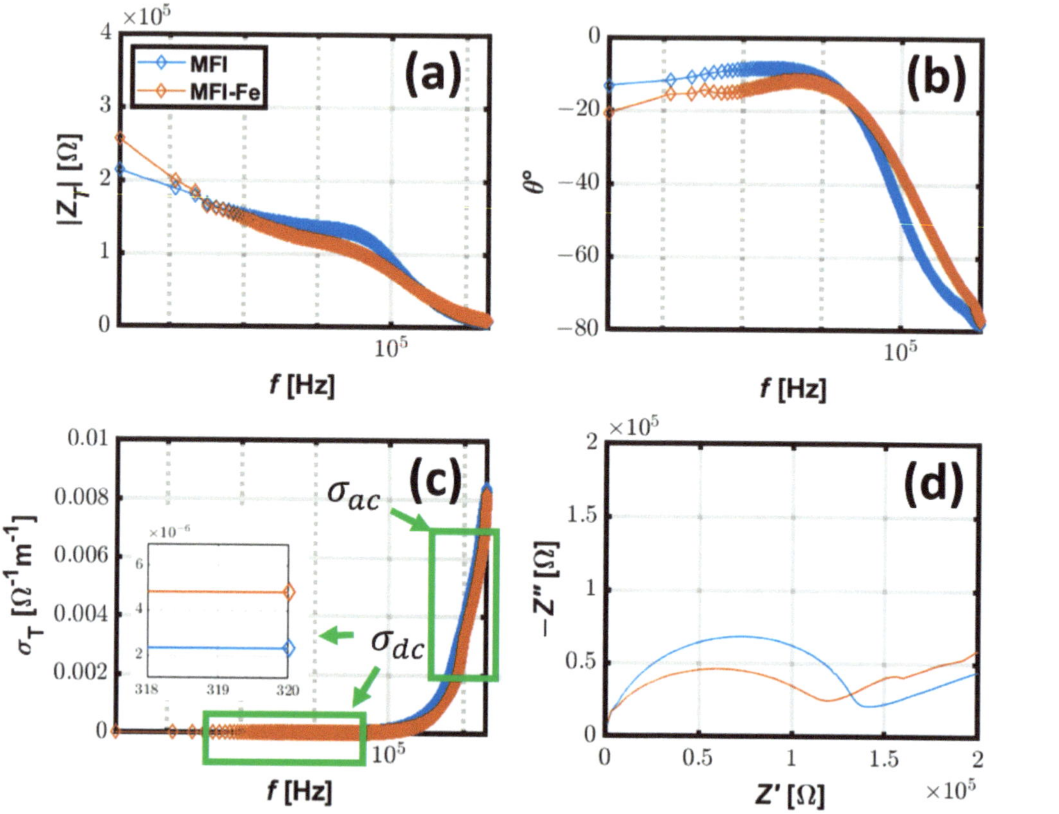

Figure 7. The magnitude of total impedance $|Z_T|$ (**a**); phase angle θ (**b**); total conductivity σ_T (**c**), and Argand diagram of the real part of impedance Z' vs. imaginary part of impedance Z'' (**d**).

The behavior of the total electrical conductivity σ_T is quite similar, this is illustrated by the parameters of Joshcher's power law (Table 4). In particular, the value of n for both samples is equal to 1, which is typical for ionic compounds. In addition, the value of n defines the transition from DC to AC conductivity, which in this case, is observed to be the same. The differences in σ_T can be attributed to σ_{ac} and how the frequency ω is escalated by A_j for each sample. The Argand diagram shows the relationship between the real and imaginary parts of the impedance. Given that

$$Z_T = Z' + jZ' \tag{4}$$

and

$$Z' = |Z_T| \cos\theta, \tag{5}$$

$$Z'' = |Z_T| \sin\theta, \tag{6}$$

Table 4. Values obtained from fitting Jonscher's power law.

Sample	σ_{dc}	A_j	n
MFI	1.5×10^{-6}	3.86×10^{-9}	1
MFI-Fe	2.5×10^{-6}	3.37×10^{-9}	1

It can be noted for both samples that at low frequencies (Figure 7d), the value of Z' is greater, there is a positive slope, and after a local minimum, there is an increment describing a semicircle. In the case of MFI, there is an absolute maximum, and in the case of MFI-Fe, there is a local maximum. Both samples have a decrement with a final value at the origin. Since Equation (4) apparently defines total impedance in terms of a resistive Z' and reactive Z'' parts, it could be considered that in the Argand diagram, the reactive effects have a greater influence in the MFI sample at higher frequencies. Alike in MFI zeolite, after modification with Fe, MFI-Fe shows more reactive effects at low frequencies. These observations can be attributed to a similar ionic conductivity mechanism but with a different contribution of each element involved in such a process.

3.8. Magnetic Characterization

Figure 8 shows the hysteresis loops at 300 K for MFI and MFI-Fe samples. For the MFI sample in Figure 8a, an anti-S-type hysteresis curve is observed, which indicates a principal diamagnetic behavior caused by the matrix and the sample holder. The inset shows the ferromagnetic-like behavior after subtracting the diamagnetic component. Meanwhile, the Fe-modified MFI sample, see Figure 8b, exhibits an S-type hysteresis curve, i.e., presents weak ferromagnetic (FM) behavior with a coercivity around 600 Oe and very small magnetization values without reaching saturation, according to the response and magnetic behavior for this type of materials [21]. This FM behavior can be associated with the presence of oxy-hydroxides; however, it can also be related to an existing ferrimagnetic order. However, both samples exhibit small magnetization values, which may be due to the influence of preferred orientation in a single axis or the presence of a small phase fraction causing a magnetic dilution effect leading to the rather low magnetization, which is also consistent with the ICP analysis.

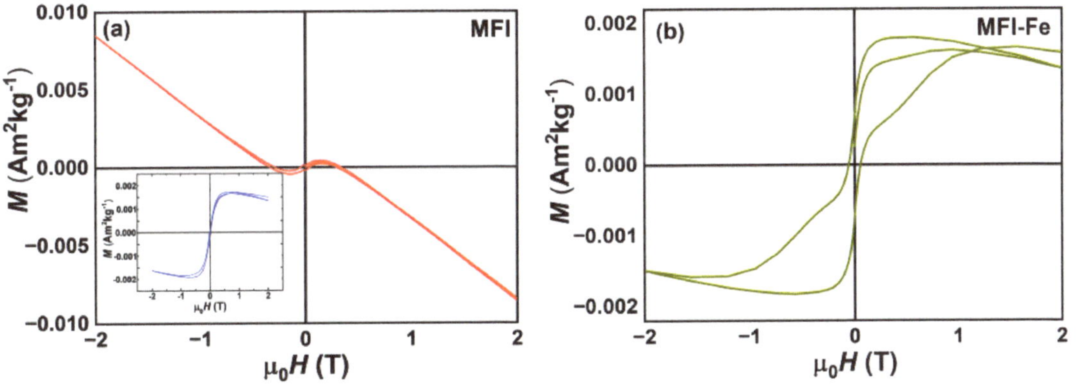

Figure 8. Hysteresis loops of both zeolites at room temperature; (**a**) MFI, inset: corrected for diamagnetic background, and (**b**) Fe-modified MFI sample.

4. Conclusions

This article reports a solvent-free approach to the rapid one-step synthesis of zeolites modified with iron and/or iron oxide particles. The modification of the MFI zeolite with iron was carried out by the mechanochemical method. As a result, it was observed that the zeolite, after modification, retains its original crystalline structure; however, there is a variation in physicochemical properties, such as the band gap. Moreover, as a result of the introduction of Fe into the composition of the sample, a weak ferromagnetic behavior was observed. This effect causes a change in the electrical properties of the zeolite, which is observed as a decrease in zeolite reactance.

Author Contributions: Conceptualization, F.N.M.-R., V.P. and J.A.-G.; methodology, F.N.M.-R.; validation, V.P. and R.I.Y.-G.; formal analysis, F.N.M.-R., V.P., J.A.-G. and J.Z.; investigation, F.N.M.-R., V.P., J.A.-G., R.I.Y.-G., J.Z., A.R.-S. and A.P.; writing—original draft preparation, F.N.M.-R., V.P., J.A.-G., R.I.Y.-G., J.Z., A.R.-S. and A.P.; writing—review and editing, F.N.M.-R., V.P., J.A.-G., R.I.Y.-G., J.Z., A.R.-S. and A.P.; funding acquisition, V.P., A.R.-S. and A.P. All authors have read and agreed to the published version of the manuscript.

Funding: This research was funded by CONACYT, grant number "Basic Science Project A1-S-33492", and the APC was funded by MDPI, Sevastopol State University Research grant 42-01-09/169/2021-4.

Data Availability Statement: Not applicable.

Acknowledgments: The authors would like to thank Jorge Barreto and Arturo Martínez (IF-UNAM), and David Dominguez (CNYN-UNAM) for their valuable technical assistance.

Conflicts of Interest: The authors declare no conflict of interest.

References

1. Derbe, T.; Temesgen, S.; Bitew, M.A. Short Review on Synthesis, Characterization, and Applications of Zeolites. *Adv. Mater. Sci. Eng.* **2021**, *2021*, 6637898. [CrossRef]
2. Yue, B.; Liu, S.S.; Chai, Y.C.; Wu, G.J.; Guan, N.J.; Li, L.D. Zeolites for separation: Fundamental and application. *J. Energy Chem.* **2022**, *71*, 288–303. [CrossRef]
3. Villa, C.C.; Ortega-Toro, R.; Ahmed, S.; Gutierrez, T.J.; Valencia, G.A.; Cordoba, A.L. Zeolites for food applications: A review. *Food Biosci.* **2022**, *46*, 101577. [CrossRef]
4. Li, Y.; Yu, J.H. Emerging applications of zeolites in catalysis, separation and host-guest assembly. *Nat. Rev. Mater.* **2021**, *6*, 1156–1174. [CrossRef]
5. Sun, Q.M.; Wang, N.; Yu, J.H. Advances in Catalytic Applications of Zeolite-Supported Metal Catalysts. *Adv. Mater.* **2021**, *33*, 2104442. [CrossRef] [PubMed]
6. Van der Mynsbrugge, J.; Bell, A.T. Challenges for the theoretical description of the mechanism and kinetics of reactions catalyzed by zeolites. *J. Catal.* **2021**, *404*, 832–849. [CrossRef]
7. Murrieta-Rico, F.N.; Petranovskii, V.; Galván, D.H.; Antúnez-García, J.; Sergiyenko, O.; Lindner, L.; Rivas-Lopez, M.; Grishin, M.; Sarvadii, S. Basic Aspects in the Application of QCMs as Sensors: A Tutorial. *IEEE Sens. J.* **2022**, *22*, 10163–10172. [CrossRef]
8. Database of Zeolite Structures. Available online: http://www.izastructure.org/databases (accessed on 31 August 2021).
9. Lei, Q.F.; Wang, C.; Dai, W.L.; Wu, G.J.; Guan, N.J.; Li, L.D. Multifunctional heteroatom zeolites: Construction and applications. *Front. Chem. Sci. Eng.* **2021**, *15*, 1462–1486. [CrossRef]
10. Yabushita, M.; Osuga, R.; Muramatsu, A. Control of location and distribution of heteroatoms substituted isomorphously in framework of zeolites and zeotype materials. *CrystEngComm* **2021**, *23*, 6226–6233. [CrossRef]
11. Zhang, J.; Tang, X.; Yi, H.; Yu, Q.; Zhang, Y.; Wei, J.; Yuan, Y. Synthesis, characterization and application of Fe-zeolite: A review. *Appl. Catal. A Gen.* **2022**, *630*, 118467. [CrossRef]
12. Loiola, A.R.; Bessa, R.A.; Oliveira, C.P.; Freitas, A.D.; Soares, S.A.; Bohn, F.; Pergher, S.B. Magnetic zeolite composites: Classification, synthesis routes, and technological applications. *J. Magn. Magn. Mater.* **2022**, *560*, 169651. [CrossRef]
13. Oliveira Guidolin, T.D.; Cechinel, M.A.P.; Arcaro, S. Iron-Based Nanomaterials for Fenton Reaction. In *Environmental Applications of Nanomaterials*; Springer: Berlin/Heidelberg, Germany, 2022; pp. 133–152.
14. Parthasarathy, M.; Mohan, C.N. Catalytic applications of iron oxide nanoparticles. In *Iron Oxide Nanoparticles and Their Applications*; Nova Science Publishers: Hauppauge, NY, USA, 2021; pp. 257–287.
15. Pereira, M.C.; Oliveira, L.C.A.; Murad, E. Iron oxide catalysts: Fenton and Fenton-like reactions—A review. *Clay Miner.* **2012**, *47*, 285–302. [CrossRef]
16. Liu, Q.; Bian, C.; Ming, S.; Guo, L.; Zhang, S.; Pang, L.; Liu, P.; Chen, Z.; Li, T. The opportunities and challenges of iron-zeolite as NH_3-SCR catalyst in purification of vehicle exhaust. *Appl. Catal. A Gen.* **2020**, *607*, 117865. [CrossRef]
17. Heinrich, F.; Schmidt, C.; Löffler, E.; Grünert, W. A highly active intra-zeolite iron site for the selective catalytic reduction of NO by isobutane. *Catal. Commun.* **2001**, *2*, 317–321. [CrossRef]
18. Yocupicio-Gaxiola, R.I.; Petranovskii, V.; Antúnez-García, J.; Zepeda, T.A.; Fuentes, S. Effect of alkalinity variation in gel composition developed for hierarchical ZSM-5 growth: Conversion of ZSM-5 to mordenite. *Rev. Mex. Ing. Quím.* **2018**, *17*, 1159–1172. [CrossRef]
19. Jones, J.B. Al–O and Si–O tetrahedral distances in aluminosilicate framework structures. *Acta Crystallogr. Sect. B Struct. Crystallogr. Cryst. Chem.* **1968**, *24*, 355–358. [CrossRef]
20. García-Sosa, A.T.; Castro, M. Density functional study of FeO_2, FeO_2^+ and FeO_2^-. *Int. J. Quantum Chem.* **2000**, *80*, 307–319. [CrossRef]
21. Belviso, C.; Agostinelli, E.; Belviso, S.; Cavalcante, F.; Pascucci, S.; Peddis, D.; Varvaro, G.; Fiore, S. Synthesis of magnetic zeolite at low temperature using a waste material mixture: Fly ash and red mud. *Microporous Mesoporous Mater.* **2015**, *202*, 208–216. [CrossRef]

materials

MDPI

Article

Thermal Properties of Porous Mullite Ceramics Modified with Microsized ZrO_2 and WO_3

Ludmila Mahnicka-Goremikina [1,*], Ruta Svinka [1], Visvaldis Svinka [1], Liga Grase [1], Inna Juhnevica [1], Maris Rundans [1], Vadims Goremikins [2], Sanat Tolendiuly [3] and Sergey Fomenko [4]

1 Institute of Materials and Surface Engineering, Faculty of Materials Science and Applied Chemistry, Riga Technical University, Paula Valdena St. 3/7, LV-1048 Riga, Latvia
2 Institute of Structural Engineering and Reconstruction, Riga Technical University, Kipsalas St. 6A, LV-1048 Riga, Latvia
3 Space Engineering Department, AUPET Named G. Daukeev, Baitursynov St., 126/1, Almaty 050013, Kazakhstan
4 Institute of Combustion Problems, Bogenbay Batyr St. 172, Almaty 050012, Kazakhstan
* Correspondence: mahnicka@inbox.lv

Abstract: Mullite ceramics are well known as materials with a high temperature stability, strength and creep resistance. In this research, the effect of a modification with magnesia-stabilized zirconia and yttria-stabilized zirconia, separately, as well as in a mixture with WO_3, in 1:1 and 1:2 ratios on the thermal properties of porous mullite ceramics was investigated. The porous mullite-containing ceramics were prepared by a slip casting of the concentrated slurry of raw materials with the addition of a suspension of Al paste for the pore formation due to the H_2 evolution as a result of the reaction of Al with water. The formed samples were sintered at 1600 °C and the holding time was 1 h. The materials were characterized using X-ray diffractometry, scanning electron microscopy, mercury porosimetry, the laser flash contactless method, thermal shock resistance testing and the non-destructive impulse excitation method for determining the elasticity modulus. The modification of the porous mullite ceramic with a mixture of ZrO_2 and WO_3 oxides had a positive effect by decreasing the thermal conductivity, due to the increased porosity, in comparison to the undoped samples and samples with only ZrO_2. The doubling of the WO_3 amount in the modifying oxide mixtures improved the ceramic thermal shock resistance. The porous mullite ceramics which were modified with magnesia-stabilized zirconia (2.8 mol% MgO) and WO_3 had a lower thermal conductivity and improved thermal shock resistance than the samples with yttria-stabilized zirconia (8 mol% Y_2O_3) and WO_3.

Keywords: mullite; porous ceramic; zirconia; tungsten oxide; thermal conductivity; thermal shock

Citation: Mahnicka-Goremikina, L.; Svinka, R.; Svinka, V.; Grase, L.; Juhnevica, I.; Rundans, M.; Goremikins, V.; Tolendiuly, S.; Fomenko, S. Thermal Properties of Porous Mullite Ceramics Modified with Microsized ZrO_2 and WO_3. *Materials* **2022**, *15*, 7935. https://doi.org/10.3390/ma15227935

Academic Editors: Andres Sotelo, Zhifeng Wang, Weiqing Zhang and Yichao Wang

Received: 20 October 2022
Accepted: 7 November 2022
Published: 10 November 2022

1. Introduction

Special materials are needed to save thermal energy in the working space of thermal high-temperature units and prevent it from flowing into the environment. Such materials are called high-temperature thermal insulation materials. Thermal energy losses during high-temperature processes often exceed its theoretical need by several times. Climate-neutral manufacturing aims to boost the efficient use of fossil and energy resources by reducing air and water pollution, reducing heat losses and slowing down climate change. The investigation and use of porous high-temperature ceramics for thermal insulation help to achieve the goals of climate-neutral manufacturing and decrease the environmental degradation [1–4].

Ceramic materials such as corundum, cordierite, zirconia, mullite and multiphase composite ceramics are used for high-temperature engineering applications. In order to apply these ceramic materials, it is important to take into account their thermal properties such as their thermal conductivity, specific heat capacity, thermal expansion coefficient, resistance to sudden changes in temperature, thermal shock cracking and thermal shock induced fracture, as well as their mechanical properties [5–13].

Several layers of ceramic refractory material with different thermal resistances and thermal conductivities or ceramic composite materials with gradient properties are applied often in attempt to meet the aim of limiting the heat loss of certain equipment. For example, the cement kiln body can be divided into three layers, the working layer, the thermal-preservation layer and the thermal-insulation layer [6,7,14,15]. Layers can be made from different ceramic materials. Each layer has different thermal properties and its own role. The main properties of the working layer are its high thermal shock resistance, low thermal conductivity, high strength and long lifetime. The thermal-preservation layer must also have a high strength and lower thermal conductivity. The thermal-insulation layer has an ultra-low thermal conductivity to prevent heat loss and ensure the thermal protection of the other layers. The same situation is encountered in the thermal insulation of a spacecraft, where the thermal insulation consists of several layers [8–12,14]. Chen al. investigated the specific facing from composite material–multilayer mullite-based brick and porous plates, in which the thermal properties change along a gradient [6]. It is difficult to choose layers of materials or form a composite material such that the materials' components are physically and chemically compatible with one another. In addition, layering increases the load on the construction. It is important that there is no peeling or destruction during the operation under elevated temperature conditions [6–9]. In the case where thermal-insulating ceramics are used at high temperatures or extreme conditions, such as rapid increase and decrease in temperature, the introduced thermal shock may produce microcracks, whose growth and development can cause the structural failure of the component [10,11].

In terms of the economy and technology, it is beneficial when the insulation part of the thermal technical equipment and spaceships consists of one type of material with a low bulk density that combines a lot of functions and properties. There are difficulties in searching for ceramic candidates for thermal insulators. The first is the selection of a ceramic material oxide or composition of oxides that will have a low thermal conductivity. The second difficulty is achieving a low bulk density due to the certain porosity. The presence of pores with an effective pore size leads to a significant reduction in the thermal conductivity compared to a dense material. The third difficulty is obtaining ceramic materials that will simultaneously have low parameters, such as linear thermal expansion coefficients and thermal conductivity coefficients [13,16]. The use of multi-phase polycrystalline ceramics can solve this problem. It is also important to take into account the specific heat capacity of heat insulation ceramics. The specific heat capacity is defined as the quantity of heat absorbed per unit mass of the material when its temperature increases by 1 K [17]. More heat energy is required to increase the temperature of materials with a high specific heat capacity than ones with a low specific heat capacity. Depending on the purpose of the ceramics, the high or low specific heat capacity and thermal diffusivity of the ceramics must be taken into account [17–19].

Porous mullite-containing refractory ceramics are good candidates for a high-temperature thermal insulation and for reducing heat losses. Porous mullite-containing ceramics have a significant share in the field of technical ceramics for industrial and space fields [1,2,5]. The catalyst supports, filters for hot gases and molten metals, parts of burners, heat insulators of industrial furnaces, technical equipment and spaceships are produced from this type of ceramic. The choice of a refractory material for an application will be determined by the type of required functionality of the furnace, heating unit or refractory insulator component and the prevailing conditions, e.g., the gaseous atmosphere, the presence of slags and the type of metal charge [1,15,16,20–22].

In order to improve the mullite-containing ceramics, researchers investigated the influence of modifying such ceramics with different oxides on the thermal properties. Xu et al. sintered a cordierite–mullite–corundum composite with added Sm_2O_3 and achieved an increase in the thermal shock resistance and a decrease in the thermal conductivity (6.81 W/mK) and thermal expansion ($5.96 \times 10^{-6}\ ^{\circ}C^{-1}$) [23]. Li et al. synthesized columnar self-reinforced mullite porous ceramics by adding V_2O_5 at 1350–1550 °C and achieved a porosity of about 63% and a thermal conductivity of about 1.04 W/mK [24]. The use of Ho_2O_3 [25], Gd_2O_3 [26] and HfO_2 [27] improved the thermal shock resistance and thermal durability of mullite-containing ceramics. There is still ongoing research related to the investigation of and improvement in the thermo-mechanical properties of mullite and alumina–mullite ceramic composites with zirconia or zirconia–zircon components [28–33]. The thermal conductivity, thermal diffusivity and specific heat capacity of ceramic samples could be obtained using different international test standards, provided in Table 1.

Table 1. Thermal analysis tests.

Thermal Analysis Tests	References
Standard test method for thermal conductivity and thermal diffusivity by modulated temperature differential scanning calorimetry.	[34,35]
Standard test method for determining specific heat capacity by differential scanning calorimetry.	[36,37]
Hot-plate system. Guarded hot-plate systems are used to measure steady-state heat flow through materials with low thermal conductivity (insulators).	[38–42]
Heat flow system. Guarded Comparative–Longitudinal Heat Flow Technique.	[42,43]
The laser flash method and laser pulse: Laser Flash Thermal Conductivity.	[42,44,45]

The application of WO_3 as an additive or raw component for the formation of an additional crystalline phase is relevant in different fields such as gas sensing, chromogenic, photocatalytic and emerging applications (biomedical, antibiotic and artificial intelligence) [46]. The use of WO_3 for a ceramic modification has not been extensively investigated. The main aim of the research work was the formation of the thermal-insulating mullite ceramic material with a high porosity, a high thermal shock resistance, at the same time as low linear thermal expansion coefficients, low thermal conductivity coefficients and a low specific heat capacity. The main tasks of the investigation were the modification of porous mullite ceramic with ZrO_2 and WO_3 oxides and the analysis of the thermal properties depending on the chemical compositions, structural features, porosity and pore morphology. Such ceramic materials will help reduce the heat loss and withstand the rapid temperature fluctuations [9,12–16].

2. Materials and Methods

2.1. Materials

Two types of aluminas, α-Al_2O_3 (d_{50} = 2 µm) and γ-Al_2O_3 (d_{50} = 80 µm), were purchased from Nabalox, Nabaltec AG, Schwandorf, Germany. Kaolin (d_{50} = 1.5 µm; SiO_2—56.5 wt.%, Al_2O_3—31.0 wt.%) was purchased from MEKA, Amberger Kaolinwerke, Hirschau, Germany. The magnesia-stabilized zirconia (2.8 mol% MgO) with d_{50} = 0.8 µm was obtained from Goodfellow, Huntingdon, UK. The yttria-stabilized zirconia (8 mol% Y_2O_3) with d_{50} = 0.5 µm, SiO_2 amorphous with d_{50} = 3–5 µm and WO_3 with d_{50} = 5 µm were acquired from GetNanoMaterials, Saint-Cannat, France. Aluminum paste (solid content of 70 ± 2%) with d_{50} = 12 µm was purchased from Aquapor-9008, Schlenk Metallic Pigments GmbH, Roth, Germany.

2.2. Material Proportions

The base of the mullite ceramic was prepared from two types of aluminas (α-Al_2O_3 and γ-Al_2O_3), amorphous SiO_2 and kaolin. The ratio of Al_2O_3 to SiO_2 was 2.57:1 due to the mullite stoichiometric composition. The quantity of γ-Al_2O_3 was three times more than the quantity of α-Al_2O_3. Kaolin was used at 30 wt.%. Oxides such as yttria-stabilized zirconia

(8 mol% Y_2O_3), magnesia-stabilized zirconia (2.8 mol% MgO) and WO_3 were used for the ceramic modifications. Both oxides were also used together in a mixture for modifying the mullite ceramics. The ratios of the different stabilized zirconia and tungsten oxide were 1:1 and 1:2.

2.3. Sample Preparation Methods

A slip casting of the concentrated slurry of raw materials was used for the sample preparation. The water content of the concentrated slurry was 38–40 wt.%. First, the dry raw materials were mixed in a dry state. Then, its suspension was created with distilled water and mixed for 10 min with a mechanical mixer to obtain a uniform particle distribution in the slurry. The suspension of aluminum paste was added into the raw material suspension and mixed for about 7–10 min. Then, the raw material slurry was slip casted into the mold. The porosity of the mullite ceramics was obtained due to the hydrogen gas evolution as a result of the reaction between the aluminum paste and water. The formed initial pores became visually noticeable 15–20 min after the slip casting. The pore formation took 1 to 3 h. After that, the samples were dried for 24 h at 20–25 °C and then for 24 h at 100 °C. The dried samples were sintered at 1600 °C with a 250 °C/h (4.2 °C/min) heating rate, and the holding time at the maximum temperature was 1 h. The cooling process of the fired samples was as slow as the heating process.

2.4. XRD Analysis

The phase compositions of the sintered materials were characterized by X-ray diffraction analysis (XRD; Rigaku Ultima + (Japan)) with CuK_α radiation, a voltage on the Cu anode of 30 kV, a current intensity of 20 mA, a range of the measurement angle of 5–60 $2\theta^\circ$ and a speed of the goniometer of 2°/min).

2.5. SEM Analysis

The morphology of the prepared samples was observed by using scanning electron microscopes: a TableTop SEM Hitachi TM3000 (Japan) at an electron beam energy of 5 keV and 15 keV, and a high-resolution SEM FEI Nova NanoSEM 650 (the Netherlands) at an electron beam energy of 10 keV. Metal coating sputtering was not used because the structures were observed in the low vacuum mode.

2.6. Apparent Porosity

The apparent porosity was mathematically calculated and based on the Archimedes' principle (European standard EN 623-2) after soaking the samples in distilled water. The apparent porosity P is the ratio of the total volume of the open pores in a porous body to its bulk volume. The apparent porosity P was calculated as a percentage using the following Equation (1):

$$P = ((m_3 - m_1)/(m_3 - m_2)) \times 100, \quad (1)$$

where m_1 is the mass of the dry test piece in grams, m_2 is the apparent mass of the immersed test piece in grams and m_3 is the mass of the soaked test piece in grams.

2.7. Hg Porosimetry

The pore size distribution of the porous mullite ceramic was analyzed by a mercury porosimeter (Quantachrome, Pore Master 33, USA). Mercury intrusion porosimetry makes it possible to obtain data on the pores in a material in a limited range from 0.006 μm to 1000 μm.

2.8. Thermal Analysis

The specific heat capacity and thermal conductivity measurements and the calculation of the thermal diffusivity of the samples were carried out using the laser flash contactless method using the universal equipment table-top instrument Netzsch LFA 457 MicroFlash,

Germany. The measurements were carried out in the temperature range from 25 °C to 1100 °C. The sample dimensions were 10 mm × 10 mm and the thickness was 3 mm.

Thermal diffusivity (α with the unit mm^2/s) is a material-specific property for characterizing an unsteady heat conduction. This value describes how fast heat diffuses through the material. The thermal diffusivity was calculated from the following Equation (2):

$$\alpha = \kappa/(\rho Cp), \quad (2)$$

where κ is the thermal conductivity (W/(m·K)), ρ is the density (kg/m^3) and Cp is the specific heat capacity (J/kg/K) [18,19].

2.9. Thermal Shock Resistance Testing

The thermal shock resistance of the ceramics was determined during 10 cycles of thermal shock corresponding to the scheme 20 °C → 1000 °C → 20 °C with an exposure for 1 h at 1000 °C.

2.10. Elasticity Modulus Determination

The change in the elasticity modulus was measured before and after the 1st, 2nd, 5th and 10th thermal shock test by a non-destructive impulse excitation method (equipment Buzz-O-Sonic 5.0; BuzzMac International, LLC, USA). The non-destructive method provides an opportunity to examine the same sample after each thermal shock cycle, which allowed for more precise results.

3. Results and Discussion

3.1. Mineralogical Phase Composition

The XRD patterns of the prepared materials are shown in Figures 1 and 2. Comparing the experimental diffraction pattern with those from the International Centre of Diffraction Data (ICDD), the crystalline phases of the undoped samples correspond to mullite and corundum (Figure 1a).

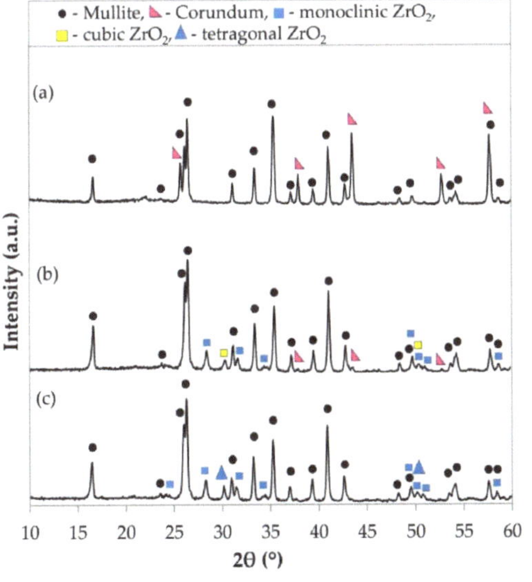

Figure 1. X-ray diffraction data of the sintered porous ceramic materials: (**a**) undoped samples, (**b**) samples with YSZ and (**c**) samples with MSZ.

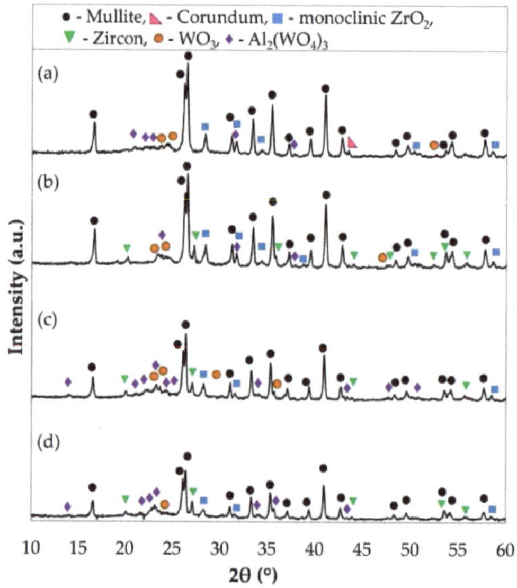

Figure 2. X-ray diffraction data of the sintered porous ceramic materials: (**a**) with YSZ:WO_3 (1:1), (**b**) with MSZ:WO_3 (1:1), (**c**) with YSZ:WO_3 (1:2) and (**d**) with MSZ:WO_3 (1:2).

The samples with yttria-stabilized zirconia and magnesia-stabilized zirconia contained mullite and monoclinic and tetragonal ZrO_2 (Figure 1b,c). The corundum phase remained in the samples with YSZ. The XRD patterns (Figure 2a,b) of the samples with a mixture of YSZ or MSZ and WO_3 in a 1:1 ratio show the presence of mullite, monoclinic ZrO_2, WO_3 and aluminum tungstate in both the materials of such compositions. Zircon was formed additionally in samples with MSZ and WO_3 (1:1).

The samples with a mixture of YSZ or MSZ and WO_3 in a 1:2 ratio have the main phase, the mullite phase (Figure 2c,d), as well as such phases as monoclinic ZrO_2, zircon and aluminum tungstate. The presence of WO_3 in the compositions with MSZ and WO_3 (1:2) was less pronounced than in the samples with YSZ and WO_3 (1:2). A detailed description of the crystalline phase formation was considered in a previous investigation, which was described in the article [47].

3.2. Macrostructure

The images of the porous mullite ceramic surface were obtained using a Table Top SEM and ×20 magnification (Figure 3). The pores have a spherical shape and approximately the same cross section as in the samples without the addition of modifying oxides (Figure 3a).

The pores of the modified samples only with YSZ or MSZ as well as with these oxides and WO_3 mixtures do not have the strong spherical form with a uniform diameter. All samples with modifying oxide additives are characterized by an elongated pore shape in the ceramic structure. It is important to note that these elongated pores are characterized by an expressed orientation in the ceramic structure. Such elongation and orientation occurred in the direction parallel to the base of the molds, respectively, parallel to the horizontal plane of the samples or along the sample length. When analyzing the photo of the surface of the modified samples along the pores and across the pores, it can be seen that to a greater extent, the pores are isolated from each other. The pores have to slit shapes with a partial networking because they have a joint intersection and branching, as well as a stomach and dead ends (Figure 3b–g). The length of the pores is 2–3 times greater than their width and height. The walls of the pores are not smooth but uneven due to the fact that large pores are formed from the merging of several smaller pores.

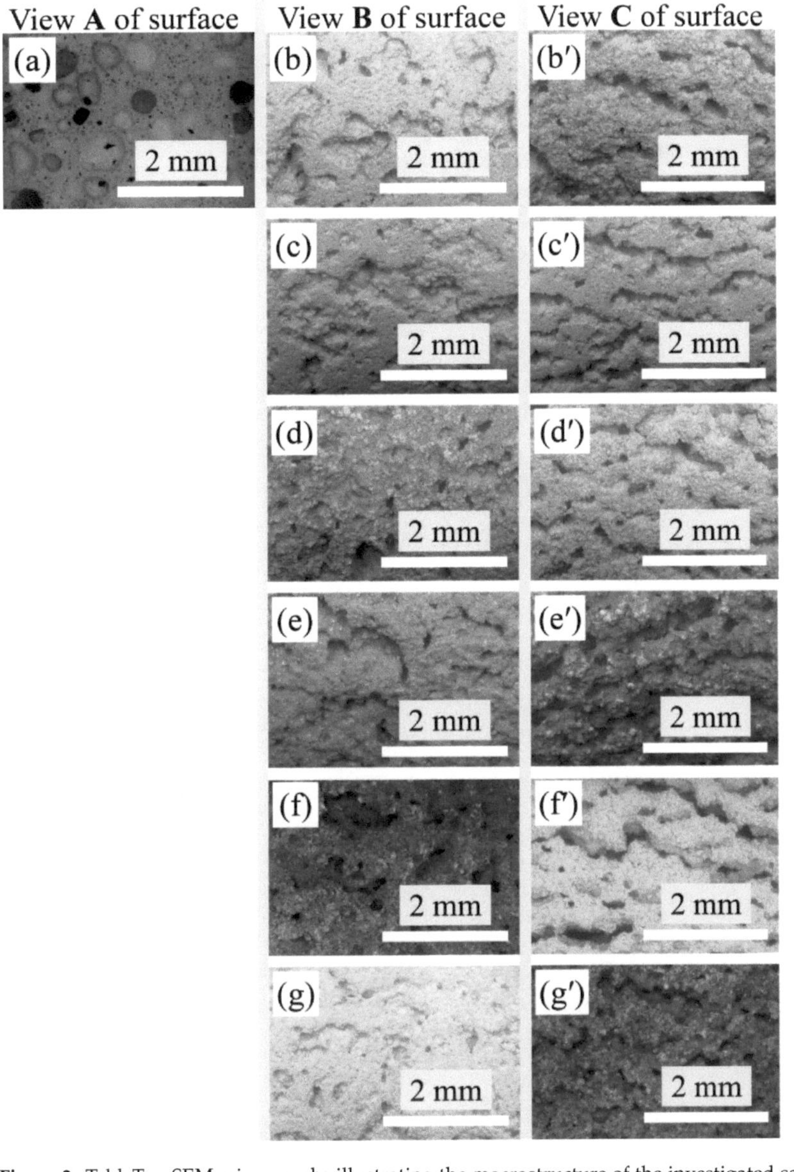

Figure 3. TableTop SEM micrographs illustrating the macrostructure of the investigated samples, magnification ×20: (**a**) view A of the undoped sample surface; (**b–g**) view B of the modified sample surface perpendicular to the base of the molds; (**b′–g′**) view C of the modified sample surface parallel to the base of the molds; (**b**,**b′**) samples with YSZ; (**c**,**c′**) with MSZ; (**d**,**d′**) with YSZ:WO_3 (1:1); (**e**,**e′**) with MSZ:WO_3 (1:1); (**f**,**f′**) with YSZ:WO_3 (1:2); and (**g**,**g′**) with MSZ:WO_3 (1:2).

3.3. Microstructure

The microstructure of the porous mullite ceramic without a modifying additive and only with yttria-stabilized zirconia or magnesia-stabilized zirconia formed from densely packed and closely bordered crystals is shown in Figure 4.

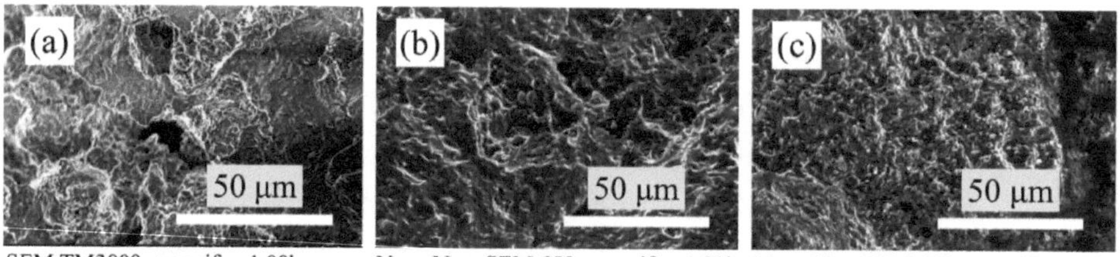

SEM TM3000, magnif.: x1.00k   Nova NanoSEM 650, magnif.: x1.00k   Nova NanoSEM 650, magnif.: x1.00k

Figure 4. SEM micrographs of the microstructure of sintered samples: (**a**) undoped samples, (**b**) with YSZ and (**c**) with MSZ.

The use of the ZrO_2 and WO_3 mixtures in a 1:1 and 1:2 ratio for modifying porous mullite ceramics caused the formation of elongated needle-shaped mullite crystals that were randomly located in relation to each other (Figures 5 and 6). Mullite crystals slightly border each other, therefore the structure of the samples is not dense.

Nova NanoSEM 650, magnif.: x1.00k   Nova NanoSEM 650, magnif.: x1.00k   SEM TM3000, magnif.: x1.00k

Figure 5. SEM micrographs of the microstructure of sintered samples: (**a**) with YSZ:WO_3 (1:1), (**b**) with MSZ:WO_3 (1:1) and (**c**) with YSZ:WO_3 (1:2).

SEM TM3000, magnif.: x1.00k   SEM TM3000, magnif.: x2.00k   SEM TM3000, magnif.: x5.00k

Figure 6. SEM micrographs of the microstructure of sintered samples with MSZ:WO_3 (1:2): (**a**) magnification ×1.00k, (**b**) magnification ×2.00k and (**c**) magnification ×5.00k.

The samples with a mixture of YSZ:WO_3 and MSZ:WO_3 in a 1:2 ratio have relatively thinner mullite crystals with a distinct acicular or needle-shaped mullite crystals–crystals with a narrow thin ending. The samples with a mixture of MSZ:WO_3 in a 1:2 ratio differ from other sintered materials in that they contain the porous mullite crystals. On the SEM micrographs (Figure 6c), such porous mullite crystals are displayed as elongated needle-shaped crystals with a round hole at the and. It can be assumed that such mullite crystals have internal hollow or voids along the length of the crystals. The formation of hollow mullite crystals occurred due to the formation of the $Al_2(WO_4)_3$ at 1075–1100 °C [48,49] and the presence of the Al–Si–O agglomerations on the surface of the $Al_2(WO_4)_3$ particles. This coincides with the results of the investigations of Liu et al. [49]. The mullitization of such Al–Si–O agglomerations occurred at about 1200 °C, with a simultaneous decomposition of the aluminum tungstate from within. Thus, the porous crystals consisted of a mullite shell and a hollow inner part [49].

3.4. Porosity and Pore Size Distributions

The apparent porosity of the sintered samples is shown in Figure 7. The samples without a modifying additive and samples with a different stabilized zirconia have a similar apparent porosity of about 40 ± 2%. The samples with mixtures of modifying additives have an apparent porosity higher than 59 ± 2%.

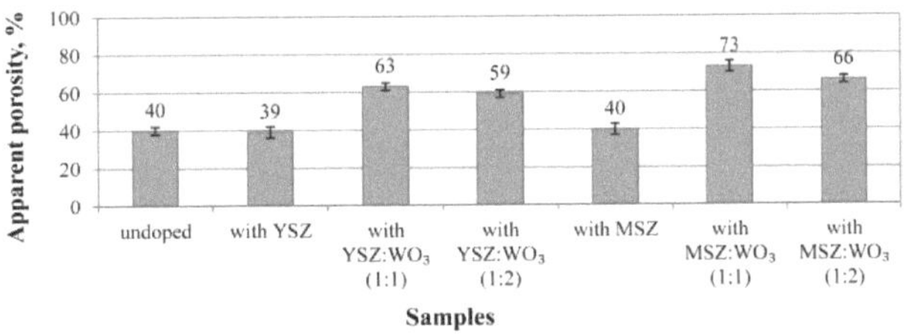

Figure 7. Apparent porosity of the investigated samples.

The use of WO_3 in the equivalent ratio to zirconia noticeably increases the apparent porosity of the samples in comparison with the undoped samples. Porous mullite ceramics modified with magnesia-stabilized zirconia and WO_3 in a 1:1 ratio have the highest apparent porosity (73 ± 2%). The doubling of WO_3 in the MSZ:WO_3 mixture slightly decreases the porosity (66 ± 2%) of such samples in comparison with the samples with MSZ and WO_3 in a 1:1 ratio.

The graphs of the pore size distribution after the mercury porosimetry are shown in Figure 8. The undoped samples have three ranges of pore size distributions: 0.05–0.2 μm, 0.3–5 μm and 7–1000 μm, with the most pronounced size of ≈0.08 μm, ≈0.15 μm and ≈150 μm in these ranges (Figure 8a).

Two ranges of pore size distributions are observed for the samples with yttria-stabilized zirconia and magnesia-stabilized zirconia (Figure 8a). The biggest pores of these samples are in range from 20 μm to 500 μm, in which ≈150 μm pores are more intensely expressed. Such large pores occupy approximately the same percentage for both compositions. The smaller pores of the samples with the YSZ additive are in the range from 1 μm to 20 μm. For the samples with MSZ, the range of the smaller pores narrows and occupies from approximately 3 μm to 20 μm, and such pores occupy a smaller percentage than in the case of the samples with YSZ. The 5 μm pores are more strongly expressed in the smaller pore size range for the samples of both compositions.

Additionally, the two pore size distribution ranges are in the samples with the YSZ:WO_3 and MSZ:WO_3 mixture in a 1:1 ratio (Figure 8b). The range of the smaller pores expanded with the addition of WO_3, which takes them from 1 to 20 μm, with a pronounced predominance of a 6–7 μm pore size. It can be seen from the graphs in Figure 8b that pores of a smaller diameter (1–20 μm) predominate over the larger sized pores (20–500 μm). The smaller pores of the samples with YSZ:WO_3 (1:1) occupy a slightly bigger percentage as for the samples with MSZ:WO_3 (1:1). The 150 μm pores are more intense in the large size pores range samples with an additional oxide mixture in a 1:1 ratio.

The pore size distributions of the samples with the YSZ:WO_3 and MSZ:WO_3 mixture in a 1:2 ratio are shown in Figure 8c. The doubling of WO_3 decreases the formation of large pores. The range of the large pores takes from 20 μm to 1000 μm. The doubling of WO_3 caused the formation of small pores. The samples with YSZ:WO_3 (1:2) have pores from ≈1.5 μm to 20 μm, with primarily 6 μm pores. The pores from 4 μm to 10 μm occupy a significant volume of the samples with MSZ:WO_3 (1:2). The 3–7 μm pores predominate in the case of the samples with MSZ:WO_3 (1:2).

Pores larger than 1000 μm (or 1 mm) are not shown on the pore distribution graphs, although they are shown in the SEM pictures in Figure 3.

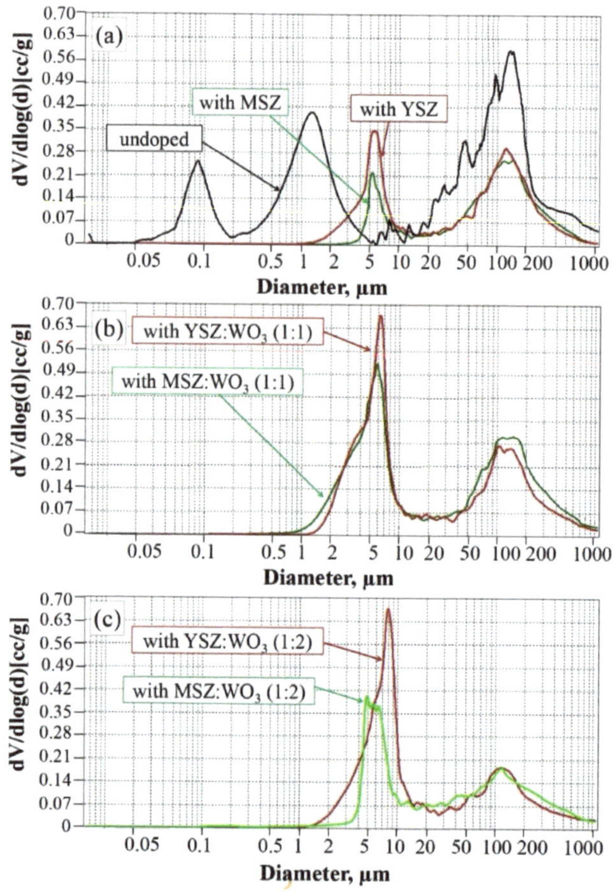

Figure 8. Pore size distributions of the samples: (**a**) undoped samples, with YSZ and with MSZ; (**b**) samples with YSZ:WO_3 (1:1) and with MSZ:WO_3 (1:1); and (**c**) samples with YSZ:WO_3 (1:2) and with MSZ:WO_3 (1:2).

3.5. Specific Heat Capacity

Figure 9 shows the temperature dependence of the specific heat capacity of the investigated samples. The approximately 1200–1300 °C specific heat capacity curves of the mullite single crystals and polycrystalline mullite show an anomalous sigmoidal increase [50], which also corresponds to the specific heat capacity lines obtained for the undoped samples and samples modified with YSZ and MSZ, as well as for YSZ:WO_3 in a 1:1 ratio and MSZ:WO_3 in a 1:2 ratio. These samples have a similar temperature dependence of the specific heat capacity from 25 °C to 1100 °C and an anomalous sigmoidal increase. The specific heat capacity values of the samples with a mixture of YSZ:WO_3 in a 1:2 ratio and a mixture of MSZ:WO_3 in a 1:1 ratio rapidly increase at about 1000 °C and then decrease due to the presence of WO_3 with a characteristic temperature dependence of the specific heat capacity [51–55]. This can be explained by the fact that the phase compositions of these samples have the pronounced presence of the WO_3 phase and its phase transitions during heating. Han et al. determined that the WO_3 phase transitions tetragonal(t_1) $\rightarrow$ tetragonal(t_2) at 1200 K (927 °C) occur with a large increase in the specific heat capacity and with a subsequent decrease [52,56–59]. In these works, such an increase in the specific heat capacity of the porous mullite ceramics occurs closer to 1000 °C perhaps due to the presence of other phases that can affect the phase transition temperature.

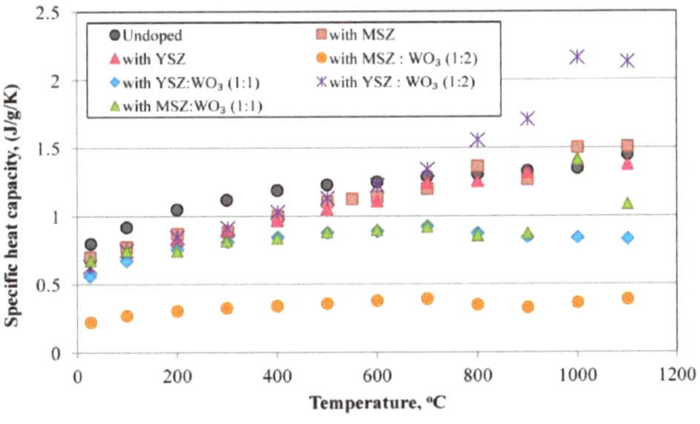

Figure 9. Specific heat capacity of the samples.

The samples with a mixture of MSZ:WO_3 in a 1:2 ratio have the lowest specific heat capacity, which does not exceed ≈0.40 J/g/K at all temperature ranges due to the presence of zircon with a relatively low specific heat capacity.

3.6. Thermal Conductivity

Figure 10 shows the thermal conductivity of the sintered porous mullite ceramic samples after the laser flash contactless method measurements. From the point of view of thermal conduction, the porous ceramics can be regarded as a two-phase system [53]. The first phase is the ceramic material skeleton or matrix. The second phase varies as the porosity, pore size and form, as well as the roughness of the pore walls. The heat transfer through such phases forms and describes the common thermal conductivity of the porous ceramic material. The mullite materials are the matrices of the investigated porous ceramics.

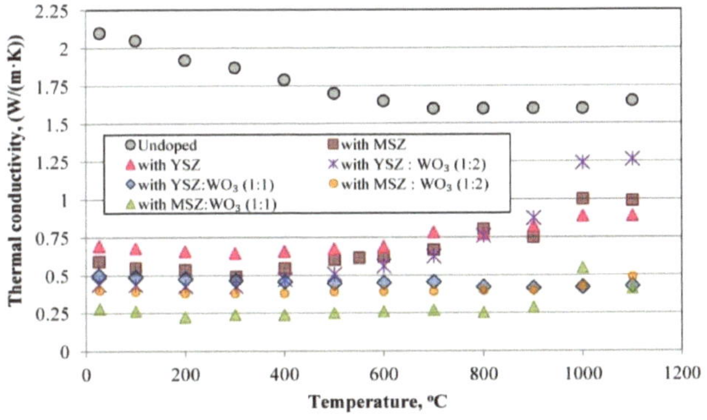

Figure 10. Thermal conductivity of the samples.

The thermal conductivity of the dense fully mullite ceramic is 5.1 W/mK and the thermal conductivity of the air is 0.026 W/mK [53]. The thermal conductivity of the porous mullite ceramics with a porosity of about 75%, according to the published data, is 0.31–0.42 W/mK at room temperature [53]. The thermal conductivity is directly proportional to the porosity of the investigated samples. Samples with a lower porosity of about 39–40% have the highest thermal conductivity, and porous ceramic samples with a porosity of about 63–73% have a lower thermal conductivity.

The undoped samples with a 40% porosity have a higher thermal conductivity in the temperature range of 25–1100 °C compared to all the modified samples. The thermal conductivity of the undoped mullite porous ceramics with a porosity of ≈40% is 2.1 W/mK at room temperature. The undoped samples also have a high thermal conductivity due to the predominance of isolated, large-sized spherical pores, as well as due to the presence of a corundum phase with a high $\lambda_{T=25°C}$ = 41.9 W/mK [53,55] that increases the thermal conductivity of the ceramic matrix. The change in the slope on the temperature dependence of the thermal conductivity at temperatures greater than about 1000 °C is due to a radiative contribution to the measured thermal conductivity.

The thermal conductivity of the samples modified with YSZ and MSZ and with a porosity of ≈40% at room temperature is 0.69 W/mK and 0.59 W/mK, respectively. The thermal conductivity of such samples with these compositions increased to 0.82 and 0.74 W/mK at 900 °C. Such parameters as the porosity, microstructure and presence of mullite and monoclinic ZrO_2 in the samples only with YSZ or with MSZ are similar, but the thermal conductivity is higher for the samples with YSZ due to the presence of a corundum phase with a high thermal conductivity coefficient. The intensity of the corundum phase of these composition samples is much less than that for the undoped samples, therefore, its influence on the thermal conductivity is less than in the case of the undoped samples.

The thermal conductivity of the samples with a porosity of about 59%, 63% and 66% at room temperature is 0.50 W/mK, 0.40 W/mK and 0.28 W/mK, respectively, for samples with a mixtures of YSZ:WO_3 in a 1:1 and 1:2 ratio and samples with a mixture of MSZ:WO_3 in a 1:2 ratio. Its λ at 900 °C became 0.41 W/mK, 0.40 W/mK and 0.88 W/mK. The thermal conductivity of the samples with a mixture of YSZ:WO_3 in a 1:2 ratio increases with an increasing temperature and became about 0.88 W/mK at 900 °C. The thermal conductivity of the samples with a YSZ:WO_3 mixture in a 1:1 ratio and with MSZ:WO_3 in a 1:2 ratio does not change intensively with an increasing temperature and remains within 0.40 ± 0.02 at 900 °C.

The sintered samples with a MSZ:WO_3 mixture in a 1:1 ratio and with a porosity of 73% have the lowest thermal conductivity of the investigated samples. The thermal conductivity of these samples does not change with an increasing temperature until 900 °C. It is 0.28 W/mK at room temperature and 0.29 W/mK at 900 °C.

Considering the point of view of the so-called mullite ceramic skeleton or matrix, the thermal conductivity is lower for those samples with a looser texture and its structure consists of relatively short, thin, randomly located and loosely packed mullite crystals in the structure. The inclusions of other phase's crystal grains with a low thermal conductivity reduce the overall thermal conductivity of the porous mullite ceramics. This is clearly noticeable in the case of an undoped sample and in the case of a sample with a mixture of MSZ:WO_3 in a 1:1 ratio, respectively, for the samples with the highest and lowest thermal conductivity. The dependence of the grain-size and crystal size on the thermal conductivity cannot be ignored. The thermal conductivity is lower for the samples with smaller and thin crystals, respectively, for the samples with a mixture of MSZ:WO_3 in a 1:1 and 1:2 ratio. This is due to the decrease in the phonon mean free path. As a result, the anharmonic phonon scattering within the grain is dominated and this decreases the thermal conductivity [60].

In its turn, the thermal conductivity is a little higher for the samples with MSZ:WO_3 in a 1:2 ratio than for the samples with MSZ:WO_3 in a 1:1 ratio. The thermal conductivity of the samples with MSZ:WO_3 in a 1:2 ratio is the same at different temperatures, due to the presence of different phases, that prevent the increase in the thermal conductivity with an increasing temperature. The stability of the thermal conductivity of the samples MSZ:WO_3 (1:2) can also be explained by the chaotic arrangement of mullite crystals, which are weakly adjacent to each other. The porous mullite crystals of such samples also decreases the thermal conductivity.

The Increase in the thermal conductivity of the ceramic samples with only YSZ and MSZ or with YSZ:WO_3 in a 1:2 ratio with an increasing temperature from 600 °C to 1000 °C could be explained by the expressed content of the monoclinic ZrO_2. Such an elevation of the thermal conductivity with an increasing temperature is a characteristic property of zirconium ceramics [61]. For samples with YSZ:WO_3 in a 1:2 ratio, such an increase in the thermal conductivity is also caused by the expressed presence of WO_3. The specific heat capacity and thermal conductivity of WO_3 increases with an increasing temperature [52]. Accordingly, it is less pronounced for the samples with a mixture of YSZ:WO_3 in a 1:1 ratio and MSZ:WO_3 in a 1:1 and 1:2 ratio due to additional phases such as zircon and aluminum tungstate that reduce this effect.

The change in the slope on the temperature dependence of the thermal conductivity at temperatures greater than about 1000 °C is due to a radiative contribution to the measured thermal conductivity.

3.7. Thermal Diffusivity

Figure 11 shows the temperature dependence of the thermal diffusivity of the investigated samples. The undoped samples have the larger thermal diffusivity in comparison with the modified porous mullite ceramics, due to a higher thermal conductivity and the similar specific heat capacity in comparison with the modified samples. These undoped samples potentially faster propagate the heat into the medium.

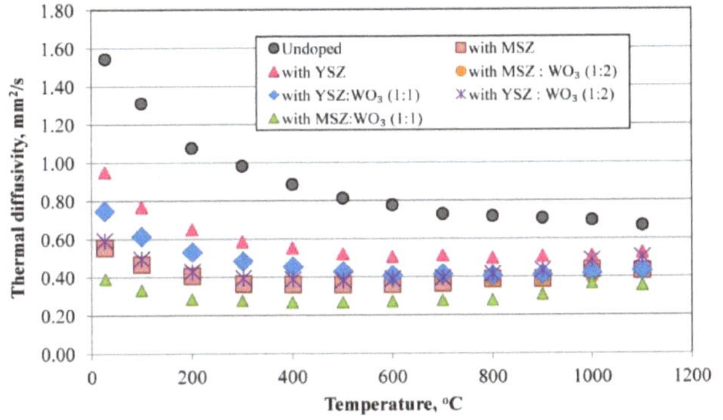

Figure 11. Thermal diffusivity of the samples.

The modified samples have a relatively similar thermal diffusivity. The samples modified with a mixture of MSZ:WO_3 in a 1:1 ratio have a lower thermal diffusivity due to the lower thermal conductivity and relatively low specific heat capacity. The low thermal diffusivity of these samples means that heat is mostly absorbed by the material and a small amount of heat is conducted farther.

3.8. Thermal Shock Resistance

The dependence of the relative change in the elastic modulus on the cycle number of the thermal shock is shown in Figure 12. The undoped samples and samples modified with YSZ and MSZ already have a lower resistance to thermal shock after the first cycle of the temperature change corresponding to the scheme 20 °C → 1000 °C → 20 °C with exposure for 1 h at 1000 °C. Such samples lose more than 15% of the elastic modulus after the 10th cycle due to the induced thermal stress after the rapid temperature change. The reading of the elastic modulus of the samples with YSZ and MSZ improved between the 2nd and 5th cycles due to the process of the crack healing of the ceramic materials. Samples with YSZ and MSZ have the ZrO_2 mainly as a monoclinic (m-ZrO_2) modification but also as tetragonal (t-ZrO_2). With a temperature change from 1000 °C to 20 °C, tetragonal ZrO_2 grains transform into monoclinic grains, thus, the martensitic phase transformation occurs. This phase transformation is accompanied by a volume expansion of 3–4%, which is directed opposite to the crack propagation. The growth of the crack and propagation were prevented due to the induced compressive stress [21,22,61–64]. However, a further exposure to thermal shock leads to a decrease in the elastic modulus.

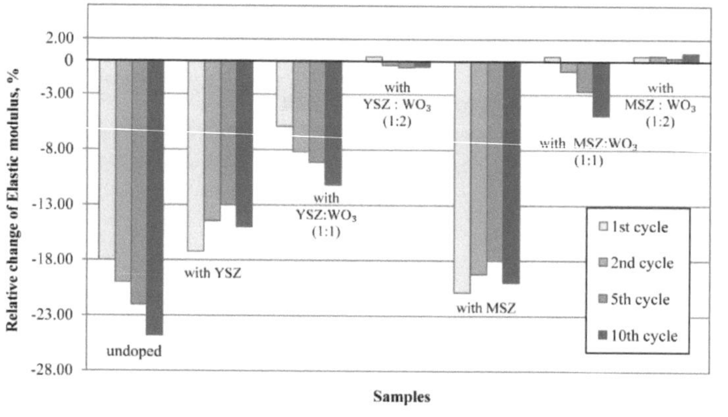

Figure 12. Relative change in the elastic modulus of samples after the thermal shock tests.

The thermal shock resistance of the investigated samples improves with the use of WO_3 in the additive mixture. In the case of the samples with YSZ:WO_3 in a 1:1 ratio, the relative change in the elastic modulus is negative already after the first thermal shock test and proportionally reduces with the increase in the thermal shock tests' numbers in comparison with the undoped samples and samples with YSZ and MSZ. The relative change in the elastic modulus after the 10th thermal shock cycles is ≈11% for the samples with YSZ:WO_3 in a 1:1 ratio.

The samples doped with a mixture of MSZ:WO_3 in a 1:1 ratio demonstrate a high resistance to the thermal shock. The elastic modulus of these samples increases by ≈0.5% after the 1st thermal shock cycle and decreases within 5% after the 10th cycle.

The samples with a mixture of YSZ:WO_3 and MSZ:WO_3 in a 1:2 ratio have a high thermal shock resistance. The elastic modulus of these samples also increases by ≈0.4% after the 1st thermal shock cycle, but further behavior is different. In its turn, the elastic modulus of the samples with a mixture of YSZ:WO_3 in a 1:2 ratio decreases after the 2nd cycle and decreases by 0.5% after the 10th cycle. The elastic modulus of the samples with a mixture of MSZ:WO_3 in a 1:2 ratio does not decrease. The elastic modulus of such samples increases by 0.5% after the 1st thermal shock test and does not decrease after the 2nd and 5th cycle, but it becomes greater than the initial value by ≈0.8% after the 10th cycle.

For samples with a mixture of MSZ:WO_3 in a 1:1 ratio and a mixture of YSZ:WO_3 and MSZ:WO_3 in a 1:2 ratio, the increased thermal shock resistance can be explained by the presence of such crystalline phases as zircon and aluminum tungstate [11,65]. The presence of the aluminum tungstate crystalline phase with a negative linear thermal expansion ($\alpha_{alumin\cdot\ tungst\cdot} = -1.5 \times 10^{-6}\ ^{\circ}C^{-1}$) [48] has an influence on the thermal shock resistance of the investigated ceramics. In the case of the aluminum tungstate phase, it shrinks and allows for the expansion of other the crystalline phases of the investigated ceramic without the formation of the internal stresses in the structure at the thermal shock time. Zircon with its low thermal expansion ($\alpha_{zircon} = 4.1 \times 10^{-6}\ ^{\circ}C^{-1}$ from room temperature to 1400 °C) [65] does not increase the expansion of the ceramic samples.

The samples modified with a mixture of MSZ:WO_3 in a 1:2 ratio together with the thermal insulating ability show the best thermal shock resistance. The polycrystalline structure with chaotic mullite crystals of these samples and a comparable high porosity of ≈63%, as well as the presence of small pores with a pore size range from 4 to 10 μm, do not cause the localization of stresses at the moment of the thermal shock. The relatively small branched pores can deflect, slow down or stop the propagation of cracks by its pinning [66].

The elastic modulus does not decrease as a result of a rapid temperature change due to the compensation of stresses and a reduction in the probability of crack formation and growth.

4. Conclusions

The influence of the porous mullite ceramic modification with different microsized stabilized ZrO_2 and WO_3 on the thermal properties was investigated. The porous mullite ceramic with a simultaneously low thermal conductivity and high thermal shock resistance was achieved. The following conclusions were reached:

(a) The use of the microsized ZrO_2 and WO_3 additive promotes the formation of elongated partially networked pores with an orientation in a direction parallel to the base of the molds.
(b) The thermal conductivity decreases with an increasing sample porosity and the randomness of the ceramics structure, as well as with the decreasing mullite crystal thickness.
(c) The formation of the hollow mullite crystals decreases the thermal conductivity of the ceramics and stabilizes its temperature dependence.
(d) The increase in the zircon content in the phase compositions of the porous mullite ceramic causes the decrease in the specific heat capacity of these ceramics.
(e) The presence of zircon and aluminum tungstate in the phase compositions of the porous mullite ceramic improves the thermal shock resistance of the investigated ceramics.

Porous mullite ceramics from raw material compositions with a mixture of magnesia-stabilized zirconia and WO_3 in a 1:2 ratio can be used as a potential thermal-insulating material in conditions of sharp temperature changes.

Author Contributions: Conceptualization, L.M.-G. and V.G.; methodology, L.M.-G., R.S. and V.S.; validation, L.M.-G., M.R. and I.J.; formal analysis, L.M.-G., S.T. and S.F.; investigation, L.M.-G.; resources, L.G. and I.J.; data curation, L.M.-G. and M.R.; writing—original draft preparation, L.M.-G.; writing—review and editing, L.M.-G.; visualization, L.M.-G. and V.G.; supervision, R.S. and V.S.; All authors have read and agreed to the published version of the manuscript.

Funding: This research was funded by the European Regional Development Fund within the Activity 1.1.1.2 "Post-doctoral Research Aid" of the Specific Aid Objective 1.1.1 "To increase the research and innovative capacity of scientific institutions of Latvia and the ability to attract external financing, investing in human resources and infrastructure" of the Operational Programme "Growth and Employment, (No.1.1.1.2/VIAA/1/16/121).

Institutional Review Board Statement: Not applicable.

Informed Consent Statement: Not applicable.

Data Availability Statement: The data presented in this study are available on request from the corresponding author.

Conflicts of Interest: The authors declare no conflict of interest.

References

1. Dedinec, A.; Dedinec, A.; Taseska-Gjorgievska, V.; Markovska, N.; Kanevce, G. Energy transition of a developing country following the pillars of the EU Green Deal. *Therm. Sci.* **2022**, *26*, 1317–1329. [CrossRef]
2. Salomão, R.; Oliveira, K.; Fernandes, L.; Tiba, P.; Prado, U. Porous refractory ceramics for high-temperature thermal insulation Part 1: The Science Behind Energy Saving. *Interceram Int. Ceram. Rev.* **2021**, *70*, 38–45. [CrossRef]
3. Medvedovski, E. Alumina–mullite ceramics for structural applications. *Ceram. Int.* **2006**, *32*, 369–375. [CrossRef]
4. Kurovics, E.; Kulkov, A.S.; Ibrahim, J.-E.F.M.; Kashin, A.D.; Pala, P.; Nagy, V.; Kulkov, S.N.; Gomze, L.A. Mechanical properties of mullite reinforced ceramics composite produced from kaolin and corn starch. *Epitőanyag–JSBCM* **2021**, *73*, 149–153. [CrossRef]
5. Krenzel, T.F.; Schreuer, J.; Laubner, D.; Cichocki, M.; Schneider, H. Thermo-mechanical properties of mullite ceramics: New data. *J. Am. Ceram. Soc.* **2019**, *10*, 416–426. [CrossRef]
6. Chen, S.; Wang, J.; Yuan, L.; Li, Q. Composite design of low thermal conductivity mullite brick for application to cement kiln. *MATEC Web Conf.* **2018**, *142*, 02008. [CrossRef]
7. Salomão, R.; Oliveira, K.; Fernandes, L.; Tiba, P.; Prado, U. Porous refractory ceramics for high-temperature thermal insulation—Part 2: The technology behind energy saving. *Interceram Int. Ceram. Rev.* **2022**, *71*, 38–50. [CrossRef]

8. Wu, D.; Ren, H.; Wang, F.; Wang, H. High temperature thermal insulation performance of light nanomaterials for aerospace craft. *Hangkong Xuebao/Acta Aeronaut. Astronaut. Sin.* **2018**, *39*, 221636.
9. Clarke, D.R. Materials selection guidelines for low thermal conductivity thermal barrier coatings. *Surf. Coat. Technol.* **2003**, *163–164*, 67–74. [CrossRef]
10. Tavangarian, F.; Hui, D.; Li, G. Crack-healing in ceramics. *Compos. Part B Eng.* **2018**, *144*, 56–87. [CrossRef]
11. Rendtorff, N.M.; Garrido, L.B.; Aglietti, E.F. Thermal shock resistance and fatigue of Zircon–Mullite composite materials. *Ceram. Int.* **2011**, *137*, 1427–1434. [CrossRef]
12. Deng, Z.-Y.; Ferreira, J.M.F.; Tanaka, Y.; Isoda, Y. Microstructure and thermal conductivity of porous ZrO_2 ceramics. *Acta Mater.* **2007**, *55*, 3663–3669. [CrossRef]
13. Lee, W.J.; Cho, Y.J.; Lee, H.S.; Park, I.M.; Park, Y.H. Effect of pore morphology on elastic, heat conduction and thermal shock fracture behaviors of porous ceramics. *Proc. Eng.* **2011**, *10*, 2459–2463. [CrossRef]
14. Gong, L.; Wang, Y.; Cheng, X.; Zhang, R.; Zhang, H. A novel effective medium theory for modelling the thermal conductivity of porous materials. *Int. J. Heat Mass Transf.* **2014**, *68*, 295–298. [CrossRef]
15. Belogurova, O.A.; Grishin, N.N. Heat-insulating mullite-cordierite materials made from staurolite. *Refract. Ind. Ceram.* **2013**, *54*, 64–68. [CrossRef]
16. Angle, J.P.; Wang, Z.; Dames, C.; Mecartney, M.L. Comparison of two-phase thermal conductivity models with experiments on dilute ceramic composites. *J. Am. Ceram. Soc.* **2013**, *96*, 2935–2942. [CrossRef]
17. Feidt, M. *Finite Physical Dimensions Optimal Thermodynamics 1, From Thermostatics to Non-Equilibrium Thermodynamics*; Elsevier: Amsterdam, The Netherlands, 2017; ISBN 978-1-78548-232-8.
18. See, A.; Hassan, J.; Hashim, M.; Wahab, Z.A. Thermal diffusivity of kaolinite–mullite ceramic matrix composite with silicon nitride nanoparticle filler. *Thermochim. Acta* **2014**, *593*, 76–81. [CrossRef]
19. Jannot, Y.; Degiovanni, A. *Thermal Properties Measurement of Materials*; John Wiley & Sons: Hoboken, NJ, USA, 2018; p. 342, ISBN 978-1-119-47505-7.
20. Ilić, S.; Ivanovski, V.N.; Radovanović, Ž.; Egelja, A.; Kokunešoski, M.; Šaponjić, A.; Matović, B. Structural, microstructural and mechanical properties of sintered iron-doped mullite. *Mater. Sci. Eng. B* **2020**, *256*, 114543. [CrossRef]
21. Aksel, C. The influence of zircon on the mechanical properties and thermal shock behaviour of slip-cast alumina–mullite refractories. *Mater. Lett.* **2002**, *57*, 992–997. [CrossRef]
22. Kenawy, S.H.; Awaad, M.; Awad, H. In-situ mullite-zirconia composites from kaolin. *Am. Ceram. Soc. Bull.* **2006**, *85*, 9401–9406.
23. Xu, X.; Xu, X.; Wu, J.; Lao, X.; Zhang, Y.; Li, K. Effect of Sm_2O_3 on microstructure, thermal shock resistance and thermal conductivity of cordierite-mullite-corundum composite ceramics for solar heat transmission pipeline. *Ceram. Int.* **2016**, *42*, 13525–13534. [CrossRef]
24. Li, K.; Ge, S.; Yuan, G.; Zhang, H.; Zhang, J.; He, J.; Jia, Q.; Zhang, S. Effects of V_2O_5 addition on the synthesis of columnar self-reinforced mullite porous ceramics. *Ceram. Int.* **2021**, *47*, 11240–11248. [CrossRef]
25. Wu, J.; Ding, C.; Xu, X.; Liu, Y.; Wang, Y. Microstructure and performances of corundum−mullite composite ceramics for heat transmission pipelines: Effects of Ho_2O_3 additive content. *Ceram. Int.* **2021**, *47*, 34794–34801. [CrossRef]
26. Wu, J.; Ding, C.; Xu, X.; Zhou, S.; Zhou, Y.; Zhang, Q. Microstructure and performance of Cd_2O_3 added corundum-mullite ceramics composites for concentrated solar power application. *Ceram. Int.* **2021**, *47*, 17177–17185. [CrossRef]
27. Gai, K.; Guan, B.; Liang, L.; Li, J.; Wang, Q.; Zhao, T. Continuous aluminum oxide-mullite-hafnium oxide composite ceramic fibers with high strength and thermal stability by melt-spinning from polymer precursor. *J. Eur. Ceram. Soc.* **2022**, *42*, 5911–5921. [CrossRef]
28. Khorsand, A.; Majidian, H.; Farvizi, M. Wear behavior and microstructure of alumina-mullite- zirconia composites prepared by a novel method: Coating of zircon powder by aluminium alkoxide. *Ceram. Int.* **2022**, *48*, 33594–33603. [CrossRef]
29. Reinders, L.; Pfeifer, S.; Kröner, S.; Stolpmann, H.; Renfftlen, A.; Greiler, L.C.; Clauß, B.; Buchmeiser, M.R. Development of mullite fibers and novel zirconia-toughened mullite fibres for high temperature applications. *J. Eur. Ceram. Soc.* **2021**, *41*, 3570–3580. [CrossRef]
30. Weinberg, A.V.; Goeuriot, D.; Poirier, J.; Varona, C.; Chaucherie, X. Mullite–zirconia composite for the bonding phase of refractory bricks in hazardous waste incineration rotary kiln. *J. Eur. Ceram. Soc.* **2021**, *41*, 995–1002. [CrossRef]
31. Kumar, P.; Nath, M.; Ghosh, A.; Tripathi, H.S. Thermo-mechanical properties of mullite—Zirconia composites derived from reaction sintering of zircon and sillimanite beach sand: Effect of CaO. *Trans. Nonferrous Met. Soc. China* **2016**, *26*, 2397–2403. [CrossRef]
32. Bouchetou, M.L.; Poirier, J.; Morales, L.A.; Chotard, T.; Joubert, O.; Weissenbacher, M. Synthesis of an innovative zirconia-mullite raw material sintered from andalusite and zircon precursors and an evaluation of its corrosion and thermal shock performance. *Ceram. Int.* **2019**, *45*, 12832–12844. [CrossRef]
33. Rendtorff, N.; Garrido, L.; Aglietti, E. Mullite-zirconia-zircon composites: Properties and thermal shock resistance. *Ceram. Int.* **2009**, *35*, 779–786. [CrossRef]
34. *ASTM Standards: E 1952–01*; Standard Test Method for Thermal Conductivity and Thermal Diffusivity by Modulated Temperature Differential Scanning Calorimetry. ASTM: West Conshohocken, PA, USA, 2017.
35. Camirand, C.P. Measurement of thermal conductivity by differential scanning calorimetry. *Thermochim. Acta* **2004**, *417*, 1–4. [CrossRef]

36. *ASTM Standards: E1269*; Standard Test Method for Determining Specific Heat Capacity by Differential Scanning Calorimetry. ASTM: West Conshohocken, PA, USA, 2018.
37. Robertis, E.D.; Cosme, E.H.H.; Neves, R.S.; Kuznetsov, A.Y.; Campos, A.P.C.; Landi, A.M.; Achete, C.A. Application of the modulated temperature differential scanning calorimetry technique for the determination of the specific heat of copper nanofluids. *Appl. Therm. Eng.* **2012**, *41*, 10–17. [CrossRef]
38. *ASTM Standards: C177-13*; Standard Test Method for Steady-State Heat Flux Measurements and Thermal Transmission Properties by Means of the Guarded-Hot-Plate Apparatus. ASTM: West Conshohocken, PA, USA, 2019.
39. *ASTM Standards: C1044*; Standard Practice for Using a Guarded-Hot-Plate Apparatus or Thin-Heater Apparatus in the Single-Sided Mode. ASTM: West Conshohocken, PA, USA, 2020.
40. *ASTM Standards: C1114*; Standard Test Method for Steady-State Thermal Transmission Properties by Means of the Thin-Heater Apparatus. ASTM: West Conshohocken, PA, USA, 2019.
41. *ASTM Standards: C1113/C1113M-09*; Standard Test Method for Thermal Conductivity of Refractories by Hot Wire (Platinum Resistance Thermometer Technique). ASTM: West Conshohocken, PA, USA, 2019.
42. Codorniu, D.M.; Moyano, J.J.; Belmonte, M.; Osendi, M.I.; Miranzo, P. Thermal conduction in three-dimensional printed porous samples by high resolution infrared thermography. *Open Ceram.* **2020**, *4*, 100028. [CrossRef]
43. *ASTM Standards: E1225-13*; Standard Test Method for Thermal Conductivity of Solids Using the Guarded-Comparative-Longitudinal Heat Flow Technique. ASTM: West Conshohocken, PA, USA, 2013.
44. *ASTM Standarts: E1461*; Standard Test Method for Thermal Diffusivity by the Flash Method. ASTM: West Conshohocken, PA, USA, 2022.
45. Rezaee, S.; Ranjbar, K. Thermal conductivity of porous Alumina-20 wt% zirconia ceramic composites. *Ceram. Int.* **2020**, *46*, 16564–16571. [CrossRef]
46. Zhang, M.; Yang, C.; Zhang, Z.; Tian, W.; Hui, B.; Zhang, J.; Zhang, K. Tungsten oxide polymorphs and their multifunctional applications. *Adv. Coll. Int. Sci.* **2022**, *300*, 102596. [CrossRef]
47. Mahnicka-Goremikina, L.; Svinka, R.; Svinka, V.; Grase, L.; Goremikins, V. The formation of phases with low or negative linear thermal expansion coefficient in porous mullite ceramics. *Epitőanyag–JSBCM* **2020**, *72*, 91–98. [CrossRef]
48. Achary, S.N.; Mukherjee, G.D.; Tyagi, A.K.; Vaidya, S.N. Preparation, thermal expansion, high pressure and high temperature behaviour of $Al_2(WO_4)_3$. *J. Mater. Sci.* **2002**, *37*, 2501–2509. [CrossRef]
49. Liu, H.; Xiong, X.; Li, M.; Wang, Z.; Wang, X.; Ma, Y.; Yuan, L. Fabrication and properties of mullite thermal insulation materials with in-situ synthesized mullite hollow whiskers. *Ceram. Int.* **2020**, *46*, 14474–14480. [CrossRef]
50. Girolamo, G.D.; Blasi, C.; Pilloni, L.; Schioppa, M. Microstructural and thermal properties of plasma sprayed mullite coatings. *Ceram. Int.* **2010**, *36*, 1389–1395. [CrossRef]
51. Chase, M.W., Jr. *NIST-JANAF Thermochemical Tables*, 4th ed.; American Institute of Physics: University Park, ML, USA, 1998; pp. 1–1951.
52. Han, B.-y.; Khoroshilov, A.V.; Tyurin, A.V.; Baranchikov, A.E.; Razumov, M.I.; Ivanova, O.S.; Gavrichev, K.S.; Ivanov, V.K. WO_3 thermodynamic properties at 80–1256 K revisited. *J. Therm. Anal. Calorim.* **2020**, *142*, 1533–1543. [CrossRef]
53. Gong, L.; Wang, Y.; Cheng, X.; Zhang, R.; Zhang, H. Porous mullite ceramics with low thermal conductivity prepared by foaming and starch consolidation. *J. Porous Mater.* **2014**, *21*, 15–21. [CrossRef]
54. Manga, M.; Jeanloz, R. Thermal conductivity of corundum and periclase and implications for the lower mantle. *J. Geophys. Res.* **1997**, *102*, 2999–3008. [CrossRef]
55. Dietrich, B.; Kind, M.; Martin, H. Axial two-phase thermal conductivity of ceramic sponges—Experimental results and correlation. *Int. J. Heat Mass Transf.* **2011**, *54*, 2276–2282. [CrossRef]
56. Mardare, C.C.; Hassel, A.W. Review on the versatility of tungsten oxide coatings. *Phys. Status Solidi A* **2019**, *216*, 1900047. [CrossRef]
57. Rao, M.C. Structure and properties of WO_3 thin films for electrochromic device application. *J. Non-Oxide Glasses* **2013**, *5*, 1–8.
58. Zheng, H.; Zhen, O.J.; Strano, M.S.; Kaner, R.B.; Mitchell, A.; Kalantar-zadeh, K. Nanostructured tungsten oxide—Properties, synthesis, and applications. *Adv. Funct. Mater.* **2011**, *21*, 2175–2196. [CrossRef]
59. Ning, S.; Huberman, S.C.; Ding, Z.; Nahm, Y.-H.; Kim, H.-S.; Chen, G.; Ross, C.A. Anomalous defect dependence of thermal conductivity in epitaxial WO_3. *Adv. Mater.* **2019**, *31*, 1903738. [CrossRef]
60. Wang, H.; Du, X.; Shi, Y.; Deng, M.; Wang, J.; Qi, J.; Huang, Z. Grain-size dependent thermal conductivity of $Gd_2Zr_2O_7$ ceramics. *Ceram. Int.* **2022**, *48*, 16444–16448. [CrossRef]
61. Giordano, R. Machinable zirconia. A study on the building blocks of restorative dentistry. *Inside Dent. Technol.* **2012**, *3*, 1–3.
62. Žmak, I.; Ćorić, D.; Mandić, V.; Ćurković, L. Hardness and indentation fracture toughness of slip cast alumina and alumina-zirconia ceramics. *Materials* **2020**, *13*, 122. [CrossRef]
63. Banerjee, S.; Mukhopadhyay, P. Chapter 4—Martensitic Transformations. In *Phase Transformations. Examples from Titanium and Zirconium Alloys*; Elsevier: Amsterdam, The Netherlands, 2007; Volume 12, pp. 1–813, ISBN 9780080548791.
64. Zanelli, C.; Dondi, M.; Raimondo, M.; Guarini, G. Phase composition of alumina–mullite–zirconia refractory materials. *J. Eur. Ceram. Soc.* **2010**, *30*, 29–35. [CrossRef]

65. Wei, J.; Bingqiang, H.; Wei, Y.; Li, N.; Miao, Z. Influence of phase evolution and thermal decomposition kinetics on the properties of zircon ceramic. *Ceram. Int.* **2021**, *47*, 27285–27293. [CrossRef]
66. Cao, M.C.; Jin, X.X.; Dong, L.M.; Li, X.Y.; Zhang, X.Y. Effect of pore composition on thermal shock resistance of porous ceramics. *Dig. J. Nanomater. Biostructures* **2021**, *16*, 1–9. [CrossRef]

MDPI AG
Grosspeteranlage 5
4052 Basel
Switzerland
Tel.: +41 61 683 77 34

Materials Editorial Office
E-mail: materials@mdpi.com
www.mdpi.com/journal/materials

www.ingramcontent.com/pod-product-compliance
Lightning Source LLC
Chambersburg PA
CBHW041345210926
51850CB00013B/2995